Advances in Marine Bioprocesses and Bioproducts

Series Editor

Wei Zhang, New York, USA

The new book series "Advances in Marine Bioprocesses and Bioproducts" reviews current scientific and technological advances and progresses in industry applications in relation to developing marine bioprocesses for the production of bioproducts from marine bioresources. To acknowledge the multidisciplinary and cross-disciplinary nature of marine biotechnology, the aim of this book series is to establish a fresh look at bioprocessing of marine bioresources that is oriented to bioproduct development and their industrial applications. A further aim is to synthesise the new knowledge and expertise developed at the disciplinary interfaces and boundaries that will result in new disciplines or new industries. It will cover the achievements in both fundamental science and applied technology that lead to commercial developments of marine bioproducts and their applications in medicine, healthcare, agriculture, environment, energy, material, manufacturing and food. Topics selected will have a direct relevance in an industrial context.

Topical volumes will be written and edited by eminent experts, as guest editors in the field. The guest editors will invite reputed authors for contributions to their volumes. Each volume will be dedicated to a selected topic related to specific bioprocessing technologies, marine bioproducts and their applications. They are expected to provide a valuable, informative, comprehensive and updated account of the selected field of research innovation towards industry applications. In order to meet the criteria of industrial relevance, an expert from the respective industries is expected to act as a reviewer.

The book series is written for academics, scientists, engineers, postgraduate students, industry professionals and policymakers who work in public and private sectors, academic institutions and industries. It provides a unique forum to link the fundamentals to applications by timely dissemination of the topical R&D achievements in the field of marine bioprocesses and bioproducts.

Changhu Xue
Editor

Advances in Sea Cucumber Processing Technology and Product Development

Springer

Editor
Changhu Xue
College of Food Science and Engineering
Ocean University of China
Qingdao, China

ISSN 2731-4421 ISSN 2731-443X (electronic)
Advances in Marine Bioprocesses and Bioproducts
ISBN 978-3-031-16511-5 ISBN 978-3-031-16512-2 (eBook)
https://doi.org/10.1007/978-3-031-16512-2

This Springer imprint is published by the registered company Springer Nature Switzerland AG
The registered company address is: Gewerbestrasse 11, 6330 Cham, Switzerland

Contents

Diversity, Distribution, and Biology of Sea Cucumber

Hongying Liu, Changhu Xue, and Zhaojie Li

Abstract The sea cucumber is a benthic invertebrate with great nutritional and commercial value. There are about 900 species of sea cucumber encountered all over the world. Most sea cucumbers are inedible, whereas only about 40 species are edible. This requires consumers and trade managers to distinguish based on their knowledge of sea cucumber species. The fishery and trade of sea cucumber cover more than 70 countries and regions in the world. Sea cucumbers are exploited and utilized in polar regions, temperate regions, and tropical regions in the form of industrialized, semi-industrialized, and artisanal workshop (small-scale) fisheries. The purpose of this chapter is to introduce the distribution of sea cucumber species and provide identification guidelines for industrial workers to distinguish different sea cucumber species being exploited, utilized, and traded worldwide.

Keywords Sea cucumber · Species · Distribution · Biological characteristics

1 Introduction

The sea cucumber is a member of marine echinoderm belonging to the class *Holothuroidea* or *Holothuroidea* of the phylum *Echinodermata*. It was named *Holothuria* by Lamarck from France in 1801 (Liao 1997). There are many species of sea cucumbers, among which with around 900 varieties recorded in the world. While most sea cucumbers are inedible, only a few species are not. There are approximately 40 species of edible sea cucumbers in the world (Liao 2001).

In China, the sea cucumber is valued not only as precious food but also as a valuable traditional Chinese medicinal material. For this reason, it is as famous as

H. Liu
College of Ocean, Hebei Agricultural University, Qinhuangdao, P. R. China

C. Xue · Z. Li (✉)
College of Food Science and Engineering, Ocean University of China, Qingdao, P. R. China
e-mail: xuech@ouc.edu.cn; lizhaojie@ouc.edu.cn

C. Xue (ed.), *Advances in Sea Cucumber Processing Technology and Product Development*, Advances in Marine Bioprocesses and Bioproducts,
https://doi.org/10.1007/978-3-031-16512-2_1

ginseng, bird's nest, and shark's fin and is known as one of the eight treasures in the world. The sea cucumber is rich in nutrients, especially arginine (Arg) (Guo et al. 2015). The sea cucumber contains a variety of biologically active compounds such as active polypeptides, polyunsaturated fatty acids, saponins, lectins, and sulfated polysaccharides, featuring anti-oxidation, anticancer, and anti-inflammatory effects, among others (Bordbar 2011). The medicinal effect of sea cucumber has been recorded in the *Compendium of Materia Medica* by Li Shizhen of the Ming Dynasty in China. In *A Supplement to the Compendium of Materia Medica* written by Zhao Xuemin of the Qing Dynasty, it was recorded that "The sea cucumber tastes sweet and salty, and its effect of tonifying kidney essence is equivalent to that of ginseng, hence the name sea cucumber."

2 Classification of Sea Cucumbers

The traditional classification of sea cucumbers is mainly based on morphology, including external morphology, such as tentacle and tubular feet, as well as micro-morphological features, such as ossicle. Since the ossicle is very stable and has significant species specificity, it has been, and still is, used as the main basis for species identification of sea cucumbers (Li et al. 2008).

With the continuing discovery of sea cucumber's biological characteristics and new sea cucumber varieties, the classification system for sea cucumber has gradually evolved. Generally, there exist Selenka (1867), Semper (1868), Theel (1886), Ludwig (1892), Mortensen (1927), and Pawson et Fell (1965) classification systems. Fell (1965) believed that the earliest sea cucumber probably evolved from animals similar to the Cambrian fossil *Helicoplacus* or the Ordovician fossil *Eothuria*, which can be traced back to 500 million years ago. According to the classification system of Pawson et Fell (1965), the sea cucumber belongs to six orders: *Dendrochirotida*, *Dactylochirotida*, *Aspidochirotida*, *Elasipodida*, *Molpadiida*, and *Apodida*. The specific classification system is as follows (Table 1):

3 Distribution and Main Economic Types of Sea Cucumber

3.1 Distribution of Sea Cucumber Resources in the World

The sea cucumber is widely distributed in all oceans, but mainly in tropical and temperate seas, at depths ranging from the intertidal zone to the deepest parts of the ocean. Except for a few floating and roaming *Elasipodida*, most of them are benthic and can be found on rocky banks, coral reefs, and various sand bottoms (Pawson et al. 2001). Chin's manned submersible "striver" found a kind of relatively small sea cucumber at the bottom of the Mariana Trench on November 10, 2021.

Table 1 Main classification of sea cucumber

Class	Order	Family
Dendrochirotacea	*Dendrochirotida*	*Placothuriidae*
		Paracucumariidae
		Psolidae
		Heterothyonidae
		Phyllophoridae
		Sclerodactylidae
		Cucumariidae
	Dactylochirotida	*Ypsilothuriidae*
		Vaneyellidae
		Rhopalodinidae
Aspidochirotacea	*Aspidochirotida*	*Holothuriidae*
		Stichopodidae
		Synallactidae
	Elasipodida	*Deimatidae*
		Laetmogonidae
		Elipidiidae
		Psychropotidae
		Pelagothuriidae
Apodacea	Apodida	*Synaptidae*
		Chiridotidae
		Myriotrochidae
	Molpadida	*Molpadiidae*
		Caudinidae
		Eupyrgidae

The population of sea cucumber in tropical sea areas is diverse and is largely concentrated in the tropical Pacific and the Indian Oceans. In this regard, the major countries of origin are Indonesia, the Philippines, Malaysia, as well as New Caledonia, Fiji, and Barbuda in the South Pacific. Among them, the Indo-West Pacific region is the region with the largest number of species and the greatest resources of sea cucumber in the world. The sea cucumber resources in the temperate zones are monotypic and are mostly distributed in the temperate zones on the east and west sides of the Pacific Ocean (Xu 2009), being dominated by *Stichopodidae* (Conand and Byrne 1993). Among them, the most highly valued varieties are mainly distributed along the Pacific coast of the northern hemisphere, the coast of Latin America and the coast of the Arctic Ocean. The southern hemisphere boasts the most species of sea cucumber, but most of them have low food value.

According to records, there are 287 species of sea cucumber in the shallow water areas (no deeper than 20 m) of the Indo-West Pacific region Clark and Rowe (1971), which are mainly sea cucumbers of *Aspidochirotida*, *Bohadschia*, *Stichopus*, and *Actinopyga*, including *B. marmorata, B. argus, A. lecanora, A. echinites, A. mauritiana, A. miliaris, H. cinerascens, H. atra, H. edulis, H. difficilis, H. impatiens, H. arenicola, H. hilla, H. leucospilota, H. pervicax, H. pardalis,*

H. scabra, *H. nobilis*, *Stichopus chloronotus*, *S. variegatus*, *T. ananas*, and *T. anax*. Most of the 41 species of sea cucumber found near China's Xisha Islands and reported by Liao Yulin are Indo-West Pacific species. The species of deep-sea cucumber are mainly *Elasipodida*. According to Hansen (1975), there are 171 species of *Elasipodida* sea cucumbers in the world.

Sea cucumbers in the temperate zone mainly include *Stichopus Japonicus* and *Apostichopus japonicus* distributed in the Yellow Sea and the Bohai Sea areas of China, the Japanese archipelago, and the Korean Peninsula; *Paralichopus californicus* and *Paralichopus parvimensis* are distributed along the coasts of North America; *Isostichopus fuscus* and *Isostichopus badionotus* are distributed along the coasts of Latin America, the Caribbean, and Mexico; *Isostichopus fuscus* is distributed from the California peninsula to continental Ecuador; *Holothuria tubulosa* is distributed in the Mediterranean and the eastern Atlantic (Liu et al. 2019). They are mainly distributed in the northern hemisphere; while there exists a significant diversity of sea cucumbers in the temperate waters of the southern hemisphere, most of them have lower economic value.

3.2 Sea Cucumbers in China

Zhang recorded 15 species of edible sea cucumber in China in 1954 (Zhang 1954). Liao and Zhang recorded 21 species of edible sea cucumber in the book *Annelid (Polychaeta), Echinoderms and Protozoans of the Economic Fauna of China* in 1963. Fifty-nine species of sea cucumber in China were recorded in the *Atlas of Animals in China: Echinoderma* in 1964; Liao published *Fauna of China: Echinoderma: Holothurian* in 1977, which recorded 41 species of *holothurian echinoderms* in the Xisha Islands of China. Liao recorded 101 species of sea cucumber in his book *Echinoderms in South China (English edition)* coauthored with A. M. Clark in 1995. Liao recorded 147 species of sea cucumber native in China (Liao 2001), covering 21 species of edible and medicinal value, which mainly include *Apostichopus japonicus* of *Apostichopus* (*Stichopodidae*); *Stichopus chloronotus*, *Stichopus horrens* and *Stichopus variegatus* of *Stichopus* (*Stichopidae*); *Holothuria atra*, *Holothuria cinerascens*, *Holothuria edulis*, *Holothuria fuscocinerea*, *Holothuria mobili*, *Holothuria nobilis*, *Holothuria pervicax*, *Holothuria leucospilota*, *Holothuria impatiens*, and *Holothuria scabra* of *Holothuria* (*Holothuriidae*); *Bohadschia argus* and *Bohadschia marmorata* of *Holothuria* (*Holothuriidae*); *Actinopjga echinites*, *Actinopyga lecanora*, *Actinopyga mauritiana*, and *Actinopyga miliaris* of *Actinopyga* (*Holothuroidea*); and *Thelenota ananas* of *Thelenota*.

3.3 Distribution and Utilization Status of Edible Sea Cucumbers in the World

Common large edible sea cucumbers are in a thick and cylindrical shape, with parapodia on the back and ambulacral feet on the abdomen. Depending on whether there are conical prickly parapodia on the back, sea cucumbers can be classified either as *Stichopus japonicus* or *Cucumaria japonica*. *Stichopus japonicus* mainly includes *Apostichopus japonicus*, *Thelenota ananas*, *Stichopus chloronotus*, and *Stichopus variegatus* of *Stichopodidae*; the *Cucumaria japonica* mainly includes *B. marmorata*, *B. argus*, *A. lecanora*, *A. mauritiana*, *A. miliaris*, *H. atra*, *H. leucospilota*, *H. nobilis*, *H. scabra*, etc. of *Holothuriidae*, *Peniacia quadrangularis*, *Acolochirus inornatus* and *Spiny Sea Cucumber* of *Cucumariidae*, and *Acaudina leucoprocta* and *Paracaudina chinens var.* of *Molpadida*. Conand et al. (2006) believed that there are approximately 40 species of sea cucumber harvested by commercial fishery, while Toral-Granda et al. (2008) listed at least 47 species to this end. Later, Purcell et al. (2012) listed 66 species that are currently being widely exploited and utilized all over the world. According to the estimates of those who have been long engaged in sea cucumber trade, the 23 kinds of sea cucumber with the largest trade volume account for 90% of the total volume of global sea cucumber trade (Table 2).

Different species of sea cucumber vary significantly in size. Small species are only a few centimeters long, mainly including *Elasipodida*, *Dendrochirotida*, and *Leptosynapta* of *Apodida*. *Molpadida* is generally of medium size. In general, *Aspidochirotida* are large, including varieties such as *Holothuria*, *Actinopyga*, and *Stichopus*. The length of these is usually 30 to 50 cm. The largest of *Thelenota* can reach 100 cm. The longest sea cucumber is *Apodida*, with a length of 1 m or even 2 m.

3.3.1 *Apostichopus Japonicus* Selenka

Commonly known as Japanese sea cucumber (FAO), *Apostichopus japonicus Selenka* is generally about 200 mm in body length and 40 mm in diameter. Its body is cylindrical, with a bulge on the back, featuring 4–6 rows of irregularly arranged conical parapodia (meat spines) of various sizes. The body color varies significantly, with the back generally being tawny, or chestnut brown, and its ventral surface being light tawny or russet. In addition, it may also be in green, russet, purple-brown, gray-white, or pure white color.

Apostichopus japonicus is mainly distributed in Northeast Asian areas such as China, Japan, Russia, South Korea, and North Korea. However, its wild resources have declined drastically in recent decades. In the past 30 to 50 years, the biological resources of wild *Apostichopus japonicus* have dropped by 30% in Japan, 40% in South Korea, 80% in Russia, and more than 95% in China (Liao 2001). *Apostichopus japonicus* has been included in the *Red List of Threatened Species*

Table 2 Biological classification and geographical distribution of 23 sea cucumber species with the largest trade volume (Jiang et al. 2021)

Order	Family	Genus	Species	Geographical distribution
Aspidochirotida	*Stichopodidae*	*Apostichopus*	*Apostichopus japonicus* (Selenka)	Shallow waters along the North Pacific coasts, offshore waters of Japan, North Korea, and Russia; Liaoning Peninsula, Shandong Peninsula, Jiangsu, and other coastal areas in China
			Apostichopus parvimensis	The Gulf of California to Point Cape at the northern tip of California
		Stichopus	*Stichopus horrens*	The markets involving *Stichopus monotremulatus* and *Stichopus naso* are collectively referred to as *Stichopus horrens*, which is widely distributed in the Indo-Pacific region, the Red Sea, East Africa, Maldives, the East Indies, northern Australia, the Philippines, China and southern Japan, and Pacific Islands
			Stichopus badionotus	Widely distributed in the whole Caribbean, from subtropical Atlantic, Brazil, Venezuela, Colombia, Panama, Yucatan, and Mexico to southern Florida, Bahamas, South Carolina, Ascension Islands in the middle of the Atlantic, and the Gulf of Guinea on the west coast of Africa

(continued)

Table 2 (continued)

Order	Family	Genus	Species	Geographical distribution
			Stichopus variegatus	The shallow sea along the coasts of the Xisha Islands, Hainan Island, and Leizhou Peninsula in China
			Stichopus chloronotus	Xisha Islands, southern Hainan Island, and Beihai and Dongzhou Island in Guangxi, China
		Thelenota	*Thelenota ananas* (Jaeger)	Tropical Indo-West Pacific. The Red Sea, Mascarene Islands, Maldives, the East Indies, northern Australia, the Philippines, Indonesia, China, and southern Japan, as well as China's Pacific Islands; as far east as French Polynesia
			Thelenota anax (Clark)	Tropical Indo-West Pacific. The Red Sea, Mascarene Islands, Maldives, the East Indies, northern Australia, the Philippines, Indonesia, China, and southern Japan, as well as China's Pacific Islands; as far east as French Polynesia
		Parastichopus	*Parastichopus californicus*	Distributed along the east coast of the Pacific Ocean in North America, from Aleutian Islands and Alaska to the Gulf of California
			Australostichopus mollis	Throughout New Zealand and Southeast Australia
	Holothuriidae	*Bohadschia*	*Bohadschia argus* (Jaeger)	Seychelles, Tahiti, northern Australia; southern Taiwan, southern tip of

(continued)

Table 2 (continued)

Order	Family	Genus	Species	Geographical distribution
				Hainan Island, and the Xisha Islands in China
			Bohadschia marmorata (Jaeger)	Indo-West Pacific region; southern tip of Hainan Island and the Xisha Islands in China
		Actinopyga	*Actinopyga lecanora* (Jaeger)	Indo-West Pacific region, from Mauritius to the Ryukyu Islands, Tongatapu, and northern Australia; the Xisha Islands
			Actinopyga mauritiana (Quoy & Gaimard)	Indo-West Pacific region; southern Taiwan, southern Hainan Island, and the Xisha Islands in China
			Actinopyga echinites (Jaeger)	Indo-West Pacific region; Taiwan, Guangdong, Hainan Island, and the Xisha Islands in China
		Holothuria	*Holothuria mammata*	Mainly distributed in the Mediterranean region, including *Holothuria tubulosa* and *Holothuria poli*
			Holothuria hilla	Most of the Western and Central Pacific regions, to the Pitcairn Islands and eastern Central America; also found in Southeast Asia, East Africa, and the Indian Ocean
			Holothuria mexicana (Ludwig)	Widely distributed in the Florida Keys, Bahamas, Cuba, Puerto Rico, Jamaica, Barbados, Tobago, Aruba, Yucatan Peninsula, Belize, Bonaire Island,

(continued)

Table 2 (continued)

Order	Family	Genus	Species	Geographical distribution
				Venezuela, and islands of Columbia; *Holothuria floridana* is distributed in similar waters, and the two species are mixed for sale
			Holothuria nobilis (Selenka)	Indo-West Pacific region; Taiwan, Hainan Island, and the Xisha Islands in China
			Holothuria scabra (Jaeger)	From Natal to the Red Sea in the west, the Caroline Islands and Fiji Islands in the east, Japan in the north, and Australia in the south; Guangxi, Guangdong, and Hainan in China
Dendrochirotida	*Cucumariidae*	*Cucumaria*	*Cucumaria frondosa*	The North Atlantic, from the North Pole to Cape Cod, and the northern part of the United Kingdom of Great Britain and Northern Ireland (Scotland and Orkney Islands); also distributed in Iceland, Barents Sea along the coast of Russia, and Greenland
		Mensamaria	*Mensamaria intercedens* (Lampert)	Indonesia, northern Australia; Fujian and Hainan Island in China
Molpadiida	*Caudinidae*	*Acaudina*	*Acaudina leucoprocta* (Clark)	Iran, Northwest Australia; Zhoushan Islands in Zhejiang Province and Hainan Island, China

by the International Union for Conservation of Nature and is categorized as endangered (EN) (Liao et al. 2021). In order to solve the problem of this decline in numbers, the aquaculture industry of *Apostichopus japonicus* has come into being and gradually evolved into an emerging cultivation industry (Han et al. 2016). *Stichopus japonicus* is a pillar variety of sea cucumber that is artificially bred, multiplied, and artificially cultivated in China, and it is also the most expensive variety on the market. According to statistics, the breeding output *of Apostichopus japonicus* is about 200,000 tons, equivalent to more than 4000 tons of dried sea cucumber in China.

The processed products of *Apostichopus japonicus* mainly include various dried, salted and ready-to-eat products, and capsule and oral liquid products processed from its by-products. The main consumer markets of *Apostichopus japonicus* are China mainland and Southeast Asia, where a lot of Chinese people live.

3.3.2 Cucumaria Frondosa

Commonly known as orange-footed sea cucumber, northern sea cucumber, phenix sea cucumber, pumpkin (Canada), and Atlantic sea cucumber (United States and Russia). The color of *Cucumaria frondosa* is usually light or dark brown, tending to yellowish near the mouth and tentacles. Its body is cylindrical, stout, slightly curved on the back, and tapering at both ends. When touched, the body of a live *Cucumaria frondosa* becomes almost spherical.

Cucumaria frondosa is mainly distributed in the cold water along the Atlantic coast of the northern hemisphere and is commonly found on the seafloor at a depth of 30–300 m. Although its fishing history is merely 30 years, it is possible to catch *Cucumaria frondosa* mechanically by trawling. It takes 10 years for a *Cucumaria frondosa* to reach commercial grade (25 ~ 30 cm). *Cucumaria frondosa* are also in danger of depletion. For instance, the catch of *Cucumaria frondosa* in Maine, United States, reached a record high of 4362 tons in 2004, while the catch in 2018 was almost zero (Gianasi et al. 2021). The fundamental biological and ecological traits of *Cucumaria frondosa* (Gianasi et al. 2021; Nelson et al. 2012; Dvoretsky and Dvoretsky 2021) and the effects of environmental factors (Couillard et al. 2021; Sun et al. 2020) and bait (Yu et al. 2020) on its growth have been studied in depth. The research findings indicate that *Cucumaria frondosa* is a variety of sea cucumber with significant breeding potential.

Cucumaria frondosa is mainly processed through deheading, gutting, cleaning, cooking, freezing, etc. and then traded in the form of frozen products. While this process has been partially mechanized (Gianasi et al. 2021), *C. frondosa* is processed into dried or ready-to-eat sea cucumber, or eaten raw after thawing (Feng 2013).

3.3.3 Holothuria Scabra

British sandfish (FAO). The body color of *Holothuria scabra* is generally dark greenish-brown, scattered with sparse black markings, and the base of the parapodia is white. Its body length is 30 ~ 40 cm, while the largest can reach 70 cm, and the body width of *H. scabra* is about 1/4 of body length. The body color of *H. scabra* varies significantly, tending to have a dark greenish-brown dorsally mingled with a few black markings and being darker along the dorsal line but lighter on the sides and white on the ventral surface.

Holothuria scabra is mainly distributed in the Indian Ocean. In India, *H. scabra* has a fishing history of 1000 years (Feng 2013). As a result of overfishing, the natural population resources of *Holothuria scabra* are declining sharply, and the resources of *H. scabra* in the coastal areas of Guangxi and Guangdong in China have been exhausted. India promulgated measures to ban the harvesting of *Holothuria scabra* in 2001. *H. scabra* is considered an excellent species for tropical sea cucumber culture (Yang et al. 2011; Hair et al. 2022; Kurnianto and Indriana 2021; Rahantoknam et al. 2021). In 2014, Hainan University in China achieved a breakthrough in large-scale breeding technology of *Holothuria scabra* seedlings. In 2017, more than two million *Holothuria scabra* seedlings were bred, laying the foundation for artificial breeding in tropical marine areas. Several tropical countries and regions have started their *Holothuria scabra* aquaculture programs, while small-scale cultivation of *Holothuria scabra* has begun in Hainan, China (Feng 2013).

Holothuria scabra is being exploited and utilized in large quantities through manual or industrialized fishery. The major form of consumption is the dried product of its body wall, which has been developed commercially for the preparation of traditional medicinal products in Vietnam (Purcell et al. 2012).

3.3.4 Parastichopus Californicus

Giant red sea cucumber (FAO). The body of *Parastichopus californicus* is in dark red, brown, or yellow color, and it is covered with red parapodia on the back. Its belly is in light cream color. *Parastichopus californicus,* 25–40 cm in body length, is the leading sea cucumber species on the Pacific Northwest coast, and is mainly produced in California, Alaska, and Oregon in the United States.

Fishery of *Parastichopus californicus* began to be harvested in the 1970s, mainly through sailor diving or trawling (Purcell et al. 2012). In the coastal waters of California, *Parastichopus californicus* accounts for about 11.4% of the total biomass estimate (Carson 2016). Technology for the commercial cultivation of *Parastichopus californicus* has been developed, and the effects of different microalgal diets on the quality of *Parastichopus californicus* (Ren et al. 2016; Haddad et al. 2018) and the effects of stocking density and temperature on the growth rate of *P. californicus* (Ren et al. 2018) have been systematically studied, laying the foundation for its commercialization.

Parastichopus californicus is mainly exported to China in the form of frozen or salted products, and a small part is exported in the form of dried products.

3.3.5 *Isostichopus Fuscus* (Ludwig 1875)

Brown sea cucumber or giant sea cucumber (FAO). The body of *Isostichopus fuscus* has a dark brown dorsally with many stout, sharp, yellow parapodia, which are randomly arranged. Its body is 20–24 cm in length, weighing 294–497 g. Mexico and Ecuador are the major fishing countries for *Isostichopus fuscus*, but the body length and weight of *Isostichopus fuscus* from Ecuador are smaller than those from Mexico.

The major form of consumption for *Isostichopus fuscus* in Asia is the dried product of its body wall. As a result of overfishing, it has been listed twice in Annex III of the *Convention on International Trade in Endangered Species of Wild Fauna and Flora (CITES)* in 2009 and 2019; *Isostichopus fuscus* is prohibited for commercial purposes. In 2019, CITES also listed *Holothuria fuscogilva* (Cherbonnier 1980), *Holothuria nobilis*, and *Holothuria whitmaei* (Bell 1887) in the Annex II, the protection of which took effect since August 28, 2020.

3.3.6 *Thelenota Ananas* Jaeger

Commonly known as prickly redfish (FAO) or *Holothurie ananas* (FAO), *Thelenota ananas Jaeger* is one of the large-size varieties of sea cucumber. It is generally 70 cm in length, 10 cm in diameter, and 1000–3000 g in wet weight, while the length after drying is 20–25 cm. The back color of *Thelenota ananas Jaeger* is variable, from orange-red to brown or purplish red. The parapodia on its back are large and thorn-like, and the bases of every 3–11 thorns are connected, like a "plum blossom." The appearance of *T. ananas Jaeger* is also somewhat like a pineapple, so it is sometimes called the "pineapple sea cucumber."

Thelenota ananas Jaeger is exploited as one of the most valuable species in the Indo-Pacific region. The major countries and regions of *Thelenota ananas Jaeger* fishery are the Maldives, Northern Australia, the Philippines, Indonesia, and China. The fishing methods include manual fishing, semi-industrialized fishing, and industrialized fishing. The major form of consumption for *Thelenota ananas Jaeger* in Asia is its dried body wall products. In some places in the Pacific region, *Thelenota ananas Jaeger* as a whole or its intestines and/or gonads are consumed as a delicacy. It is also consumed as traditional food and as a protein supplement in hard times. The products from Singapore and Vietnam are reexported to China.

3.3.7 *Stichopus Chloronotus* Brandt

Commonly known as greenfish (FAO) or Trepang vert (FAO), *Stichopus chloronotus* Brandt is in tetragonal prism shape. Its maximum size is 350 mm in length, with most species being around 200 mm under adequate growth conditions. Along the sides of the body and the ambulacrum of its back are each two alternating rows of large conical parapodia. The body is in dark green or slightly bluish-black color, and the ends of its parapodia are orange or orange red.

Stichopus chloronotus Brandt is being developed through artisanal or semi-industrialized fishing methods. The major form of consumption for *Stichopus chloronotus* Brandt in Asia is its dried body wall products. In some Pacific islands, it is consumed as part of the traditional food. The main export markets are Singapore and Hong Kong Special Administrative Region in China.

3.3.8 *Stichopus Horrens* Selenka

Commonly known as Selenka's sea cucumber (FAO), *Stichopus horrens Selenka* is generally 200 mm in body length and about 40 mm in diameter. Its body is cylindrical, with huge conical parapodia on the back, arranged in four irregular longitudinal rows along with the two ambulacra on the back and the ventral ambulacrum.

The major form of consumption for this species in Asia is its dried body wall products. Its intestines and gonads are also consumed as traditional food. It is also used in the preparation of traditional medicinal products in Malaysia.

3.3.9 *Actinopyga Mauritiana* Quoy et Gaimard

Commonly known as surf redfish (FAO) or *Holothurie brune des brisants* (FAO), *Actinopyga mauritiana Quoy et Gaimard* is generally 200 mm in length, or 350 mm maximum. Its back end is often thicker, about 8 cm wide. It has about 25–27 large tentacles, and these are arranged in irregular inner and outer circles.

Actinopyga mauritiana Quoy et Gaimard is a kind of quality edible sea cucumber. Dried products are the major form of consumption in Asia, and its intestines and gonads can also be eaten.

4 Biological Characteristics of Sea Cucumber

4.1 *External Form*

Most sea cucumbers are cylindrical with slightly thinner ends, with dorsal and ventral sides, and with radial and left-right symmetrical structures (Fig. 1), but the thickness, shape, and size vary significantly by species. Common large edible sea cucumbers are relatively sturdy and cylindrical, with parapodia on the back and ambulacral feet on the ventral surface. The parapodia on the back have sensory functions, while the ambulacral feet on the abdomen are the attachment and locomotive organs of the sea cucumber. The mouth is at the front end of the sea cucumber's body, while the anus is at the back end. The sea cucumber has 10–30

Fig. 1 Main external structural features of sea cucumber (Xue 2015)

tentacles, usually in multiples of 5. The shape of the tentacles of sea cucumbers is an important basis for classification. Branch-shaped tentacles are mainly found in *Dendrochirotida* sea cucumbers, shielded-shaped tentacles are mainly found in *Aspidochirotida* and *Elasipodida* sea cucumbers, feather-shaped tentacles are mainly found in *Apodida* sea cucumbers, and finger-shaped tentacles are mainly found in *Dactylochirotida* sea cucumbers (Liao 2001).

The body color of sea cucumbers is mostly dull, ranging from gray or brown to green or black, but the color of ventral surface is generally lighter. The body of *Elasipodida* living in the deep sea is often in purple, reddish-brown, or violet color. The body of small *Elasipodida* is mostly in gray or flesh red color, whereas the body of large *Elasipodida* is mostly in grayish brown color. The body of *Molpadiida* is often in light brown color with red wine like brown spots. The body of *Dendrochirotida* is mainly yellow, red, or orange color.

4.2 Internal Structure

The body wall of sea cucumber is flexible and rich in connective tissue. The thickness of a sea cucumber's body wall varies by species. Most species with thick body walls can be eaten, while those with thin body walls lose their edible value. The body wall of sea cucumbers is composed of cortex, muscle layer, and coelom membrane (Chang 2004). After Van Gieson staining for body wall sections of fresh sea cucumber, it could be seen that, except for the edge of parapodia on the inner wall that was stained with yellow, the rest was stained with red, indicating that the sea cucumber body wall is mainly composed of collagen fibers and only a short amount of muscle fibrils (Fig. 2).

The endoskeleton of sea cucumber is mostly underdeveloped, and many tiny calcareous ossicles that can be seen only with a microscope are buried under the skin, about which the amount is enormous. The shape of sea cucumber ossicles is the most important basis for the classification of sea cucumbers. Common ossicles include table-shaped, cingulate-shaped, rod-shaped, perforated plate-shaped, pattern-shaped, wheel-shaped, and anchor-shaped bodies (Fig. 3).

The pharynx of sea cucumber is surrounded by a ring of calcareous bony plate, which is called the lime ring. The lime ring is an organ unique to sea cucumbers and is also an important basis for the classification of varieties of sea cucumbers.

The internal organs of sea cucumber are composed of the pharynx, esophagus, stomach, and intestines. The structures of the pharynx, esophagus, and stomach are short, but the intestines are much longer (Fig. 4). The intestines are composed of three parts, which form a loop structure that first descends, then ascends, and finally descends again. This structure is connected with the rectum and cloaca, and finally leads to the outside of the body through the anus opening. In the taxa with respiratory tree, which is connected to the cloaca, oxygenated water is brought into the body through the respiratory tree. This structure is found in all orders except *Apodida*. Ductus Cuvieri are found in some species of *Aspidochirotida*, and these are generally considered to be defensive structure. These Ductus Cuvieri are sticky tubules attached to the base of the respiratory tree, and some species of *Holothuria* and *Bohadschia* can eject such tubules through the cloaca towards the stimulus.

In contrast to other echinoderms, the reproductive system of sea cucumber consists of a single gonad or gonad (Fig. 4). The sea cucumbers are usually hermaphroditic, and there is little difference between males and females in nonsexual maturity. The gonads of sea cucumbers are located on the dorsal side, with the gonads being composed of two clusters of tubules (*Stichopus japonicus*) or one cluster of tubules (*Holothuriidae*) in *Aspidochirotida*. The gonads are attached to the dorsal mesentery through which the gonoduct or genital stolon opening passes, leading to the outside by the gonopore (genital orifice) or genital papilla. In most species, mature gametes of *Aspidochirotida* are released freely into seawater. Through the observation of many species of *Aspidochirotida*, it has been found that sea cucumbers maintain an upright posture while spawning, with the bodies of

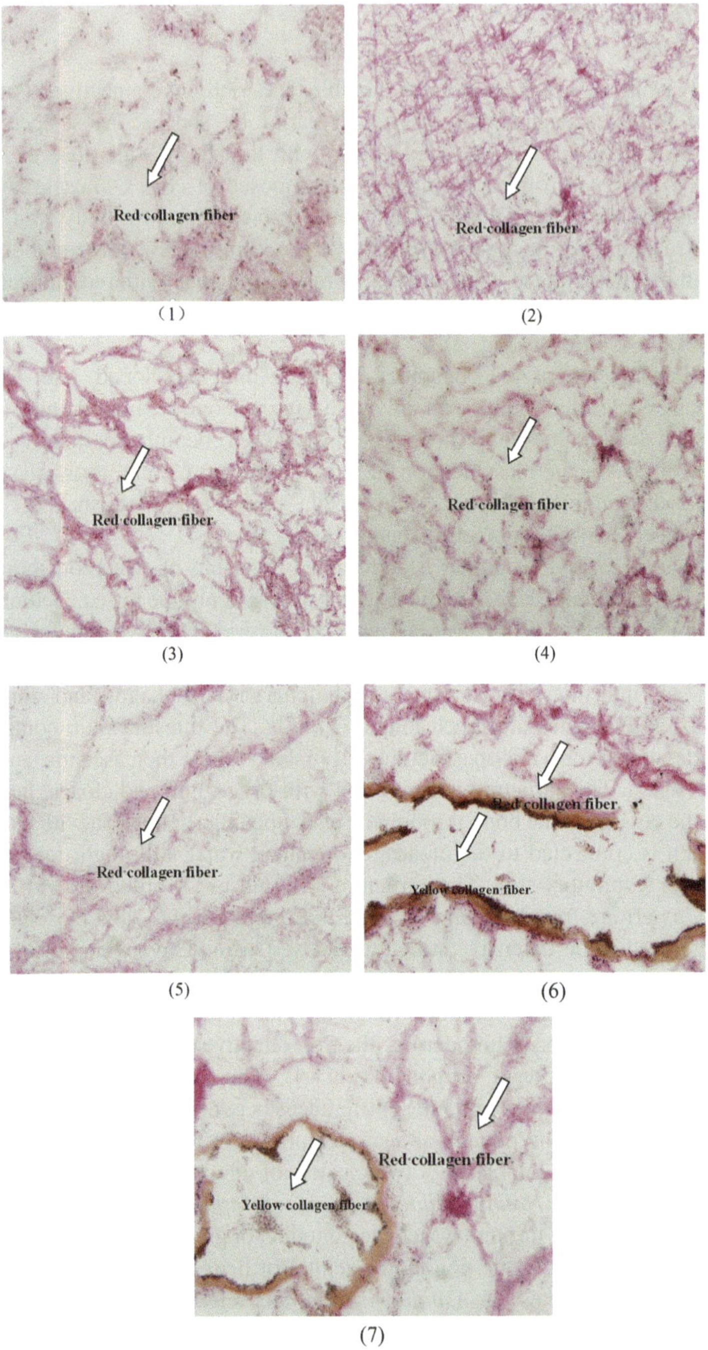

Fig. 2 Light microscope pictures of different parts of fresh *Apostichopus japonicus* (Xue 2015) (**a**) Epidermis, longitudinal section. (**b**) Inner wall, longitudinal section. (**c**) Inner wall, transverse section. (**d**) Epidermis parapodium, transverse section. (**e**) Epidermal parapodium, longitudinal section. (**f**) Inner wall parapodium, longitudinal section. (**g**) Inner wall parapodium, transverse section

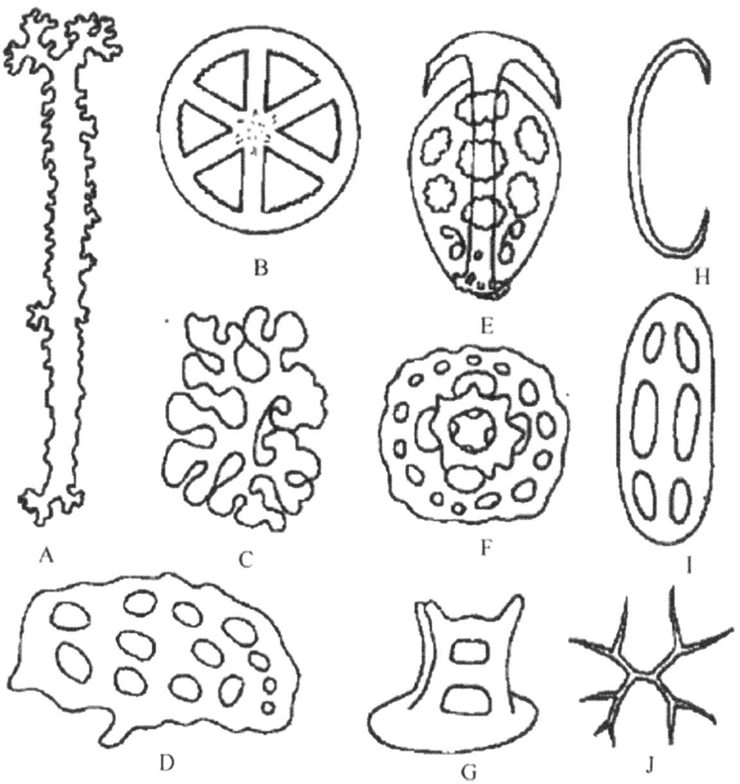

Fig. 3 Basic types of sea cucumber ossicles (Xue 2015). (**a**). Rod body. (**b**). Wheel body. (**c**). Pattern body. (**d**). Perforated plate body. (**e**). Anchor and anchor plate bodies. (**f**). The top of the table body. (**g**). The side of the table body. (**h**). C-shaped body. (**i**). Cingulate body. (**j**). X-shaped body

male and female sea cucumbers swaying back and forth at the same time, eventually releasing gametes.

The water vascular system of sea cucumbers is a body cavity space surrounded by body cavity epithelial cells (Fig. 4). It consists of the lumens of buccal tentacles and of ambulacral feet, and a water ring surrounds the esophagus, the radial canals, the madreporic canal, and the Polian vesicles. The perivisceral coelom of sea cucumber is a large coelom that contains watery proteinaceous coelomic fluid and cells of different morphologies (coelomocytes). The haemal system is well developed, composing of a large number of blood vessels distributed along the digestive tract, sinus, and lacunae. The blood vessels connected to the digestive tract form a complex meshwork, also known as the rete mirabile. The rete mirabile is closely associated with the left breathing tree, realizing different transmission functions for nutrients and gases.

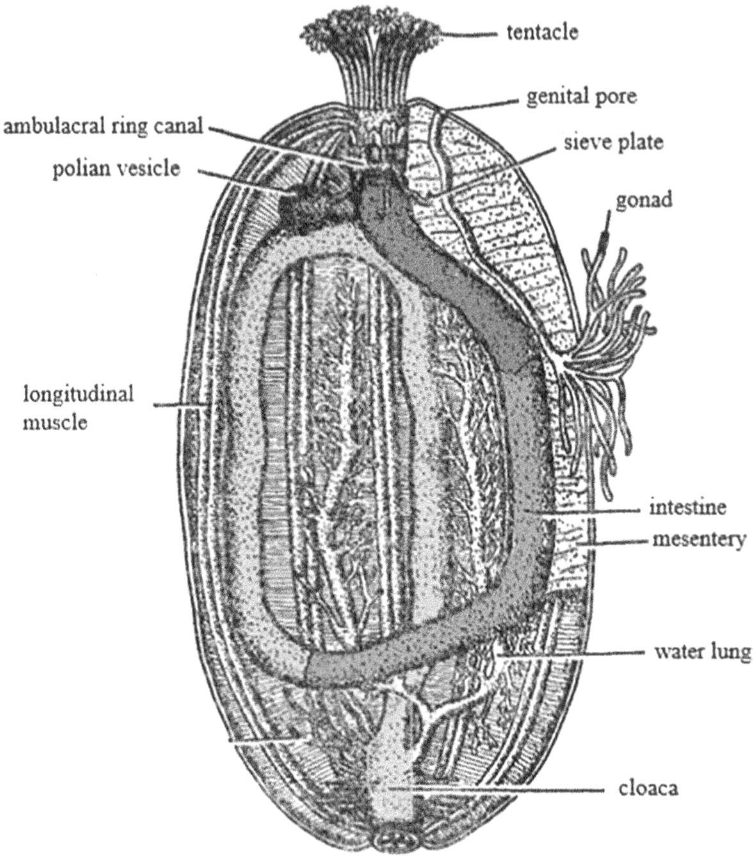

Fig. 4 Anatomical drawing of sea cucumber (Xue 2015)

5 Conclusions

This chapter reviews the main species of sea cucumber and introduces the distribution and utilization status of major commercial edible sea cucumbers in the world. Sea cucumbers with important commercial value are distributed all over the world. At present, there are about 900 species recorded globally. However, some restrictions on the harvesting of sea cucumbers have been imposed as a result of overfishing, such as in the case of *Isostichopus fuscus*. In addition, the description for ecological characteristics of sea cucumbers will help to distinguish the origins of sea cucumbers.

References

Bell FJ (1887) Studies in the Holothuroidea - VI. Descriptions of new species. Proc Zool Soc London:531–534

Bordbar S, Anwar F, Saari N (2011) High-value components and bioactives from sea cucumbers for functional foods—a review. Mar Drugs 9(10):1761–1805

Carson HS, Ulrich M, Lowry D, Pacunski RE, Sizemore R (2016) Status of the California sea cucumber (*Parastichopus californicus*) and red sea urchin (*Mesocentrotus franciscanus*) commercial dive fisheries in the San Juan Islands, Washington State, USA. Fish Res 179:179–190

Chang YQ (2004) Biological research and breeding of sea cucumber and sea urchin. China Ocean Press, Qingdao

Cherbonnier G (1980) Holothuries de Novelle-Calédonie. Bull Mus natn Hist nat 3:615–667

Clark AM, Rowe F (1971) Monograph of Shallow-Water Indo-West Pacific Echinoderms. British Museum Natural History, London

Conand C, Byrne M (1993) A review of recent developments in the world sea cucumber fisheries. Mar Fish Rev 55(4):1–13

Conand C, Muthiga N, Aumeeruddy R, Torre Castro ML, Frouin P, Mgaya Y, Mirault E, Ochiewo J, Rasolofonirina R (2006) A three-year project on sea cu-cumbers in the southwestern Indian Ocean: National and regional anal-yses to improve management. Beche-de-mer Infor Bull 23:11–15

Couillard CM, Maltais D, Rénald B (2021) Morphometrics and processing yield of *Cucumaria frondosa* (*Holothuroidea*) from the St. Lawrence Estuary, Canada. PLoS One 16(1):e0245238

Dvoretsky AG, Dvoretsky VG (2021) Cucumaria in Russian waters of the Barents Sea: Biological aspects and aquaculture potential. Front Mater Sci 8:613453

Feng YP (2013) Word important commercial sea cucumber species in China. Ocean University of China Press, Qingdao

Gianasi BL, Hamel JF, Montgomery EM, Sun J, Mercier A (2021) Current knowledge on the biology, ecology, and commercial exploitation of the sea cucumber *Cucumaria frondosa*. Rev Fish Sci Aquac 29(4):582–653

Guo Y, Ding Y, Xu F (2015) Systems pharmacology-based drug discovery for marine resources: An example using sea cucumber (*Holothurians*). J Ethnopharmacol 165:61–72

Haddad CA, Cross SF, Da Vies HL, Duff SD, Fortune AC, Pearce CM (2018) Assessing natural substrates for restricting movement of California sea cucumbers (*Parastichopus californicus*). N Am J Aquacult 80:347–354

Hair C, Militz TA, Daniels N, Southgate PC (2022) Performance of a trial sea ranch for the commercial sea cucumber, *Holothuria scabra*, in Papua New Guinea. Aquaculture 547:737500

Hansen B (1975) Systematic and biology of the deep-sea Holothurians. Part 1. Elasipoda. Galathea Reports. Scientific Results of the Danish Deep-Sea Expedition Round World (1950–1952). 13:1–262

Han QX, Keesing JK, Liu DY (2016) A review of sea cucumber aquaculture, ranching, and stock enhancement in China. Rev Fish Sci Aquac 24(4):326–341

Jiang BX, Zhang XM, He XX, Zhang JK, Zhang HW, Li ZJ, Xue CH, Feng TY, Jiang XM (2021) Recent advances in analytical techniques for the traceability of commercial sea cucumber. Food Sci 42(13):309–318

Kurnianto D, Indriana LF (2021) Investment feasibility study of grow out sea cucumber Holothuria scabra cultured in East Lombok. IOP Conf Ser: Earth Environ Sci 860(1):012054

Li Y, Fei LH, Chen JX (2008) The morphology of ossicles of 15 commercial holothurians. Period Ocean Univ China 8(2):211–216

Liao MJ, Wang YG, Li B, Rong XJ, Fan RT, Liu RY, Liu QB (2021) Present situation, existing problems and countermeasures of sea cucumber breeding industry in China. Sci Fish Cult:24–27

Liao YL (1997) Chinese fauna *Phylum Echinodermata* sea cucumber class. Science Press, Beijing

Liao YL (2001) Sea cucumber in China. Biol Bull 35(9):1–3

Liu SN, Cao R, Zhao L, Sun HH, Liu Q (2019) Analysis on nutrients of Iso-stichopus fuscus. J Food Saf Qual 10(08):2263–2267

Ludwig HL (1875) Beiträge zur Kenntniss der Holothurien. Arbeiten aus dem Zoologischen zootom. Institut in Würzburg 2(2):77–120

Nelson EJ, MacDonald BA, Robinson SMC (2012) A Review of the Northern Sea Cucumber Cucumaria frondosa (Gunnerus, 1767) as a Potential Aquaculture Species. Rev Fish Sci 20 (4):212–219

Pawson DL, Vance DJ, Ahearn C (2001) Western Atlantic sea cucumbers of the order Molpadiida (Echinodermata: Holothuroidea). P Biol Soc Wash 10:311–327

Purcell S, Samyn Y, Conand C (2012) Commercially important sea cucumbers of the world. FAO Species Catalogue for Fishery Purposes No. 6, Rome

Rahantoknam SPT, Beruatjaan MY, Rahantoknam MA, Ohoiwutun MK (2021) Environmental factors for *holothuria scabra* sea cucumber cultivation. IOP Conf Ser: Earth Environ Sci 755(1): 012058

Ren Y, Liu W, Dong S, Pearce CM (2016) Effects of mono-species and bi-species microalgal diets on the growth, survival and metamorphosis of *auricularia* larvae of the California sea cucumber, *parastichopus californicus* (Stimpson 1857). Aquac Nutr 22:304–31422(2)

Ren Y, Liu W, Pearce CM (2018) Effects of stocking density, ration and temperature on growth, survival and metamorphosis of *auricularia* larvae of the California sea cucumber, *Parastichopus californicus*. Aquac Res 49:517–525

Sun J, Hamel JF, Gianasi BL, Graham M, Mercier A (2020) Growth, health, and biochemical composition of the sea cucumber *Cucumaria frondosa* after multi-year holding in effluent waters of land-based salmon culture. Aquacult Environ Interact 12:139–151

Toral-Granda V, Lovatelli A, Vasconcellos M (2008) Sea cucumbers. A global review on fishery and trade. Beche-de-mer Infor Bull 27:4–6

Xu WD (2009) A review of sea cucumber fishery and aquaculture abroad. J Dalian Fish Univ 4: 172–174

Xue CH (2015) Intensive processing of sea cucumber theory and technology. Science Press, Beijing

Yang XM, Zhang L, Li YN (2011) Primary test on artificial spawning induction and larvae rearing of *Holothuria scabra*. South China Fish Sci 7(1):39–44

Yu Z, Robinson S, Macdonald B, Lander T, Smith C (2020) Effect of diets on the feeding behavior and physiological properties of suspension-feeding sea cucumber *cucumaria frondosa*. J Oceanol Limnol 38(3):883–893

Zhang FY (1954) Edible sea cucumber in the Xisha Islands of China. Biol Bull 06:21–22

Nutritional Components of Sea Cucumber and the Biochemical Characteristics of Autolytic Enzymes

Dayong Zhou, Yanchao Wang, Jie Xu, Sang Moo Kim, and Yaoguang Chang

Abstract The sea cucumber is a valuable marine resource with high protein and low cholesterol contents. It is considered as a fine food supplement and an ideal tonic for strengthening the body in China and Southeast Asia. The protein content of dried sea cucumber is generally more than 50%, mainly in the form of collagen; although the lipid content is low, it mainly consists of phospholipids rich in EPA/DHA, while it is also rich in sphingolipids such as cerebroside and ganglioside with outstanding functional activity. The sea cucumber is rich in sulfated polysaccharides, with a content of up to more than 10%, mainly including fucosylated chondroitin sulfate and fucoidan. However, due to the presence of an autolytic enzyme system, the sea cucumber is prone to autolysis during processing and storage, resulting in the loss of nutrients and serious deterioration of quality, and subsequent severe economic losses. This chapter focuses on the three main nutritional components of sea cucumber collagen, sea cucumber lipid, and sea cucumber sulfate polysaccharide and discusses the research progress of such main nutritional components of sea cucumber from the perspective of content distribution, structural type, and biochemical characteristics. At the same time, the research progress of sea cucumber autolytic enzymes is introduced from the perspective of the enzymatic properties of serine

D. Zhou
School of Food Science and Technology, Dalian Polytechnic University, Dalian, China

National Engineering Research Center of Seafood, Dalian, China

Y. Wang · J. Xu
College of Food Science and Engineering, Ocean University of China, Qingdao, R.P. China

S. M. Kim
Department of Marine Food Science and Technology, Gangneung-Wonju National University, Gangneung, South Korea

Y. Chang (✉)
College of Food Science and Engineering, Ocean University of China, Qingdao, R.P. China

Laboratory for Marine Drugs and Bioproducts, Pilot National Laboratory for Marine Science and Technology (Qingdao), Qingdao, China
e-mail: changyg@ouc.edu.cn

C. Xue (ed.), *Advances in Sea Cucumber Processing Technology and Product Development*, Advances in Marine Bioprocesses and Bioproducts,
https://doi.org/10.1007/978-3-031-16512-2_2

protease, cysteine protease, matrix metalloproteinase, and cathepsin and their influence on sea cucumber quality.

Keywords Sea cucumber · Collagen · Sulfated polysaccharide · Lipid · Autolytic enzyme

1 Introduction

The sea cucumber is considered to be an animal food without cholesterol, which is rich in a variety of nutrients and functional components. The main nutritional components of sea cucumber include protein, sulfated polysaccharides, and active lipids. Dried sea cucumber has the highest content of protein, which is more than 50%. The sea cucumber is marine benthos, which is distributed in various sea areas around the world, mainly in tropical and temperate regions. There are many kinds of sea cucumber resources in the tropical area, accounting for about 86% of the total output; the resources of sea cucumbers in the temperate zones are limited. At present, more than 1500 kinds of sea cucumber have been found in the world, of which approximately 100 are available for human consumption. There are abundant sea cucumber resources in the eastern waters of China, with more than 140 kinds of sea cucumber present, of which around 20 have high economic value. Among them, *Stichopus japonicus* is the most precious. It is found that the main nutritional composition and biochemical characteristics of sea cucumbers are related to species of sea cucumber, age, origin, production mode, harvest season, and some other factors.

The sea cucumber has a very strong capacity for autolysis, and it shows autolysis phenomena such as the softening of the body wall under the action of the autolytic enzyme system very easily. The sea cucumber is rich in autolytic enzymes. At present, a variety of endogenous enzymes have been extracted and purified from sea cucumber body walls, intestines, or viscera, and their enzymatic properties and effects on sea cucumber quality have been studied to varying degrees. Because *Stichopus japonicus* is the species with the highest economic value at present, *Stichopus japonicus* is generally the most studied on the structural composition and biochemical characteristics of the main nutrients of sea cucumbers. This chapter will focus on the progress on research into the main nutrients of sea cucumbers and the biochemical characteristics of autolytic enzymes.

2 Biochemical Characteristics of Main Nutrients in Sea Cucumber

2.1 Sea Cucumber Collagens

Collagen, comprising at least one triple-helix domain for its molecules, is the major structural protein of sea cucumber body wall. Collagen is the basic unit for the formation of collagen fibrils, which further aggregate to form collagen fibers. There is a repetitive sequence Gly-X-Y in collagen molecules, where Gly represents glycine, X is often proline, and Y is often hydroxyproline or hydroxylysine, which is the basic characteristic of collagen molecules.

2.1.1 Contents of Sea Cucumber Collagen

Protein is the most abundant component in sea cucumber body wall on a dry weight basis, and the protein contents of different sea cucumber species show certain interspecific specificity. The sea cucumber (*Stichopus japonicus* and *Holothuria scabra*) body walls have relatively higher content of proteins, while the protein content of sea cucumber (*Holothuria*) is relatively lower, only about 50% (Table 1), which has also been substantiated by the research of Wang et al. (2010). Identical sea cucumber species also show some seasonal differences in protein content (Li et al. 2006a, b). Collagen, accounting for more than 70% of the total proteins, is the primary protein in sea cucumber's body walls (Saito et al. 2002; Xue 2006). Sea cucumber collagens aggregate into collagen fibers through dense cross-linking, and these play an important role in the physiological activities of sea cucumbers (Trotter et al. 1995). In addition, the quality of processed sea cucumber products could be affected by the biochemical characteristics of collagen, which is one of the most important functional components influencing sea cucumber processing and storage.

Table 1 Total protein contents in different sea cucumbers (%, on a dry weight basis) (Dong 2008)

Sea cucumber species	Total protein content (%)
Stichopus japonicus (Russia)	74.24
Stichopus japonicus (Japan)	75.03
Stichopus japonicus (Australia)	65.70
Thelenota ananas	55.42
Holothuria nobilis	56.24
Holothuria nobilis	46.50
Holothuria scabra	74.86
Actinopyga miliaris	62.67
Actinopyga echinites	63.04
Pearsonothria graeffei	52.02
Holothuria mexicana	53.76

Table 2 Protein and collagen contents in the sea cucumber (*Stichopus japonicus*) body wall from different production regions and harvesting seasons

Region	Total protein content (%)		Collagen content (%)	
	Spring	Autumn	Spring	Autumn
Dalian	38.80	51.20	21.60	22.70
Yantai	46.9	52.4	23.60	25.20
Qingdao	44.3	49.9	22.50	19.60
Weihai	41.3	47.4	19.20	23.20

Protein contents in sea cucumber body wall depend on the production region and harvesting season, and the harvesting season shows a greater impact on the total protein content than the production region does. For example, Feng et al. (2021) reported protein contents in sea cucumber (*Stichopus japonicus*) body wall from four different production regions and seasons and demonstrated that *Stichopus japonicus* in the autumn from Yantai showed the highest protein content while *Stichopus japonicus* in the spring from Dalian possessed the lowest protein content. For sea cucumbers from the same production region, the protein content of those in the spring was much lower than that of those in the autumn (Table 2). The collagen contents in sea cucumber body wall also fluctuate by production region and harvesting season. The collagen contents of sea cucumber in the autumn from Yantai are at relatively higher levels, while those in the spring from Weihai are relatively lower (Table 2). The contents of certain collagen molecules (A0A1I9W676, A0A2G8LKB1, A0A2G8LKB8, A0A2G8LKB9, and A0A2G8LLA5) stayed at relatively higher levels in the sea cucumber (*Stichopus japonicus*) from Xiamen, compared with those from Dalian, Yantai, Qingdao, and Weihai (Feng et al. 2020). Therefore, exogenous environmental factors (e.g., temperature, salinity, etc.) can modulate the collagen content in sea cucumber body wall.

2.1.2 Distribution of Sea Cucumber Collagen

Isemura et al. (1973) identified some glycopeptides rich in hydroxylysine from the sea cucumber (*Holothuria forskali*) body wall, and glycosylated hydroxylysine was located at the Y position of the Gly-X-Y tripeptide and distributed throughout the peptide chain, which is different from the collagens in the vertebrate skin. Trotter et al. (1995) isolated intact collagen fibrils from the sea cucumber (*Cucumaria frondosa*) body wall and found that the collagen fibrils were covalently linked to glycosaminoglycans. Wang et al. (2018) isolated intact collagen fibrils from the sea cucumber (*Apostichopus japonicus*) body wall and illustrated that the collagen fibrils were covalently linked to fucosylated chondroitin sulfate via O-glycosidic bond.

Sea cucumber collagen fibrils showed a typical fibrillar morphology and a striated pattern of light-dark alternation under transmission electron microscopy (TEM) through phosphotungstic acid negative staining or uranyl acetate positive staining. The repeat period (D) of collagen fibrils was 69.6 ± 0.5 nm ($n = 50$), while the gap region (G) was 36.9 ± 0.5 nm ($n = 50$). The gap and overlap (O) regions accounted for 53.1% and 46.9% of the repeat cycles, respectively. Chondroitin sulfate particles were specifically distributed in the gap regions of the collagen fibrils and appeared roughly globular or elliptical with a diameter of 19.2 ± 2.6 nm ($n = 50$; based on positively stained image measurement). There were 4 ~ 6 stained particles in each gap region, indicating the presence of approximately 10 chondroitin sulfate chains or aggregates in one repeat period of collagen fibrils. The chondroitin sulfate surrounded the fibrils circumferentially and coaxially with a repeat period that matched the periodicity of the fibrils, with the shape resembling a pearl bracelet (Wang et al. 2018).

2.1.3 Structural Types of Sea Cucumber Collagen

Currently, research on the structural types of sea cucumber collagen mainly focuses on pepsin-solubilized collagen (PSC). Several researchers have investigated into the collagen composition in different sea cucumber species (Table 3). There are two types of conclusions on the subunit composition of sea cucumber collagens, including heterotrimers composed of $(\alpha 1)_2\alpha 2$ and homotrimers composed of $(\alpha 1)_3$. Wang

Table 3 Research on the pepsin-solubilized collagens from sea cucumber body wall

Species	Subunit	Structural type of sea cucumber collagen	References
Cucumaria frondosa	$(\alpha 1)_3$	Collagen composition was $(\alpha 1)_3$, amino acid composition similar to that of type I collagen	Trotter et al. (1995)
Stichopus japonicus	$(\alpha 1)_2\alpha 2$	Collagen molecular composition was $(\alpha 1)_2\alpha 2$ upon enzymolysis, rich in glutamic acid compared to the other collagens and identical to calf type I collagen	Saito et al. (2002)
Stichopus japonicus	$(\alpha 1)_3$	Collagen fibrils was type I collagen, with the molecular composition of $(\alpha 1)_3$	Cui et al. (2007)
Parastichopus californicus	$(\alpha 1)_3$	Collagens in the epidermis and connective tissue were both type I collagen, with the molecular composition of $(\alpha 1)_3$	Liu et al. (2010)
Stichopus japonicas	$(\alpha 1)_3$	Pepsin-solubilized collagen appeared as a single band in SDS-PAGE, which was designated α chain	Dong et al. (2011)
Bohadschia spp.	$(\alpha 1)_3$	Pepsin-solubilized collagen was type I collagen, with the molecular composition of $(\alpha 1)_3$	Siddiqui et al. (2013)
Stichopus vastus	$(\alpha 1)_3$	Collagen was composed of 3α1 subunits and belonged to type I collagen, with a molecular weight of 122 kDa	Abedin et al. (2013)
Holothuria parva	$(\alpha 1)_3$	Pepsin-solubilized collagen was composed of 3α1 subunits, with a molecular weight of 130 kDa, and belonged to type I collagen similar to that of calfskin	Adibzadeh et al. (2014)
Stichopus monotuberculatus	$(\alpha 1)_3$	Pepsin-solubilized collagen was $(\alpha 1)_3$, with a molecular weight of 137 kDa and characteristics and good thermal stability similar to that of type I collagen	Zhong et al. (2015)

Table 4 Five collagens identified in the collagen fibrils of sea cucumber (*Apostichopus japonicus*)

Sequence	NCBI annotation
PIK62545.1	Putative collagen α-1(V) chain isoform X2, partial
PIK60696.1	α-2 collagen
PIK60691.1	Putative collagen α-1(I) chain
PIK55424.1	Putative collagen α-2(IX) chain
PIK55422.1	Putative collagen α-1(IX) chain, partial

et al. (2020a) identified 17 collagens from the sea cucumber (*Apostichopus japonicus*) body wall and found that the collagens were distributed throughout the body wall employing the proteomics technique. Tian et al. (2020) isolated collagen fibrils from the sea cucumber (*Apostichopus japonicus*) body wall and identified five collagen molecules by proteomics analysis (Table 4).

Using the most primitive sponge fibrillar collagen sequence as a root, a phylogenetic tree is constructed utilizing all known fibrillar collagens from the SWISS-PROT database, fibrillar collagens from sea urchin, and five collagen sequences identified for the expression in the sea cucumber (*Apostichopus japonicus*) body wall. The results showed that the 1α, 5α, and 2α chains of the sea urchin are in clade A, and the 6α chain of the sea urchin is in clade B. Sea cucumber collagens PIK60696.1 and PIK60691.1 belong to the clade A fibrillar collagens, and sea cucumber collagen PIK62545.1 belongs to the clade B fibrillar collagens, which are all homologous to the fibrillar collagens of sea urchins. Since the sea cucumber and sea urchin both belong to the echinoderm phylum with a close phylogenetic relationship, the positions of sea cucumber collagen sequences and sea urchin sequences in the evolutionary tree are very close to each other. However, sea cucumber collagen sequences are not included in typical collagen branches. Compared with vertebrate collagens, invertebrate collagens are relatively primitive and generally cannot be classified into the typical collagen types, while typical collagens are defined in the context of vertebrate collagens (Exposito et al. 2010). Sea cucumber collagens PIK55424.1 and PIK55422.1 are not included in the phylogenetic tree of fibrillar collagen sequences and are recorded in the NCBI database as type IX. Therefore, the sea cucumber collagen fibrils consist of multiple collagens of different types, demonstrating a mixture of types I-like, V-like, and IX-like collagens (Tian et al. 2020).

2.1.4 Biochemical Characteristics of Sea Cucumber Collagen

For the amino acid composition of sea cucumber (*Apostichopus japonicus*) pepsin-solubilized collagens, glycine accounts for approximately 30% of the total amino acid residues, and the contents of hydroxyproline and proline are about 16%, with a ratio of 0.69. The pepsin-solubilized collagens are comprised of high contents of aspartic acid, glutamic acid, glycine, alanine, arginine, proline, and hydroxyproline; low contents of aromatic amino acids and histidine; and zero content of cystine (Cui

2007). The isoelectric points of sea cucumber collagens are generally in the acidic range. For instance, the pepsin-solubilized collagens of *Stichopus japonicus*, *Stichopus vastus*, and *Stichopus monotuberculatus* show isoelectric points of 4.14, 4.67, and 4.00, respectively (Zhu et al. 2012; Abedin et al. 2014; Zhong et al. 2015).

Compared with vertebrate collagens, sea cucumber collagens are more insoluble, and almost insoluble in 0.5 M acetic acid. However, they can be rapidly dissolved after enzymatic treatment with pepsin, reflecting the typical features of collagens from echinoderms (Kittiphattanabawon et al. 2005). The glass transition temperature of sea cucumber collagens depends on the source and the extraction and separation method. Abedin et al. (2013) isolated pepsin-solubilized collagens from sea cucumber (*Stichopus vastus*), and the denaturation temperature was 21.23 °C. Liu et al. (2010) isolated pepsin-solubilized collagens from sea cucumber (*Parastichopus californicus*) epidermis and connective tissue with denaturation temperatures of 18.5 °C and 17.9 °C, respectively. Compared with vertebrate collagens, sea cucumber collagens generally demonstrate lower hydroxyproline content and thermal denaturation temperature.

2.2 *Sea Cucumber Lipids*

The lipid with a relatively lower content in the sea cucumber is an important active component. Active lipids in sea cucumbers mainly include phospholipids, sphingolipids (cerebrosides, gangliosides, ceramides), and sterol sulfates. Phospholipids account for approximately one-third of the total lipid content in sea cucumbers. Sea cucumbers are abundant in ether phospholipids, especially plasmalogens. The most dominant polyunsaturated fatty acids esterified to phospholipids are C20:5 and C20:4. Cerebrosides are the most abundant sphingolipid components in sea cucumbers; these are mainly glucosyl cerebrosides (GlcCer). A typical predominant long-chain bases (LCB) is 2-amino-1,3-dihydroxy-4-heptadecene (d17:1). The occurrence of 2-hydroxy-tricosenoic acid (C23:1 h) is a characteristic of sea cucumber cerebrosides. Sea cucumber gangliosides include mono-, di-, and tri-sialylated gangliosides. Sea cucumber-derived sterols are mainly composed of two kinds of sulfated sterols, namely, cholest-5-en-3β-yl hydrogen sulfate and 24-methylene-cholest-5-en-3β-ol-3-sulfate.

2.2.1 Fatty Acid Composition Analysis of Total Lipids from Sea Cucumber

The lipid contents in sea cucumbers and the lipid compositions both vary sharply by origin (Liu 2014a, b; Luo et al. 2013). In general, the lipid contents of sea cucumbers stay low and are dominated by phospholipids, which account for about one-third of the total lipid contents. Saturated fatty acids in the body wall of sea cucumbers account for 9.29%–24.98%, dominated by C16:0 and C18:0. Mono-unsaturated fatty acids account for 23.03%–34.25%, mainly, C16:1, C18:1, and C20:1. Sea cucumbers also have a high content of C23:1, which is a characteristic fatty acid of sea cucumbers (Lou 2011). The content of polyunsaturated fatty acids accounts for 27.29%–43.38%, with C20:4n-6 and C20:5n-3 predominating and the content of C22:2n-6 relatively lower. Among them, C20:4n-6 (AA), C20:5n-3 (EPA), and C22:6n-3 (DHA) account for 73.1%–84.8% of the total PUFA (Liu 2014a, b). The ability to enrich C20:4n-6 and C20:5n-3 is considered to be an essential characteristic of marine echinoderms (Takagi et al. 1980). Studies have shown that tropical sea cucumbers, such as *Isostichopus badionotus*, *Halodeima atra*, *Isostichopus fuscus*, and *Holothuria edulis*, contain much higher levels of C20:4n-6 than of C20:5n-3. However, boreal sea cucumbers, such as *Cucumaria frondosa*, *Stichopus tremulus*, and *Patallus mollis*, contain much higher levels of C20:5n-3 than of C20:4n-6, suggesting that tropical sea cucumbers enrich C20:4n-6 more easily, whereas boreal sea cucumbers enrich C20:5n-3 more easily (Luo et al. 2013).

2.2.2 Sea Cucumber Phospholipids

Sea cucumber phospholipids mainly refer to glycerophosphate (PG), which can be divided into phosphatidylcholine (PC), phosphatidylethanolamine (PE), phosphatidylserine (PS), phosphatidylinositol (PI), bisphosphatidylglycerol (PG), and glycerophosphatidic acid (PA) by substituent group (X). PC and PE are the main components of sea cucumber phospholipids, which account for more than 80% of the total phospholipid contents of sea cucumbers. Among them, 1-alkyl-2-acyl-PC (PC-O) accounts for 59%–83% of the total PC content, and 1-alkenyl-2-acyl-PE (PE-P) accounts for 83%–98% of the total PE content. Phospholipids isolated from sea cucumbers contain high percentages of polyunsaturated fatty acids (PUFA), particularly arachidonic acid (AA, 20:4, n-6) and eicosapentaenoic acid (EPA, 20:5, n-3). The PUFA contents of PC-O in sea cucumbers range from 69.6% to 83.9%, and *Bohadschia marmorata* has the highest amount of PE-P molecular species, containing 20:4 and 20:5 at approximately 90% (Wang et al. 2020a, b). In addition, there are significant differences in the content (total contents, proportion) of EPA and DHA in the phospholipids of the body walls of sea cucumbers by origin, with the total amount of EPA and DHA ranging from 14.00% to 25.20%. Among them, the EPA content is 10.13%–21.94%, and the percentage of DHA is 3.26%–4.27% (Liu 2014a, b).

2.2.3 Sea Cucumber Sphingolipids

2.2.3.1 Sea Cucumber Cerebrosides

Cerebroside (Cer) is composed of ceramide and glycosyl, which can be divided into glucose cerebroside (GlcCer) and galactose cerebroside (GalCer) by glycosyl group (Chen et al. 2009). Ryuichi Higuchi's research team isolated and identified the cerebrosides in six species of fresh sea cucumbers, namely, *Cucumaria echinata* (Higuchi et al. 1994a, b), *Pentacta australis* (Higuchi et al. 1994a, b), *Holothuria pervicax* (Yamada et al. 2002), *Holothuria leucospilota* (Yamada et al. 2005), *Stichopus japonicas* (Kisa et al. 2005), and *Bohadschia argus* (Ikeda et al. 2009). Structural analysis showed that the cerebrosides in all the five species of sea cucumbers are β-GlcCer, except for the sea cucumber *Bohadschia argus,* which contains a small amount of β-GalCer. The fatty acid moieties of the sea cucumber cerebrosides are usually saturated (mostly C18:0, C20:0, C22:0, and C24:0), mono-unsaturated (mostly C20:1, C22:1, and C24:1), and α-hydroxyl fatty acids (mostly C22:0 h, C23:0 h, C23:1 h, and C24:1 h), while the structures of long-chain bases are dihydroxy (d17:1, d18:2, and d18:1) and trihydroxy (t17:0 and t16:0) (Xu et al. 2013; Jia et al. 2015; Ikeda et al. 2009). Xu et al. showed that the sphingolipid contents of sea cucumbers (*Isostichopus badionotus* and *Cucumaria frondosa*) in the boreal sea are higher than in the tropical sea. Among them, *Cucumaria frondosa* has the highest sphingolipid content (4.17 ± 0.11 mg/g); *Apostichopus japonicus* has the lowest sphingolipid content (0.35 ± 0.02 mg/g) (Xu 2011).

2.2.3.2 Sea Cucumber Gangliosides

Gangliosides (GLS) are glycosphingolipids containing mono- or poly-sialylated glycans attached to ceramide (Cer). The gangliosides of *Holothuria pervicax, Holothuria leucospilota, Cucumaria echinata, Stichopus japonicus,* and *Stichopus chloronotus* have been isolated, extracted, and structurally characterized. The purified gangliosides include mono-, di-, and tri-sialylated gangliosides as well as gangliosides CEG-3, CEG-4, CEG-5, and HLG-3 containing fucose at the terminals, most of which are new compounds (Kaneko et al. 1999, 2003; Kisa et al. 2006; Yamada et al. 1998, 2000, 2003). Wang et al. (2021) showed that the total GLS amounts in sea cucumbers are 200.1–1592.7 μg/g. GD4 is the major GLS in the sea cucumber samples, which is 27–67% of the total GLSs. GM4 is present in the sea cucumber samples at high proportions of 6–38% of the total GLSs (48.2–278.6 μg/g). Likewise, the total GD4(1S) level in sea cucumbers is high (2–20% of total GLSs). The basic sea cucumber GLSs are elucidated as NeuGc2–6Glc1-1Cer. Neu5Gc, sulfated and fucosylated Neu5Gc are prevalently in sea cucumber GLSs. The structural characteristics of gangliosides from different sea cucumber species are listed in Table 5.

2.2.3.3 Sea Cucumber Ceramides

Ceramide is the basic structural unit of sphingolipids and is the product of sphingosine aminoacylation at 2-position. The carbon chains of ceramide in *Cucumaria frondosa* are 20–24 in length, and the long-chain bases are 17–19 in length. C24:1 is the major fatty acid composition, and d18:2 and d18:3 are the main long-chain base components. The long-chain bases of ceramide in sea cucumber *Acaudina*

Table 5 Structural characteristics of gangliosides in sea cucumber

Sea cucumber species	GLS structure	GLS name	Ref.
H. pervicax	NeuAcα2–6Glcβ1-1ceramide	HPG-8	Yamada et al. (1998)
	Fucα1-8NeuGcα2–4NeuGcα2-6Glcβ1-1ceramide	HPG-1	
	NeuGcα2–4NeuAcα2–6Glcβ1-1ceramide	HPG-3	
	Fucβ1-4NeuAcα2-11NeuGcα2–4NeuAcα2–4-Glcβ1-1ceramide	HPG-7	
H. Leucospilota	NeuGcα2–6Glcβ1-1ceramide	HLG-1	Kaneko et al. (1999)
	NeuGcα2–4NeuAcα2–6Glcβ1-1ceramide	HLG-2	
	Fucβ1-11NeuGcα2–4NeuAcα-6Glcβ1-1ceramide	HLG-3	
C. echinata	8-OSO_3NeuGcα2–6Glcβ1-1ceramide	CG-1	Kaneko et al. (2003)
	4-OAcFucβ1-11NeuGcα2-6Glcβ1-1ceramide	CEG-3	
	Fucβ1-11NeuGcα2–6Glcβ1-1ceramide	CEG-4, CEG-5	
	NeuGcα2-11NeuGcα2–4NeuAcα2–6Glcβ1-1ceramide	CEG-8, CEG-9	
	Fucα1-11NeuGcα2–4NeuAcα2–6Glcβ1-1ceramide	CEG-6	
S. japonicus	NeuGCα2–6Glcβ1-1ceramide	SJG-1	Yamada et al. (2000)
	NeuAcα2–4[NeuAcα2-3]Galβ1-8NeuAcα2–3-GalNAcβ1-3Galβ1-1ceramide	SJG-2	
S. chloronotus	NeuGcα2–6Glcβ1-1ceramide	SCG-1, SCG-2	Yamada et al. (2003)
	Fucβ1-11NeuGcα2–6Glcβ1-1ceramide	SCG-3	
B. marmorata	HSO_3-8Neu5Ac2-8Neu5Ac2–6Glc1-1ceramide	GD_4(1S)	Wang et al. (2021)
I. fuscus	Neu5Ac2-8Neu5Ac2–6Glc1-1Cer	GD_4	
H. poli	Neu5Gc2-11Neu5Gc2-11Neu5Gc2–6Glc-Cer	GT_4	
H. mexicana	Fuc-Neu5Gc-Neu5Gc-Neu5Gc2–6Glc-Cer	GT_4(1Fuc)	
P. californicus	Neu5Gc(8S)-Neu5Gc-Neu5Gc2–6Glc-Cer	GT_4(1S)	
B. marmorata	Fuc-Neu5Gc(8S)-Neu5Gc-Neu5Gc2–6Glc-Cer	GT_4(1Fuc, 1S)	
P. californicus	Neu5GcMe2–6GalNAc1-3Gal1-4Glc-Cer(d18:3-C24:1 h)	GM_2	

(continued)

Table 5 (continued)

Sea cucumber species	GLS structure	GLS name	Ref.
	Neu5GcMe2-1 1Neu5GcMe2–6[Neu5GcMe2–3] GalNAc1-3Gal1-4Glc1- 1Cer	GT_2	
	Neu5GcMe2-1 1Neu5GcMe2–6[Neu5Gc2-1 1Neu5GcMe2–3]GalNAc1-3Gal1-4Glc1-1Cer	GQ_2	

molpadioides is single d17:1, with carbon chain lengths mostly C16–C18. Ceramide in *Apostichopus japonicus* has a high monounsaturated fatty acid content (75%) and long-chain base lengths of 16–19 (Yu 2013). Xu et al. identified the ceramide contents in seven dried sea cucumbers, including *Bohadschia marmorata*, *Pearsonothuria graeffei*, *Thelenota ananas*, *Cucumaria frondosa*, *Acaudina molpadioides*, *Isostichopus badionotus*, and *Holothuria atra*. The lowest content (0.23 mg/g) is found in sea cucumber *Acaudina molpadioides*; the highest content (1.58 mg/g) is found in *Isostichopus badionotus* (Xu 2011).

2.2.4 Sea Cucumber Sterol Sulfates

Sterols, as a major component of biofilms, are natural potent components that are widely found in the nature. According to the source of raw materials, sterols are divided into animal sterols, plant sterols, and fungal sterols. Unlike terrestrial sterols, marine-derived sterols, such as those from echinoderms, exhibit more diverse skeletons and branched chains, especially sulfate groups. Sea cucumber sterols have a steroidal ring structure with a polar head group (sulfate) at the C-3 site and a branched acyl chain at the C-17 site. Sea cucumber-derived sterols are mainly composed of two kinds of sulfated sterols, namely, cholest-5-en-3β-yl hydrogen sulfate and 24-methylene-cholest-5-en-3β-ol-3-sulfate (Li et al. 2021). In addition, 14α-methyl-5α-cholest-9(11)-en-3β-ol (32.2%) and 4α,14α-dimethyl-5α-cholest-9(11)-en-3β-ol (34.2%) are identified as preponderant constituents. 5α-Cholest-7-en-3β-ol (7.65%) and 24-methyl-5α-cholesta-7,24(28)-dien-3β-ol (approx. 6.9%) are presented as other conspicuous sterols in sea cucumber (Ponomarenko et al. 2001). Since two sterols have been isolated and purified from sea cucumber (*Cucumaria frondosa*), the sterols isolated from sea cucumber are a mixture of cholest-5-en-3β-yl hydrogen sulfate (51.5%) and 24-methylene-cholesterol sulfate (40.5%) (Zeng et al. 2020).

2.3 *Sulfated Polysaccharides*

Sulfated polysaccharides are one of the important active components of the sea cucumber's body wall, mainly including fucosylated chondroitin sulfate (fCS) and sulfated fucan (SF). Fucosylated chondroitin sulfate is composed of N-acetyl-D-galactosamine (GalNAc), D-glucuronic acid (GlcA), and L-fucose (Fuc), while sulfated fucan is merely composed of Fuc.

2.3.1 Content of Sulfated Polysaccharides in Sea Cucumbers

The content of sulfate polysaccharides accounts for about 4%–10% of the weight of dried sea cucumbers, with this number varying by species (Sheng 2007; Yin 2009). In addition, the difference in extraction method is also the main reason for variation in polysaccharides content.

The contents of fCS and SF in different sea cucumbers varies significantly, with the ratio of fCS:SF ranging from 8.2 to 0.72, which may affect the activities of polysaccharides and the nutritional quality of sea cucumbers (Sheng 2007; Yin 2009). Even for sea cucumbers of the same species, the difference in origin can lead to differences in sulfated polysaccharides content. It is found that there are discrepancies in the content of polysaccharides in the body wall of *Apostichopus japonicus* from different origins. Overall, there is a trend of high sulfated polysaccharides content in northern *A. japonicus* and low content in southern *A. japonicus* (Liu 2014a, b).

2.3.2 Molecular Weight of Sulfated Polysaccharides from Sea Cucumbers

2.3.2.1 Molecular Weight of fCS

The molecular weight of fCS in most sea cucumbers is between 20 and 100 kDa (Wang et al. 2021). However, considerable differences in the fCS from the same species of sea cucumber were observed in different studies, which may be caused by different extraction methods. For instance, Ustyuzhanina et al. (2016) reported that the molecular weight of fCS extracted from *A. japonicus* by papain digestion was 27 kDa, while the molecular weight of fCS extracted by papain digestion combined with alkali treatment was nearly 100 kDa (Mou et al. 2018; Yang et al. 2015). fCS from nine species of commercial dried sea cucumbers was isolated with papain digestion (0.1% papain at 60 °C for 24 h), and their molecular weights were between 12.5 and 20.6 kDa. The molecular weights of fCS extracted from different species of sea cucumbers by papain digestion are shown in Table 6.

2.3.2.2 Molecular Weight of SF

The molecular weight of SF is generally higher than that of fCS in sea cucumbers, and in most cases, the SF molecular weight is above 100 kDa. For example, the molecular weight of SF from *Acaudina leucoprocta* is 593 kDa (He et al. 2020), and

Table 6 Molecular weight of fCS prepared by enzymatic digestion with papain

Species	Papain extraction conditions	MW (kDa)	Reference
Apostichopus japonicus	Papain (0.33%) at 45–50 °C for 24 h	27.0	Ustyuzhanina et al. (2016)
Actinopyga mauritiana	Papain (0.33%) at 45–50 °C for 24 h	26.5	Ustyuzhanina et al. (2016)
Bohadschia argus	Papain (1%) at 50 °C for 6 h, then NaOH (0.25 M) for 2 h	70.1	Yin et al. (2018)
Holothuria coluber	Papain (1%) at 50 °C for 6 h, then NaOH (0.25 M) at 50 °C for 2 h	49.5	Yang et al. (2018)
Holothuria poli	Papain (0.5%) at 60 °C for 24 h	45.8	Ben-Mansour et al. (2017)
Holothuria tubulosa	Papain (0.5%) at 60 °C for 24 h	42.0	Niu et al. (2020)
Holothuria stellati	Papain digestion	46.4	Ustyuzhanina et al. (2018)
Holothuria hilla	Papain digestion	26.7	Ustyuzhanina et al. (2020)
Massinium magnum	Papain (0.32%) at 45–50 °C for 24 h	27.0	Ustyuzhanina et al. (2017)
Paracaudina chilensis	Papain digestion	28.9	Ustyuzhanina et al. (2020)
Stichopus horrens	Papain (0.32%) at 45–50 °C for 24 h	27.0	Ustyuzhanina et al. (2018)

the molecular weight of SF from *Stichopus chloronotus* is 778.7 kDa (Li et al. 2022). Cai et al. extracted SF with a molecular weight up to 2000 kDa from *Holothuria albiventer* (Cai et al. 2018). Exceptionally, researchers extracted SF with a molecular weight of less than 100 kDa from individual species of sea cucumbers. For example, SFs with molecular weights of 61.6 kDa, 42.0 kDa, and 47.6 kDa were extracted from the body wall of *Holothuria Edulis*, *Stichopus Japonicus*, and *Holothuria Nobilis*, respectively, by Luo et al. (Luo et al. 2013). In addition, and in a similar way to the case of fCS, different extraction methods can lead to significant differences in the molecular weight of SF. For example, Yu et al. extracted SF with a molecular weight of 1284 kDa from *T. ananas* by papain hydrolysis and cetylpyridinium chloride (CPC) precipitation (Yu et al. 2014a, b), while SF with a molecular weight of only 61.2 kDa was obtained through treatment with 0.5 M NaOH, papain hydrolysis, 0.5 M KOAc salting-out, and 50% ethanol precipitation (Yu et al. 2014a, b).

2.3.3 Monosaccharide Composition and Sulfate Content of Sulfated Polysaccharides

2.3.3.1 Monosaccharide Composition and Sulfate Content of fCS

fCS is mainly composed of GalNAc, GlcA, Fuc, and sulfate groups. The molar ratio of GalNAc to GlcA of most sea cucumber fCS is about 1:1 (Sheng 2007; Ustyuzhanina et al. 2016; Mou et al. 2018; Yang et al. 2015). Exceptionally, the molar ratio of GalNAc and GlcA of fCS from *Eupentacta fraudatrix* is approximately 0.79:1 (Ustyuzhanina et al. 2017). The Fuc content is slightly higher than GalNAc content. The ratio of sulfate to GalNAc is generally greater than 3, varying from 1:3.05 to 1:5.33 and accounting for about 30%–40% of fCS. In addition, a small amount of galactose and arabinose has been detected in fCS (Yin 2009).

2.3.3.2 Monosaccharide Composition and Sulfate Content of SF

Most of the SF consists of Fuc and sulfate groups. Among them, the content of sulfate group is generally 25–35%. For example, the sulfate content of SF in *H. hilla* is 32.57% (Chen et al. 2021), which is similar to that of SF from *I. badionotus* (32.9%) (Chen et al. 2012). The contents of sulfate groups of SF in *T. ananas* and *A. molpadioides* are 28.2% and 26.3%, respectively (Yu et al. 2014a, b). Exceptionally, Yu et al. found that the SF from *Cucumaria frondosa* is composed of Fuc as well as a portion of galactose (Yu et al. 2012).

2.3.4 Structure of Sulfated Polysaccharides in Sea Cucumber

2.3.4.1 Structure of fCS

The skeleton structure of fCS from sea cucumber is similar to that of chondroitin sulfate (CS) in mammals, with →4)-β-D-GlcA-(1 → 3)-β-D-GalNAc-(1 → repeat units and bearing sulfate groups at C-4 and/or C-6 of GalNAc residues and C-2 of GlcA residues (Agyekum et al. 2018). α-L-fucosyl branches are mostly attached to the O-3 site of GlcA residues in the backbone of sea cucumber FCS, creating the Fuc-(1 → 3)-GlcA linkage, and fucosylation may also occur at the O-6 site of GalNAc residues, creating the Fuc-(1 → 6)-GlcNAc linkage (Yang, Wang et al. 2018). In most cases, the branches usually consist of only one Fuc residue. Nevertheless, the presence of branches consisting of two Fuc residues has been observed in FCS extracted from *Holothuria Hilla* and *Eupentacta fraudatrix*. Sulfation may occur at the C-4, as well as C-2 and C-3 sites of Fuc residues in fCS. Sulfation of the C-3 site of GlcA residues was present in the fCS extracted from *Cucumaria frondose* (Ustyuzhanina et al. 2017), *Cucumaria japonica* (Ustyuzhanina et al. 2016), *Eupentacta fraudatrix* (Ustyuzhanina et al. 2017), and *Hemioedema spectabilis* (Ustyuzhanina et al. 2020). Such pattern of sulfation disables α-L-fucosyl branches to link to the C-3 site of the disaccharide unit, which explains why the fucose content is relatively lower in the FCS in these sea cucumbers.

2.3.4.2 Structure of SF

Most of the reported sea cucumber SF are high-molecular-weight linear polysaccharides composed of Fuc that is connected by α-1,3 glycosidic linkages and bear abundant sulfate groups at their O-2 and (or) O-4 sites. There is a discrepancy in sulfated distribution for fucosan residues in different sea cucumbers. Ribeiro et al. (Mulloy et al. 1994) clarified that the primary structure of SF from *Ludwigothurea grisea* was [3-α-L-Fuc*p*-2,4(OSO_3^-)-1 → 3-α-L-Fuc*p*-1 → 3-α-L-Fuc*p*-2(OSO_3^-)-1 → 3-α-L-Fuc*p*-2 (OSO_3^-)-1]$_n$ using methylation analysis and nuclear magnetic resonance (NMR) analysis. In addition, SF extracted from *H. poli* and *H. tubulosa* share an identical structure with that of *L. grisea* (Chang et al. 2016; Li et al. 2022). By using heterologously overexpressed endo-1,3-fucanase as the degradation tool, Chen et al. demonstrated that the SF of *H. hilla* was composed of [→3-α-L-Fuc*p*-1 → 3-α-L-Fuc*p*2,4(OSO_3^-)-1 → 3-α-L-Fuc*p*2(OSO_3^-)-1 → 3-α-L-Fuc*p*2 (OSO_3^-)-1→]$_n$, and it shared the same structure with the SF from *I. badionotus* (Chen et al. 2012, 2021). The fine structure of *Acaudina molpadioides* and *Thelenota ananas* SF was elucidated by enzymatic hydrolysis, mass spectrometry, and NMR analysis. The SF from *A. molpadioides* is mainly composed of [→3-α-L-Fuc*p*-1 → 3-α-L-Fuc*p*2,4(OSO_3^-)-1 → 3-α-L-Fuc*p*2(OSO_3^-)-1 → 3-α-L-Fuc*p*-1→]$_n$ (Yu et al. 2014a, b), and the SF from *T. ananas* is mainly composed of [→3-α-L-Fuc*p*-1 → 3-α-L-Fuc*p*2,4(OSO_3^-)-1 → 3-α-L-Fuc*p*-1 → 3-α-L-Fuc*p*2(OSO_3^-)-1→]$_n$ (Yu et al. 2014a, b). Compared with the abovementioned SF, the main structures of SF derived from S. chloronotus and S. horrens are simpler, and both are composed of [→3-α-L-Fucp2(OSO3-)-1→]n.

In particular, the α-L-fucosyl branches are present in the SF from *A. japonicus*. Chang et al. inferred that the main structure of the SF from *A. japonicus* is [→3-α-L-Fuc*p*2(OSO_3^-)-1 → 3,(α-L-Fuc*p*-1 → 4-α-L-Fuc*p*-1→)4-α-Fuc*p*2(OSO_3^-)-1 → 3-α-L-Fuc*p*2(OSO_3^-)-1→]$_n$ (Yu et al. 2015). Kariya et al. extracted and purified two SF – type A and type B – from the body wall of *A. japonicus* by chloroform-methanol method, in which type A has the branches at O-4 site (Kariya et al. 2004).

3 Enzymatic Properties of Quality-Related Proteases in Sea Cucumber (*Apostichopus Japonicas*)

Sea cucumber (*Apostichopus japonicas*) is regarded as a valuable marine resource with high protein, low fat, low sugar, and low cholesterol contents. It is deeply loved by consumers. However, autolysis frequently occurs in harvested sea cucumbers during storage and processing, resulting in the loss of nutrients and serious deterioration of quality, which in turn leads to serious economic losses. Endogenous proteases play a critical role in the autolysis of sea cucumbers and have attracted the attention of researchers. At present, a variety of endogenous proteases have been extracted and purified from the body walls of sea cucumber, gut, and viscera. The enzymatic properties and effects on the quality of sea cucumbers have been studied to varying degrees. This chapter mainly summarizes the research status of quality-related proteases in sea cucumbers.

3.1 Enzymatic Properties

At present, the endogenous proteases of sea cucumber being studied include serine protease, cysteine protease, matrix metalloproteinase, cathepsin, etc. Their key enzymatic properties are described as follows and are summarized in Table 7.

3.1.1 Serine Protease

Serine protease is a family of proteases named by serine residues, including trypsin, chymotrypsin, and elastase. The active center is the side chain of serine residue, and it is characterized by a triad catalytic structure composed of serine (Ser) residue, histidine (His) residue, and aspartic acid (Asp) residue (Rawlings and Barrett 2013).

Yan (2014) extracted a 34 kDa protease from the intestine of sea cucumber with 20 mM Tris-HCl buffer (pH 8.0) and cloned the full cDNA sequence of the protease (GenBank: KF888630). The protease can effectively hydrolyze gelatin. Peptide mass fingerprinting revealed that the peptide fragments were identical to a proprotein convertase subtilisin/kexin type 9 preproprotein from *A. japonicus* (Serine protease, GenBank: ABC87995). Serine protease inhibitors could completely inhibit the activity of the protease, while others could not. As a result, it was inferred that the protease was serine protease.

3.1.2 Cysteine Protease

The active center of cysteine protease contains sulfhydryl groups, and most cathepsins are cysteine proteases. Cysteine proteases can form covalent intermediate

Table 7 Enzymatic properties of quality-related proteases of sea cucumber

Enzyme	Source	MW (kDa)	Optimal pH	Optimal Temp (°C)	Influence of metal ions on enzymatic activity	Influence of inhibitors and activators on enzymatic activity	Literature
Serine proteinase	Sea cucumber gut	34	7.0	40	–	Pefabloc SC, benzamidine, and leupeptin strongly inhibited protease activity; EDTA triggered protease activity	Yan et al. (2014)
Cysteine protease	Sea cucumber body wall	35.5	7.0	50	Fe^{2+} and Fe^{3+} strongly inhibited protease activity; Cu^{2+}, Mn^{2+}, and Mg^{2+} inhibited protease activity; Ca^{2+} hardly inhibited protease activity	DTNB, indoleacetic acid, EDTA, and 10-phenathroline did not inhibit protease activity; PMSF and TI partially inhibited protease activity; antipain and leupeptin strongly inhibited protease activity	Qi et al. (2007)
GMP	Sea cucumber body wall	45	9.0	40–45	Fe^{2+}, Cu^{2+}, Mn^{2+}, Mg^{2+}, and Zn^{2+} inhibited protease activity; Ca^{2+} and Ba^{2+} triggered protease activity	EDTA and 1,10-phenathroline inhibited protease activity; PMSF, E-64, benzamidine, and pepstatin A hardly inhibited protease activity	Wu et al. (2013a, b)
rMMP-2	Sea cucumber body wall	40	8.5	40	Fe^{2+}, Co^{2+}, Cu^{2+}, Mn^{2+}, and Mg^{2+} inhibited protease activity; Ca^{2+}, Ba^{2+}, and Zn^{2+} triggered protease activity	1,10-phenanthroline, EDTA, and EGTA inhibited protease activity; PMSF, leupeptin, E-64, and pepstatin A did not inhibit protease activity	Yan et al. (2018)
High-alkaline protease	Sea cucumber digestive tract	20.6	13.5	37	Cu^{2+}, Ca^{2+}, and Mg^{2+} restored the protease activity inhibited by EDTA; Zn^{2+} and Hg^{2+} inhibited protease activity	EDTA and PMSF strongly inhibited protease activity; SBTI and TLCK inhibited protease activity slightly; TI and TPCK hardly inhibited protease activity	Fu et al. (2005)
Cathepsin B	Sea cucumber gut	23/26	5.5	45	Ca^{2+} and Mg^{2+} did not inhibit protease activity; Mn^{2+}, Fe^{2+}, and Fe^{3+}inhibited protease activity; Zn^{2+} and Cu^{2+} completely inhibited protease activity	E-64, iodoacetic acid, and antipain completely inhibited protease activity; leupeptin, PMSF, 1,10-phenanthroline, and TI inhibited protease activity slightly; DTT, EDETA,	Sun et al. (2011)

(continued)

Table 7 (continued)

Enzyme	Source	MW (kDa)	Optimal pH	Optimal Temp (°C)	Influence of metal ions on enzymatic activity	Influence of inhibitors and activators on enzymatic activity	Literature
						and L-cysteine triggered protease activity	
Cathepsin D	Sea cucumber gut	–	3.0	50	Fe^{3+} and Fe^{2+} inhibited protease activity; Mg^{2+}, Ca^{2+}, Ni^{2+}, Zn^{2+}, Cd^{2+}, and K^{+} triggered protease activity	Pepstatin A, E-64, and IAA inhibited protease activity; PMSF hardly inhibited protease activity	Guo et al. (2017)
Cathepsin L	Sea cucumber body wall	63	5.0	50	Ca^{2+} and mg $^{2+}$ hardly inhibited protease activity; Fe^{2+}, K^{+}, and Mn^{2+} inhibited protease activity slightly; Cu^{2+} and Zn^{2+} completely inhibited protease activity	E-64, indoleacetic acid, antipain, and leupeptin inhibited protease activity; PMSF, TI, and 1,10-phenanthroline inhibited protease activity slightly; EDTA and DTT triggered protease activity	Zhu et al. (2008)
Cathepsin L	Sea cucumber gut	30.9	5.0–5.5	50	Zn^{2+} completely inhibited protease activity; Fe^{2+} and Cu^{2+} strongly inhibited protease activity; Na^{+}, K^{+}, Mg^{2+}, Mn^{2+}, Sn^{2+}, Ba^{2+}, Ca^{2+}, and Fe^{3+} inhibited protease activity slightly	E-64, leupeptin, indoleacetic Acid, and antipain inhibited protease activity; EDTA, DTT, and L-cysteine triggered protease activity	Zhou et al. (2014)
Cathepsin K	Sea cucumber body wall	–	5.0	50	Fe^{3+}, Fe^{2+}, Cu^{2+}, and Zn^{2+} strongly inhibited protease activity; Ca^{2+} and Mn^{2+} inhibited protease activity slightly; Mg^{2+} triggered protease activity	E-64, antipain, indoleacetic acid, CA-074, ZLLL, and PMSF strongly inhibited protease activity; DTT and L-cysteine triggered protease activity	Ji et al. (2017)

complexes with substrates and can break peptide bonds. They have universal proteolytic activity and mainly exist in cytoplasm and lysosomes.

Qi et al. (2007) extracted a protease with 50 mM citric acid-citrate sodium buffer (pH 5.0) and purified the protease from sea cucumber body wall by ion-exchange chromatography and gel filtration chromatography. The best substrate of the enzyme is casein. It was also found that cysteine protease inhibitors strongly inhibited the activity of the protease and the activator L-cysteine hydrochloride could stimulate the protease activity, indicating that the proteinase was cysteine proteinase.

3.1.3 Matrix Metalloproteinase

Matrix metalloproteinase (MMP) is a family of proteases, which requires the assistance of metal ions such as Ca^{2+} and Zn^{2+}. MMP can be divided into the following six subtypes according to the specificity and structural similarity of substrates, including collagenase, gelatinase, matrix hydrolase, matrix catabolin, membrane MMP, and other MMPs.

Wu et al. (2013a, b) extracted a protease from sea cucumber body wall with 5 mM $CaCl_2$, 20 mM Tris-HCl, and 0.02% NaN_3 buffer (pH 8.0) and named it gelatinolytic metalloproteinase (GMP). The protease activity was triggered by Ca^{2+} and was strongly inhibited by metalloproteinase inhibitors, but other protease inhibitors had no significant inhibitory effect. The peptide fragments obtained by peptide mass fingerprinting had 91.4% homology with the alkaline protease from *Pseudomonas fluorescens*. It was suggested that the protease was a Ca^{2+}-dependent matrix metalloproteinase.

Yan et al. (2018) cloned the full-length cDNA sequence (GenBank: MH348178) of matrix metalloproteinase-2 (MMP-2) from sea cucumber body wall. The prediction of protein domains showed that it contained signal peptide, N-propeptide, N-telopeptide, catalytic domain, and hemopexin-like repeats region, with the typical structural characteristics of metalloproteinases. According to the prediction of structural domains, its catalytic structural domain was selected for heterologous expression, and the obtained protein was named rMMP-2. rMMP-2 effectively degraded collagen, and its activity was stimulated by Ca^{2+} and was inhibited by metalloproteinase inhibitors. It was confirmed that rMMP-2 was a matrix metalloproteinase.

Fu et al. (2005) extracted a protease from the digestive tract of sea cucumber with 50 mM Tris-HCl buffer (pH 7.5) and named it high-alkaline protease, which has high protease activity under highly alkaline conditions. The protease was significantly inhibited by metalloproteinase inhibitor EDTA and slightly inhibited by serine protease inhibitor PMSF. The addition of exogenous Cu^{2+} could significantly weaken the effect of EDTA on protease activity. It was speculated that the protease was a matrix metalloproteinase with serine protease activity and containing Cu^{2+}.

3.1.4 Cathepsin

The existence of cathepsin, first proposed in the 1920s, is a kind of proteolytic protease existing in the cells (especially lysosomes) of various animal tissues. Cathepsin C was discovered in the 1940s (Fruton 1948), and then cathepsins B, H, and L were identified and sequenced successively. Research on cathepsin continued to develop rapidly, up to the determination of the crystal structure of cathepsin B in the 1990s (Musil et al. 1991). At present, cathepsins extracted from sea cucumbers include B, D, L, K, and their analogues.

Sun et al. (2011) extracted a protease from sea cucumber guts with 50 mM NaAc-HAC, 5 mM L-Cys, 1 mM EDTA, and 0.2% Triton X-100 buffer (pH 5.0) and then obtained two protein bands with molecular weights of 23 kDa and 26 kDa, respectively. The author believed that it might be the proprotease and active monomer of cathepsin B. The specific substrate of the protease is Z-Arg-Arg-MCA and can be completely inhibited by cysteine protease and sulfhydryl inhibitor E-64, iodoacetic acid, and antipain. It was speculated that this was a cysteine protease containing sulfhydryl group.

Guo et al. (2017) used 50 mM glycine-HCl buffer (pH 3.0) to extract a crude protease solution of the protease from intestine. The protease activity can be triggered by a variety of metal ions and inhibited by Fe^{3+} and Fe^{2+}. Aspartic protease inhibitor pepstatin A has a strong inhibitory effect on the protease activity, but this activity is not in complete inhibition, indicating that the protease solution is a crude protease solution of cathepsin D dominated by aspartic protease.

Zhu et al. (2008) and Zhou et al. (2014) extracted one protease from the body wall and one from the intestine of sea cucumber using 50 mM NaAc-HAC, 5 mM L-Cys, 1 mM EDTA, and 0.2% Triton X-100 buffer (pH 5.0). The protease extracted from sea cucumber body wall can efficiently hydrolyze the specific substrate Z-Phe-Arg-NMec of cathepsin L but hardly hydrolyze any other protease substrates, indicating that the protease was cathepsin L; the protease activity was completely inhibited by Zn^{2+} and cysteine protease inhibitors, indicating that cathepsin L in sea cucumber body wall was a cysteine protease. The protease extracted from sea cucumber intestine can hydrolyze the specific substrate of cathepsin L, but it can hardly hydrolyze the specific substrates of cathepsin B, cathepsin H, and cathepsin K. The protease can be completely inhibited by Zn^{2+} and cysteine protease inhibitors and activated by thiol activators, indicating that the protease extracted from sea cucumber intestine was also a cysteine protease.

Ji et al. (2017) used 50 mM Tris-HCl buffer (pH 7.0) to extract the crude protease solution of a protease from sea cucumber body wall. A variety of metal ions can inhibit the activity of the protease. The inhibition rate of Z-Leu-Leu-Leu-H, a specific inhibitor of cathepsin K, is up to 90%, indicating that the protease was cathepsin K. The inhibition rate of cysteine protease inhibitors on the protease can reach more than 90%, and sulfhydryl activators can trigger the activity of the protease, indicating that cathepsin K was a cysteine protease containing sulfhydryl.

3.2 *Effects of Endogenous Proteases on the Quality of Sea Cucumbers*

3.2.1 Effect of Serine Protease on the Quality of Sea Cucumbers

Trypsin is a representative serine protease, which can be used to hydrolyze the collagen fibers of sea cucumber and plays an important role in the autolysis of sea cucumber. Liu (2018) found that the collagen fibers isolated from sea cucumber remained intact after incubation with buffer for 72 h. The main body remained intact after trypsin treatment, but its surface showed many decomposed collagen fibrils and fibril bundles. Compared with the control group, the release of water-soluble total matter, protein, and glycosaminoglycan (GAG) in the trypsin group increased by 1.44, 4.84, and 2.86 times, respectively. SDS-PAGE was used to analyze the soluble substances of collagen fibers treated with trypsin. It was found that there were bands less than 1 kDa. Fourier transform infrared (FTIR) analysis showed the collagen within trypsin-treated collagen fibers (72 h) still retained their triple-helical conformation. Therefore, the authors believed that trypsin decomposes collagen fibrils by degrading the core protein of proteoglycan and then participates in the autolysis of sea cucumber dermis.

3.2.2 Effect of Cysteine Protease on the Quality of Sea Cucumbers

3.2.2.1 Effect of Cysteine Protease on Collagen Fiber Structure of Sea Cucumber Dermis

Liu et al. (2017) isolated collagen fibers from sea cucumber and found that the untreated collagen fibers from sea cucumber were composed of tightly packed collagen fibrils. After cysteine protease treatment, the collagen fibers still maintain structural integrity, but there are many decomposed collagen fibrils and fibril bundles on the surface of collagen fibers, indicating that cysteine protease does degrade the collagen fibers of sea cucumber body wall. The GAG release found that the degree of degradation of the control cysteine protease treated samples was very slight, while the trypsin treatment resulted in a significantly time-dependent release. After 72 h of hydrolysis, the release of GAG increased by 2.27 times. It was speculated that cysteine proteinase contributed to the autolysis of sea cucumber by destroying the proteoglycan bridge between fibrils.

3.2.2.2 Effect of Cysteine Protease on Non-Collagen Hydrolysis of the Body Walls of Sea Cucumbers

The edible part of the sea cucumber is the body wall, of which 70% is highly insoluble collagen fiber and 30% is non-collagen (Saito et al. 2002). The autolysis of sea cucumber can lead to the hydrolysis of tissues and proteins. Compared with collagen which is difficult to degrade, the hydrolysis of non-collagen is stronger.

Wu et al. (2013a, b) found that various cysteine protease inhibitors (E-64, indoleacetic acid, and antipain) showed similar inhibitory effects on the hydrolysis of non-collagens (vitellin and actin) extracted from sea cucumber, with inhibition rates of 63.7%, 65.2%, and 54.6%, respectively. After 6 h of autolysis, E-64, indoleacetic acid, and antipain inhibited the degradation of vitellin by 39.5%, 11.9%, and 29.4%, respectively, and the inhibition rates of actin degradation were 20.0%, 11.5%, and 10.2%, respectively. The above results showed that cysteine protease was partially involved in the proteolysis of non-collagens in the body walls of sea cucumbers.

3.2.2.3 Effect of Cysteine Protease on Autolysis of Different Body Wall Layers of Sea Cucumbers

The body wall of sea cucumbers is divided into the cuticle, epidermis, outer dermis, inner dermis, and muscle layer. The fiber network of the body walls of sea cucumbers is composed of collagen fibers and microfibril networks. Compared with the inner layer of the dermis, the outer dermis was thinner and contains less fiber networks, and the damage to fiber networks in the outer layer of dermis was more obvious during autolysis (Liu et al. 2016). In fresh tissue, muscle fibers in the muscle layer were regularly arranged with a clear outline. After undergoing autolysis, the muscle fibers swelled and mixed together, and the arrangement became irregular, which means that the muscle layer was also being degraded in the process of autolysis.

Study further found that the distribution of cysteine protease in sea cucumber body wall was uneven. Compared with the inner layer of dermis, the density of cysteine protease in the epidermis and outer dermis was higher, and the protease activity of cysteine protease in per unit weight of tissue decreases from the surface of the body walls of sea cucumbers to the body cavity. As mentioned above, compared with the inner layer of dermis, the rate and degree of autolysis for the epidermal layer and outer layer of dermis were more pronounced, which were positively correlated with the density of cysteine protease. It was suggested that the heterogeneous distribution of endogenous proteases was partially responsible for this phenomenon.

3.2.3 Effect of Matrix Metalloproteinase on the Quality of Sea Cucumber

Sun et al. (2013) induced the autolysis of sea cucumber dermis at 25 °C. After the addition of metalloproteinase inhibitors EDTA Na_2 and 1,10-phenanthroline, it was found that soluble protein increased continuously after autolysis. SDS-PAGE showed obvious protein degradation with degradation products found at the same time. However, the above two indexes can be significantly inhibited by EDTA Na_2

and 1,10-phenanthroline. It was also concluded that matrix metalloproteinases might play an important role in the autolysis of sea cucumber.

Liu et al. (2019a, b) applied commercial type I collagenase (EC 3.4.24.3) to collagen fibers extracted from sea cucumber. Chemical analysis and SDS-PAGE analysis showed that GAG, soluble protein, and hydroxyproline were released continuously, and the dissolution rate of hydroxyproline reached 11.11% after 72 hours. FTIR analysis showed that the application of collagenase resulted in the decrease of intermolecular interaction and of the structural order of collagen. TEM images showed that untreated collagen fibers exhibited a ropelike texture composed of parallel and dense collagen fibrils. After collagenase treatment, the collagen fibers were dispersed and the fibril bundles were no longer tightly packed. Therefore, the author believes that collagenase participates in the autolysis of sea cucumber by destroying macromolecules and monomer collagen.

Liu et al. (2019a, b) extracted collagen fibers from fresh sea cucumber body wall and treated them with matrix metalloproteinase. Scanning electron microscopy (SEM), differential scanning calorimetry (DSC), chemical analysis, and SDS-PAGE analysis verified that matrix metalloproteinase can degrade collagen fibers. In addition, the sea cucumber body wall was treated with metalloproteinase inhibitor, serine protease inhibitor, cysteine protease inhibitor, and deionized water (Liu et al. 2020a, b). After 72 h, the sea cucumber body wall treated with metalloproteinase inhibitor did not undergo obvious autolysis. However, the sea cucumber body wall treated with the latter three treatment methods underwent autolysis and lost its original shape. It can be inferred from the above results that matrix metalloproteinase was a key endogenous protease responsible for the autolysis of sea cucumber body wall. Liu et al. (2020a, b) also treated live sea cucumber with natural metalloproteinase chelating agents oxalic acid and tea polyphenol. It was found that external stimulation releases Ca^{2+} from within the cell to the extracellular connective tissue, which stimulated the activity of matrix metalloproteinase in sea cucumber body wall. Subsequently, matrix metalloproteinases degraded the microfibril network to release water in the gap between collagen fibers and collagen fibrils, which eventually led to autolysis of the sea cucumber body wall. Natural metal ion chelators, therefore, significantly inhibit the activity of matrix metalloproteinases by chelating Ca^{2+}, which effectively slows down the autolysis of sea cucumber.

4 Conclusion

The biochemical characteristics of collagen, sulfated polysaccharide, active lipid, and other main nutrients in sea cucumbers have been gradually analyzed with respect to various sea cucumber species, areas or production, harvest season, and other factors. The collagen fibril of sea cucumber is composed of many different types of collagen, which is a mixture of type I-like, type V-like, and type IX-like. The collagen fibril of sea cucumber and FCS are covalently linked through

O-glycosidic bond. Chondroitin sulfate exists in a ring shape, periodically distributed around the collagen fibril and coaxial with the collagen fibril. Its repetition cycle is consistent with the collagen fibril, forming a shape resembling a "pearl bracelet." Sea cucumber active lipids mainly include phospholipids, sphingolipids (cerebrosides, gangliosides, ceramides), and sulfated sterols, these being mainly phospholipids. Existing studies have clarified the effect of different kinds of enzymes, such as serine protease, cysteine protease, and metal metalloproteinase on the autolysis of sea cucumber, revealing the mechanism of autolysis of sea cucumber under the action of enzymes. The results fully suggest that targeted control for the activity of key autolytic enzymes is an important approach to control the quality of sea cucumbers. However, research on the compound effect of multiple enzymes is still limited. More systematic and comprehensive research on the endogenous enzymes of sea cucumbers is therefore of great significance to the development of quality control technology for sea cucumbers.

References

Abedin MZ, Karim AA, Ahmed F, Latiff AA, Gan CY, Ghazali FC, Sarker MZ (2013) Isolation and characterization of pepsin-solubilized collagen from the integument of sea cucumber (Stichopus vastus). J Sci Food Agric 93(5):1083–1088

Abedin MZ, Karim AA, Latiff AA, Gan CY, Ghazali FC, Barzideh Z, Ferdosh S, Akanda M, Zzaman W, Karim M, Sarker MZ (2014) Biochemical and radical-scavenging properties of sea cucumber (Stichopus vastus) collagen hydrolysates. Nat Prod Res 28(16):1302–1305

Adibzadeh N, Aminzadeh S, Jamili S, Karkhane AA, Farrokhi N (2014) Purification and characterization of pepsin-solubilized collagen from skin of sea cucumber *Holothuria parva*. Appl Biochem Biotechnol 173(1):143–154

Agyekum I, Pepi L, Yu YL, Li JH, Yan LF, Linhardt RJ, Chen SG, Amster IJ (2018) Structural elucidation of fucosylated chondroitin sulfates from sea cucumber using FTICR-MS/MS. Eur J Mass Spectrom (Chichester) 24:157–167

Ben-Mansour M, Balti R, Ollivier V, Ben-Jannet H, Chaubet F, Maaroufi RM (2017) Characterization and anticoagulant activity of a fucosylated chondroitin sulfate with unusually procoagulant effect from sea cucumber. Carbohyd Polym 174:760–771

Cai Y, Yang WJ, Yin RH, Zhou LT, Li ZK, Wu MY, Zhao JH (2018) An anticoagulant fucan sulfate with hexasaccharide repeating units from the sea cucumber *Holothuria albiventer*. Carbohydr Res 464(15):12–18

Chang YG, Hu YF, Yu L, McClements DJ, Xu XQ, Liu GC, Xue CH (2016) Primary structure and chain conformation of fucoidan extracted from sea cucumber *Holothuria tubulosa*. Carbohyd Polym 136:1091–1097

Chen SG, Hu YQ, Ye XQ, Li GY, Yu GL, Xue CH, Chai WG (2012) Sequence determination and anticoagulant and antithrombotic activities of a novel sulfated fucan isolated from the sea cucumber *Isostichopus badionotus*. BBA-Gen Subjects 1820(7):989–1000

Chen Y, Lv LJ, Duan JG et al (2009) Distribution of cerebroside lipids and progress in their biological activity from a biological evolutionary perspective. Int J Pharm Res 36(2):121–126

Chen GN, Yu L, Zhang YY, Chang YG, Liu YY, Shen JJ, Xue CH (2021) Utilizing heterologously overexpressed endo-1,3-fucanase to investigate the structure of sulfated fucan from sea cucumber (*Holothuria hilla*). Carbohyd Polym 272:118480

Cui FX (2007) Study on the biochemical characterization of collagen and activity of collagen peptide extracted from the body wall of sea cucumber *Stichopus japonicus*. Dissertation. Ocean University of China

Cui FX, Xue CH, Li ZJ, Zhang YQ, Dong P, Fu XY, Gao X (2007) Characterization and subunit composition of collagen from the body wall of sea cucumber *Stichopus japonicus*. Food Chem 100(3):1120–1125

Dong P (2008) Isolation, identification, structural modification, and activity of saponin compounds from *Pearsonothria graeffei*. Dissertation. Ocean University of China

Dong XP, Zhu BW, Sun LM, Zheng J, Jiang D, Zhou DY, Wu HT, Murata Y (2011) Changes of collagen in sea cucumber (*Stichopus japonicas*) during cooking. Food Sci Biotechnol 20(4): 1137–1141

Exposito JY, Valcourt U, Cluzel C, Lethias C (2010) The fibrillar collagen family. Int J Mol Sci 11(2):407–426

Feng JH, Zhang LN, Tang X, Xia XB, Hu W, Zhou P (2021) Season and geography induced variation in sea cucumber (*Stichopus japonicus*) nutritional composition and gut microbiota. J Food Compos Anal 101:103838

Feng JH, Zhang LN, Xia XB, Hu W, Zhou P (2020) Effect of geographic variation on the proteome of sea cucumber (*Stichopus japonicus*). Food Res Int 136:109498

Fruton HRGS (1948) On the proteolytic enzymes of animal tissues: VIII. An intracellular enzyme related to chymotrypsin. J Biol Chem 174:851–858

Fu XY, Xue CH, Miao BC, Li ZJ, Yang WG, Wang DF (2005) Study of a highly alkaline protease extracted from digestive tract of sea cucumber (*Stichopus japonicus*). Food Res Int 38:323–329

Guo XK, Li AT, Han JR, Du YN, Guo TM, Yu CP, Tang Y, Wu HT (2017) Extraction and characterization of cathepsin D from sea cucumber (*Stichopus japonicus*) guts. Sci Technol Food Indus 38:135–139

He WT, Sun HH, Su LJ, Zhou DJ, Zhang X, Shanggui D, Chen Y (2020) Structure and anticoagulant activity of a sulfated fucan from the sea cucumber *Acaudina leucoprocta*. Int J Biol Macromol 164:87–94

Higuchi R, Inagaki M, Togawa K et al (1994a) Isolation and structure of three new cerebrosides, CE-2b, CE-2c, and CE-2d, from the sea cucumber *Cucumaria echinata*. Liebigs Annalen der Chemie 1:79–81

Higuchi R, Inagaki M, Togawa K et al (1994b) Isolation and structure of cerebrosides from the sea cucumber *Pentacta australis*. Liebigs Annalen der Chemie 7:653–658

Ikeda Y, Inagaki M, Yamada K et al (2009) Isolation and structure of galactocerebroside from the sea cucumber *Bohadschia argus*. Chem Pharm Bull 57(3):315–317

Isemura M, Zahn RK, Schmid K (1973) A new neuraminic acid derivative and three types of glycopeptides isolated from the cuvierian tubules of the sea cucumber *Holothuria forskali*. Biochem J 131:509–521

Ji XT, Wang L, Xue P, Nian YY, Zhou WR, Zhang YF, Zhang GL, Hou HM, Sun LM (2017) Characteristics of cathepsin K of sea cucumber *Stichopus japonicus* and its effects on autolysis. J China Agri Univ 22:72–77

Jia ZC, Li SY, Cong PX et al (2015) High throughput analysis of cerebrosides from the sea cucumber *Pearsonothria graeffei* by liquid chromatography–quadrupole-time-of-flight mass spectrometry. J Oleo Sci 64(1):51–60

Kaneko M, Kisa F, Yamada K et al (1999) Constituents of Holothuroidea, 8. Structure of neuritogenic active ganglioside from the sea cucumber *Stichopus japonicas*. Eur J Org Chem 11:46–52

Kaneko M, Kisa F, Yamada K et al (2003) Structure of a new neuritogenic-active ganglioside from the sea cucumber *Stichopus japonicus*. Eur J Org Chem 3:1004–1008

Kariya Y, Mulloy B, Imai K, Tominaga A, Kaneko T, Asari A, Suzuki K, Masuda H, Kyogashima M, Ishii T (2004) Isolation and partial characterization of fucan sulfates from the body wall of sea cucumber *Stichopus japonicus* and their ability to inhibit *osteoclastogenesis*. Carbohydr Res 339(7):1339–1346

Kisa F, Yamada K, Kaneko M et al (2005) Constituents of Holothuroidea, 14. Isolation and structure of new glucocerebroside molecular species from the sea cucumber *Stichopus japonicas*. Chem Pharm Bull 53(4):382–386

Kisa F, Yamada K, Miyamoto T et al (2006) Constituents of Holothuroidea, 17. Isolation and structure of biologically active monosialo-gangliosides from the sea cucumber *Cucumaria echinata*. Chem Pharm Bull 54(7):982–987

Kittiphattanabawon P, Benjakul S, Visessanguan W, Nagai T, Tanaka M (2005) Characterisation of acid-soluble collagen from skin and bone of big snapper (*Priacanthus tayenus*). Food Chem 89: 363–372

Li DT, Chang YQ, Chen W, He YM, Li Y, Zhong L, Zhang YY, Pan W (2006a) Analysis of nutritive composition of body wall in wild *Apostichopus japonicus Selenka* at Zhangzi Island. J Dalian Ocean Univ 21(3):278–282

Li DT, Chang YQ, Wu ZH, Chen W, Wang JY, Chen GD (2006b) Analysis of nutritive composition of body wall in wild sea cucumber *Apostichopus japonicus Selenka* at Zhangzi Island in spring and autumn. Fisheries Sci 28(7):365–369

Li Y, Li M, Xu BJ et al (2021) The current status and future perspective in combination of the processing technologies of sulfated polysaccharides from sea cucumbers: a comprehensive review. J Funct Foods 87:104744

Li Y, Wang T, Shi HH et al (2022) Absorption, pharmacokinetics, tissue distribution, and excretion profiles of sea cucumber-derived sulfated sterols in mice. J Agr Food Chem 70:480–487

Liu XF (2014a) Analysis of regional differences in nutrient composition of sea cucumber and its phospholipid activity. Ocean University of China, Dissertation

Liu XF (2014b) Analysis of regional differences in nutritional components of sea cucumber (*Apostichopus japonicus*) and the activity evaluation of phospholipids. Ocean University of China, Dissertation

Liu YX, Liu ZQ, Song L, Ma QR, Zhou DY, Zhu BW, Shahidi F (2019a) Effects of collagenase type I on the structural features of collagen fibres from sea cucumber (*Stichopus japonicus*) body wall. Food Chem 301:125302

Liu ZQ, Liu YX, Zhou DY, Liu XY, Dong XP, Li DM, Shahidi F (2019b) The role of matrix metalloprotease (MMP) to the autolysis of sea cucumber (*Stichopus japonicus*). J Sci Food Agr 99:5752–5759

Liu Z, Oliveira AC, Su YC (2010) Purification and characterization of pepsin-solubilized collagen from skin and connective tissue of giant red sea cucumber (*Parastichopus californicus*). J Agric Food Chem 58(2):1270–1274

Liu ZQ, Tuo FY, Song L, Liu YX, Dong XP, Li DM, Zhou DY, Shahidi F (2018) Action of trypsin on structural changes of collagen fibres from sea cucumber (*Stichopus japonicus*). Food Chem 256:113–118

Liu ZQ, Zhou DY, Liu YX, Liu XY, Liu Y, Liu B, Song L, Shahidi F (2020a) In vivo mechanism of action of matrix metalloprotease (MMP) in the autolysis of sea cucumber (*Stichopus japonicus*). J Food Process Pres 44(4):e14383

Liu ZQ, Zhou DY, Liu YX, Yu MM, Liu B, Song L, Dong XP, Qi H, Shahidi F (2020b) Inhibitory effect of natural metal ion chelators on the autolysis of sea cucumber (*Stichopus japonicus*) and its mechanism. Food Res Int 133:109205

Liu YX, Zhou DY, Ma DD, Liu YF, Li DM, Dong XP, Tan MQ, Du M, Zhu BW (2016) Changes in collagenous tissue microstructures and distributions of cathepsin L in body wall of autolytic sea cucumber (*Stichopus japonicus*). Food Chem 212:341–348

Liu YX, Zhou DY, Ma DD, Liu ZQ, Liu YF, Song L, Dong XP, Li DM, Zhu BW, Konno K, Shahidi F (2017) Effects of endogenous cysteine proteinases on structures of collagen fibres from dermis of sea cucumber (Stichopus japonicus). Food Chem 232:10–18

Lou QM (2011) Lipid analysis of several marine animals and study on the biological activity of EPA phospholipids in sea cucumbers. Dissertation. Ocean University of China

Luo L, Wu MY, Xu L, Lian W, Xiang JY, Lu F, Gao N, Xiao C, Wang SM, Zhao JH (2013) Comparison of physicochemical characteristics and anticoagulant activities of polysaccharides from three sea cucumbers. Mar Drugs 11(2):399–417

Mou JJ, Li Q, Qi XH, Yang J (2018) Structural comparison, antioxidant and anti-inflammatory properties of fucosylated chondroitin sulfate of three edible sea cucumbers. Carbohyd Polym 185:41–47

Mulloy B, Ribeiro AC, Alves AP, Vieira RP, Mourao PA (1994) Sulfated fucans from echinoderms have a regular tetrasaccharide repeating unit defined by specific patterns of sulfation at the 0-2 and 0-4 positions. J Biol Chem 269(35):22113–22123

Musil D, Zucic D, Turk D, Engh RA, Mayr I, Huber R, Popovic T, Turk V, Towatari T, Katunuma N, Et A (1991) The refined 2.15 a X-ray crystal structure of human liver cathepsin B: the structural basis for its specificity. EMBO J 10:2321–2330

Niu QF, Li GY, Li C, Li QY, Li J, Liu CJ, Pan L, Li SJ, Cai C, Hao JJ, Yu GL (2020) Two different fucosylated chondroitin sulfates: structural elucidation, stimulating hematopoiesis and immune-enhancing effects. Carbohyd Polym 230:115698

Ponomarenko LP, Kalinovsky AI, Moiseenko OP et al (2001) Free sterols from the holothurians *Synapta maculata, Cladolabes bifurcatus* and *Cucumaria sp.* Comp Biochem Physiol, Part B: Biochem Mol Biol 128(1):53–62

Qi H, Dong XP, Cong LN, Gao Y, Liu L, Mikiro T, Zhu BW (2007) Purification and characterization of a cysteine-like protease from the body wall of the sea cucumber *Stichopus japonicus.* Fish Physiol Biochem 33:181–188

Rawlings ND, Barrett AJ (2013) Chapter 559–introduction: serine peptidases and their clans. In: Rawlings ND, Salvesen G (eds) Handbook of proteolytic enzymes (third edition). Academic Press, London, pp 2491–2523

Saito M, Kunisaki N, Urano N, Kimura S (2002) Collagen as the major edible component of sea cucumber (*Stichopus japonicus*). J Food Sci 67:1319–1322

Sheng WJ (2007) Extraction, separation and chemical composition analysis of polysaccharides from different sea cucumbers. Dissertation. Ocean University of China

Siddiqui YD, Arief EM, Yusoff A, Suzina AH, Abdullah SY (2013) Isolation of pepsin-solubilized collagen (PSC) from crude collagen extracted from body wall of sea cucumber (Bohadschia spp.). J Pharm Pharm Sci 5:555–559

Sun LM, Wang TT, Zhu BW, Niu HL, Zhang R, Hou HM, Zhang GL, Murata Y (2013) Effect of matrix metalloproteinase on autolysis of sea cucumber *Stichopus japonicus.* Food Sci Biotechnol 22:1–3

Sun L, Zhu B, Wu H, Yu L, Zhou D, Dong X, Yang J, Li D, Ye W, Murata Y (2011) Purification and characterization of cathepsin B from the gut of the sea cucumber (*Stichopus japonicas*). Food Sci Biotechnol 20:919–925

Takagi T, Eaton CA, Ackman RG (1980) Distribution of fatty acids in lipids of the common Atlantic Sea urchin *Strongylocentrotus droebachiensis.* Can J Fish Aquat Sci 37:195–202

Tian M, Xue CH, Chang YG, Shen JJ, Zhang YY, Li ZJ, Wang YC (2020) Collagen fibrils of sea cucumber (*Apostichopus japonicus*) are heterotypic. Food Chem 316:126272

Trotter JA, Lyons-Levy G, Thurmond FA, Koobt TJ (1995) Covalent composition of collagen fibrils from the dermis of the sea cucumber, *Cucumaria frondosa*, a tissue with mutable mechanical properties. Comp Biochem Physiol 112A:463–478

Ustyuzhanina NE, Bilan MI, Dmitrenok AS, Borodina EY, Stonik VA, Nifantiev NE, Usov AI (2017) A highly regular fucosylated chondroitin sulfate from the sea cucumber *Massinium magnum*: structure and effects on coagulation. Carbohyd Polym 167:20–26

Ustyuzhanina NE, Bilan MI, Dmitrenok AS, Nifantiev NE, Usov AI (2018) Fucosylated chondroitin sulfates from the sea cucumbers *Holothuria tubulosa* and *Holothuria stellati.* Carbohyd Polym 200:1–5

Ustyuzhanina NE, Bilan MI, Dmitrenok AS, Shashkov AS, Kusaykin MI, Stonik VA, Usov AI (2016) Structure and biological activity of a fucosylated chondroitin sulfate from the sea cucumber *Cucumaria japonica.* Glycobiology 26:449–459

Ustyuzhanina NE, Bilan MI, Dmitrenok AS, Silchenko AS, Grebnev BB, Stonik VA, Usov AI (2020) Fucosylated chondroitin sulfates from the sea cucumbers *Paracaudina chilensis* and *Holothuria hilla*: structures and anticoagulant activity. Mar Drugs 18(11):540

Wang J, Chang YG, Wu FX, Xu XQ, Xue CH (2018) Fucosylated chondroitin sulfate is covalently associated with collagen fibrils in sea cucumber *Apostichopus japonicus* body wall. Carbohydr Polym 186:439–444

Wang XC, Cong PX, Chen QS et al (2020b) Characterizing the phospholipid composition of six edible sea cucumbers by NPLC-triple TOF-MS/MS. J Food Compos Anal 94:103626

Wang YC, Tian M, Chang YG, Xue CH, Li ZJ (2020a) Investigation of structural proteins in sea cucumber (*Apostichopus japonicus*) body wall. Sci Rep 10(1):18744

Wang XC, Wang XX, Cong PX et al (2021) Characterizing gangliosides in six sea cucumber species by HILIC–ESI-MS/MS. Food Chem 352:129379

Wang YH, Wang C, Guo LP, Zhu CX, Lv ZH (2010) Analysis of the nutritional components in four kinds of Holothuroidea. Period Ocean Univ China 40(7):111–114

Wu HL, Hu YQ, Shen JD, Cai QF, Liu GM, Su WJ, Cao MJ (2013a) Identification of a novel gelatinolytic metalloproteinase (GMP) in the body wall of sea cucumber (*Stichopus japonicus*) and its involvement in collagen degradation. Process Biochem 48:871–877

Wu HT, Li DM, Zhu BW, Sun JJ, Zheng J, Wang FL, Konno K, Jiang X (2013b) Proteolysis of noncollagenous proteins in sea cucumber, *Stichopus japonicus*, body wall: characterisation and the effects of cysteine protease inhibitors. Food Chem 141:1287–1294

Xu J (2011) Isolation and purification, structural analysis and bioactivity study of sea cucumber glucocerebroside. Dissertation. Ocean University of China

Xu J, Guo S, Du L et al (2013) Isolation of cytotoxic glucoerebrosides and long-chain bases from sea cucumber *Cucumaria frondosa* using high speed counter-current chromatography. J Oleo Sci 62(3):133–142

Xue DM (2006) Discussion on physical properties and processing of sea cucumbers. Dissertation. Ocean University of China

Yamada K, Hamada A, Kisa F et al (2003) Constituents of *Holothuroidea*, 13. Structure of neuritogenic active ganglioside molecular species from the sea cucumber *Stichopus chloronotus*. Chem Pharm Bull 51(1):46–52

Yamada K, Harada Y, Miyamoto T et al (2000) Constituents of *Holothuroidea*, 9. Isolation and structure of a new ganglioside molecular species from the sea cucumber *Holothuria pervicax*. Chem Pharm Bull 48(1):157–159

Yamada K, Harada Y, Nagaregawa Y et al (1998) Constituents of *Holothuroidea*, 7. Isolation and structure of biologically active glycosphingolipids from the sea cucumber *Holothuria pervicax*. Eur J Org Chem 11:447–452

Yamada K, Sasaki K, Harada Y et al (2002) Constituents of *Holothuroidea*. 12. Isolation and structure of glucocerebrosides from the sea cucumber *Holothuria pervicax*. Chem Pharm Bull 50(11):1467–1470

Yamada K, Wada N, Hiroyuki Onaka H et al (2005) Constituents of *Holothuroidea*, 15. Isolation of anti-iso type regio-isomer on long chain base moiety of glucocerebroside from the sea cucumber *Holothuria leucospilota*. Chem Pharm Bull 53(7):788–791

Yan LJ, Jin TC, Chen YL, Zhan CL, Zhang LJ, Weng L, Liu GM, Cao MJ (2018) Characterization of a recombinant matrix metalloproteinase-2 from sea cucumber (*Stichopus japonicas*) and its application to prepare bioactive collagen hydrolysate. Process Biochem 72:63–70

Yan LJ, Zhan CL, Cai QF, Weng L, Du CH, Liu GM, Su WJ, Cao MJ (2014) Purification, characterization, cDNA cloning and in vitro expression of a serine proteinase from the intestinal tract of sea cucumber (*Stichopus japonicus*) with collagen degradation activity. J Agr Food Chem 62:4769–4777

Yang WJ, Cai Y, Yin RH, Lin LS, Li ZK, Wu MY, Zhao JH (2018) Structural analysis and anticoagulant activities of two sulfated polysaccharides from the sea cucumber *Holothuria coluber*. Int J Biol Macromol 115:1055–1062

Yang J, Wang YH, Jiang TF, Lv ZH (2015) Novel branch patterns and anticoagulant activity of glycosaminoglycan from sea cucumber *Apostichopus japonicus*. Int J Biol Macromol 72:911–918

Yin LA (2009) Isolation, purification and biochemical properties of polysaccharides from different sea cucumbers. Dissertation. Ocean University of China

Yin RH, Zhou LT, Gao N, Li Z, Zhao LY, Shang FN, Wu MY, Zhao JH (2018) Oligosaccharides from depolymerized fucosylated glycosaminoglycan: structures and minimum size for intrinsic factor Xase complex inhibition. J Biol Chem 293:14089–14099

Yu JT (2013) Molecular species analysis and activity of the main sphingolipids in sea cucumbers. Dissertation. Ocean University of China

Yu L, Ge L, Xue CH, Chang YG, Zhang CY, Xu XQ, Wang YC (2014b) Structural study of fucoidan from sea cucumber *Acaudina molpadioides*: a fucoidan containing novel tetrafucose repeating unit. Food Chem 142:197–200

Yu L, Xue CH, Chang YG, Hu YF, Xu XQ, Ge L, Liu GC (2015) Structure and rheological characteristics of fucoidan from sea cucumber *Apostichopus japonicus*. Food Chem 180:71–76

Yu L, Xue CH, Chang YG, Xu XQ, Ge L, Liu GC, Wang YC (2014a) Structure elucidation of fucoidan composed of a novel tetrafucose repeating unit from sea cucumber *Thelenota ananas*. Food Chem 146:113–119

Yu L, Xue CH, Su W, Wang YC, Zhang CY, Chang YG (2012) Purification and characterization of fucoidan extracted from *Cucumaria frondosa*. Food Sci 33(19):83–87

Zeng BB, Zhang LY, Chen C et al (2020) Sea cucumber sterol alleviates the lipid accumulation in high-fat–fructose diet fed mice. J Agr Food Chem 68(36):9707–9717

Zhong M, Chen T, Hu CQ, Ren CH (2015) Isolation and characterization of collagen from the body wall of sea cucumber *Stichopus monotuberculatus*. J Food Sci 80(4):C671–C679

Zhou DY, Chang XN, Bao SS, Song L, Zhu BW, Dong XP, Zong Y, Li DM, Zhang MM, Liu YX, Murata Y (2014) Purification and partial characterisation of a cathepsin L-like proteinase from sea cucumber (*Stichopus japonicus*) and its tissue distribution in body wall. Food Chem 158: 192–199

Zhu BW, Dong XP, Zhou DY, Gao Y, Yang JF, Li DM, Zhao XK, Ren TT, Ye WX, Tan H, Wu HT, Yu CX (2012) Physicochemical properties and radical scavenging capacities of pepsin-solubilized collagen from sea cucumber *Stichopus japonicus*. Food Hydrocoll 28:182–188

Zhu B, Zhao L, Sun L, Li D, Murata Y, Yu L, Zhang L (2008) Purification and characterization of a cathepsin L-like enzyme from the body wall of the sea cucumber *Stichopus japonicus*. Biosci Biotechnol Biochem 72:1430–1437

The Functional Components of Sea Cucumber and their Nutritional and Biological Activities

Tiantian Zhang, Jie Xu, Teruyoshi Yanagita, Yuming Wang, and Changhu Xue

Abstract The sea cucumber is an important seafood that serves as both medicine and food in the Asian diet and is of great interest due to its diverse physiological regulatory functions. The nutritional efficacy and bioactivity of sea cucumber are related to the special structure of the nutrients and non-nutrient bioactive substances, also called "bioactive food components." The main bioactive food components in sea cucumbers include proteins (collagens), peptides, sea cucumber saponins, and active lipids (phospholipids, glycolipids, and sterols). This chapter systematically introduces the physiological regulatory activities and nutritional health functions of different bioactive food components in sea cucumbers based on available research reports. These mainly include the alleviation of disorders related to glucose and lipid metabolism (metabolic syndrome), antioxidation, anticancer function, improvement of uric acidemia, anti-fatigue, and prevention and improvement of neurodegenerative diseases (Alzheimer's, Parkinson's disease, etc.) associated with aging. Advances in research on the possible biological and molecular nutritional mechanisms underlying the role of bioactive components of sea cucumbers are also included. In addition, research progresses related to the differences in the structure and efficacy of different sea cucumber species and to similar bioactive components are also covered.

Keywords Bioactive food components · Functional food · Collagen · Sulfated polysaccharide · Active lipids · Sea cucumber saponin

---

T. Zhang · J. Xu · Y. Wang · C. Xue (✉)
College of Food Science and Engineering, Ocean University of China, Qingdao, R.P. China
e-mail: zhangtiantian@ouc.edu.cn; xujie9@ouc.edu.cn; wangyuming@ouc.edu.cn; xuech@ouc.edu.cn

T. Yanagita
Department of Applied Biological Science, Saga University, Saga, Japan
e-mail: yanagitt@cc.saga-u.ac.jp

C. Xue (ed.), *Advances in Sea Cucumber Processing Technology and Product Development*, Advances in Marine Bioprocesses and Bioproducts,
https://doi.org/10.1007/978-3-031-16512-2_3

1 Introduction

The sea cucumber is a traditional tonic in the Eastern food culture centered on China, and it has a long history of consumption in East and South Asia. In China, the sea cucumber is one of the eight most precious products, together with ginseng, bird's nest, and shark's fin. In the East, the dietary and medical concept of "medicine and food from the same source" has existed since ancient times, and according to this concept, the sea cucumber is not only food but also medicine. The medicinal effects of sea cucumber were recorded in the *Compendium of Materia Medica* by Li Shizhen in the Ming Dynasty and in the *Supplement to Compendium of Materia Medica* by Zhao Xuemin in the Qing Dynasty, who believed that "The sea cucumber tastes sweet and salty, and its effect of tonifying kidney essence is equivalent to that of ginseng, hence the name sea cucumber." The variety of nutritional and health benefits associated with sea cucumber are a result of its richness in the special "functional components in nutrition and health." Sea cucumber contains a variety of biologically active ingredients, such as proteins, peptides, polysaccharides, lipids, saponins, and other active ingredients. The protein content of sea cucumber is generally more than 50%, mainly in the form of collagen; the lipid content varies by species, but the sea cucumber is generally rich in EPA/DHA phospholipids, cerebroside and ganglioside, etc. The sea cucumber is also rich in polysaccharide sulfate, containing up to 10% of the dry weight or more. In addition, saponin is also an important active ingredient of sea cucumber, which is up to 3.5% of the dry weight. They possess the activities of immunopotentiation, regulation of glycolipid metabolism, prevention and improvement of neurodegenerative diseases, and anti-oxidant, antitumor, antifungal, antiaging, anti-fatigue, and other physiological functions. This chapter summarizes the functional components and biological activity of sea cucumbers, which may provide the theoretical basis for scientists researching the sea cucumber from the cellular and molecular level and developing functional foods. It also provides new ideas for the development and utilization of sea cucumber.

2 Proteins and Peptides

2.1 Sea Cucumber Collagen Peptides

2.1.1 Antioxidant Effect

More than 70% of sea cucumber protein is collagen. In recent years, with the deepening of research on the nutrition and function of sea cucumber, the preparation of antioxidant peptides has become a particular area of interest. Different types of sea cucumbers and active peptides obtained by various processes exhibit diversified antioxidant activities.

Previous studies focused on the ability of active peptides derived from *Stichopus japonicus*, *Holothuria mexicana*, and *Pearsonothria graeffei* to scavenge reactive oxygen free radicals in vitro, with the results showing that the scavenging effects of three sea cucumber collagen peptides on reactive oxygen radicals (O_2-·) and hydroxyl radicals (·OH) were enhanced with the increase of their concentrations and the dose-effect relationship was prominent (Cui 2007; Wang et al. 2010a, b). Among the three types of sea cucumbers, the *S. japonicus* collagen polypeptide had the highest scavenging activity against superoxide anion free radicals, and its IC_{50} was 2.18 mg·mL^{-1}. The *Pearsonothria graeffei* collagen polypeptide had the highest scavenging activity against hydroxyl radicals, and its IC_{50} was 0.20 mg·mL^{-1}. Liu et al. (2007a, b) studied the antioxidant activity of peptides prepared from *Stichopus japonicus* by autolysis trypsin combined hydrolysis technology and found that the sea cucumber collagen peptides with a molecular mass of 1000–3000 u had stronger scavenging ability to DPPH free radicals than V_E. Fu et al. (2016) studied the correlation between the antioxidant activity and molecular weight of *Apostichopus japonicus* collagen peptides and found that although the scavenging activity of sea cucumber peptides with Mr. $\leq$ 5 Ku was the highest, the scavenging activities of sea cucumber peptides with different molecular weights on superoxide anion free radicals were no different. In contrast, there were great differences in the scavenging activities of sea cucumber peptides with different molecular weights. Sea cucumber peptides with molecular weights of less than 5 Ku and molecular weights of more than 10 Ku had a higher activity of scavenging hydroxyl radicals. Chen et al. (2011) studied the activity of peptides from *Acaudina molpadioides* in scavenging reactive oxygen free radicals in vitro, with the results showing that the peptides hydrolyzed with papain for 4 h had the strongest ability to scavenge free radicals and the scavenging rate was more than 70%. The IC_{50} of the peptides for scavenging hydroxyl radicals and superoxide radicals were 27.8 mg/mL and 49.3 mg/mL, respectively. Peptides with molecular mass of less than 5 kDa had the strongest activity. Xiao et al. (2012) studied the free radical scavenging activity of collagen peptides from *Paracaudina chinensvar,* and the research results showed that the scavenging rate of superoxide anion free radicals by sea cucumber peptide obtained using bromelain could reach 52.20%. Chen et al. (2013) studied the antioxidative activity of polypeptides obtained by enzymatic hydrolysis of *Holothuria nobilis* and found that it had strong scavenging effects on (DPPH·), hydroxyl radical (·OH), and superoxide anion radical ($O_2·^-$), and the IC_{50} values were 5.78 mg/mL, 4.20 mg/mL, and 0.27 mg/mL, respectively. Yang et al. (2020) compared the anti-oxidative activities of enzymatic hydrolysis extracts from small Arctic sea cucumber (*Holothuria impatiens*), large Arctic sea cucumber (*Holothuria impatiens*), leaf melon sea cucumber (*Cucumaria frondosa*), flower sea cucumber (*Stichopus monotuberculatus*), Turkish sea cucumber (*Holothuria tubulosa*), Dalian jumbo dark sea cucumber (*Holothuria nobilis*), and Australian sea cucumber, and the results showed that the anti-oxidative ability of the *Cucumaria frondosa* extract was the strongest.

Zhao (2008) used H_2O_2 to establish an oxidative stress injury model of PC12 cells in vitro and studied the protective effect of three active peptides from *A. japonicus*,

Isostichopus badionotus, and *C. frondosa*. The results indicated that the three kinds of sea cucumber collagen peptide could significantly increase the survival rate of PC12 cells damaged by H_2O_2 oxidation, and this protective effect was strengthened with the increase in peptide concentration, showing a dose-response relationship. After treatment with collagen peptides, the level of ROS in PC12 cells was significantly decreased, and the enzyme activities of SOD and GSH-PX were significantly increased. The protective effect of active peptides with molecular weight lower than 6000 was greater, and the effect of low-molecular-weight active peptide prepared from *Stichopus japonicus* was significantly higher than that of high-molecular-weight active peptide.

Song et al. (2017a) used hydroxyl radical (·OH) scavenging ability, Fe^{2+} chelating ability, and Fe^{3+} reducing ability as indicators to explore the antioxidant activity of the protein peptides of *Stichopus japonicus* in vitro and the protective effect against the H_2O_2-induced oxidative damage in RAW264.7 macrophages. The results showed that the sea cucumber peptide could scavenge ·OH free radicals and reduce Fe^{3+} and chelate Fe^{2+}, and its antioxidant capacity showed a concentration-dependent effect in vitro. Sea cucumber peptide could increase the viability of oxidatively damaged cells, significantly reduce the level of reactive oxygen species in oxidatively damaged RAW264.7 macrophages, and clearly increase the mRNA expression of heme oxygenase-1 in macrophages.

Zhang et al. (2011) studied the production of active peptides and their antioxidant activities during the autolysis of the body wall and viscera of *Stichopus japonicas*. Zhou et al. (2013) prepared sea cucumber intestinal collagen peptides from sea cucumber intestines, a by-product from sea cucumber processing as raw materials, and studied their antioxidant activity. The results showed that sea cucumber intestinal collagen peptides could reduce Fe^{3+} to a remarkable extent and also had a good scavenging effect on hydroxyl radicals and DPPH. Zhao et al. (2013) studied the preparation and antioxidant activity of active peptides in the boiled liquid of *Apostichopus japonicas*. The IC_{50} values of peptides for scavenging ·OH, O_2^{-}·, and DPPH· free radicals were 8.6, 6.7, and 2.0 mg/mL, respectively, and their scavenging ability for O_2^{-}· was better than that of TBHQ.

2.1.2 Antiaging Effect on Skin

UV rays accelerate skin aging and damage due to a process, known as photoaging. This is related to UV irradiation inducing the skin to generate a variety of free radicals that attack the polyunsaturated fatty acids in the biofilm and trigger lipid peroxidation. The changes in dermal connective tissue caused by ultraviolet irradiation in animals are very similar to those in human skin photoaging, and effects on UV on animals' skin are commonly used as a model for human skin carcinogenesis and photoaging damage.

Chai et al. (2015) prepared *Stichopus japonicas* peptide by mixed enzymolysis and comparatively studied the effect of sea cucumber peptide and other anti-wrinkle ingredients on NIH/3 T3 cells and on collagen secretion by MTT method. When the mass concentration of sea cucumber peptide was 200 μg/mL, it could significantly promote the growth of NIH/3 T3 cells, and the proliferation rate was about 35%; when the mass concentration of sea cucumber peptide was 800 μg/mL, it could significantly promote the secretion of collagen in NIH/3 T3 cells.

Wang et al. (2008) studied the protective effect of *Stichopus japonicas* collagen peptide on skin of UV-induced photoaging model mice by continuous gavage. The research results indicated that the biochemical analysis of the skin showed that the collagen peptide of *Stichopus japonicas* could significantly reduce the content of MDA in the skin and serum of photoaging mice; increase the activities of antioxidant enzymes SOD, GSH-PX, and CAT; and elevate the total amount of hydroxyproline content in the skin. Histological observation results of the skin showed that the collagen peptide of *Stichopus japonicas* could significantly alleviate the structural disorder present in mouse skin as a result of exposure to ultraviolet rays. The study indicated that the active peptide of sea cucumber collagen could effectively remove active oxygen and free radicals in the skin, promote the synthesis of collagen in the dermis, as well as having a significant antiaging effect on the skin. Liu et al. (2022) used hairless mice with UV-induced skin barrier dysfunction as a model to explore the effect of a dietary supplement of *Stichopus japonicus* (SC) and its hydrolysates (SCH) on UV-induced cellular apoptosis. The findings suggested that oral administration of SC and SCH attenuated pathological changes associated with UV-induced photoaging, such as impaired skin barrier function and wrinkle formation. The protective effect of SCH was higher than that of SC, and this might be attributed to the small bioactive collagen peptide produced during the hydrolysis, which had a strong anti-photoaging effect. In addition, SC and SCH exerted anti-photoaging effects by regulating filaggrin synthesis and exfoliation in epidermis and NF-κB signaling pathway-related gene expression in dermis. These results show that SC and SCH have potential applications in photoaging nutrition, which can delay skin aging. Chen et al. (2013) found that *Holothuria nobilis* collagen polypeptide could significantly increase the antioxidant level and hydroxyproline content in the serum and skin of D-galactose model mice.

2.1.3 Inhibition of Cellular Melanin Synthesis

Melanin plays an important role in defending against UV damage to the skin. However, excessive or massive accumulation of melanin in cells will cause freckles, age spots, and other pigmentation in the skin (Iwata et al. 1990). Wang et al. (2008) used different concentrations of *Apostichopus japonicus* collagen peptides to coculture with B16 cells in vitro system and studied the effects of different molecular weight collagen peptides of *Apostichopus japonicus* (AJCP1: 6000D < Mr. < 10,000 D, AJCP2: Mr. < 6000 D) on the activity of B16 melanoma cells. The results showed that *Apostichopus japonicus* collagen peptides could significantly reduce the melanin content of B16 cells, inhibit the activity of tyrosinase, downregulate the expression of tyrosinase mRNA, increase the content of reduced glutathione (GSH), and reduce the content of cAMP with a dosage-effect relationship.

Tyrosinase is the major rate-limiting enzyme regulating melanin synthesis in mammals (Iwata et al. 1990). The synthesis of melanin is a process coupled with reactive oxygen species. Studies have shown that antioxidants can inhibit the formation of melanin. On the one hand, antioxidants directly inhibited the activity of tyrosinase by scavenging reactive oxygen species (Kim et al. 2005). On the other hand, antioxidants could simultaneously increase the level of GSH, which plays an important regulatory role in the process of intracellular melanin production, and then prevent melanin precursors from forming melanosomes. Therefore, the way that *Apostichopus japonicus* collagen peptide inhibited melanin synthesis was related to its effective scavenging of free radicals. The above studies have shown that *Apostichopus japonicus* collagen peptides could significantly inhibit the synthesis of melanin in cells, as well as have a significant whitening effect.

2.1.4 Regulation of Lipid Metabolism

Ox-LDL is a major risk factor for atherosclerosis. Zhao et al. (2008) used ox-LDL to establish an oxidative damage model of vascular endothelial cells and studied the protective effect of three sea cucumber active peptides from *A. japonicus*, *I. badionotus*, and *C. frondosa* on artery and on vascular endothelial cells. Results showed that sea cucumber collagen peptide could promote the proliferation of oxidatively damaged endothelial cells. Collagen peptides from *A. japonicus* and *I. badionotus* could significantly promote the proliferation of normal endothelial cells, while the collagen peptides from *C. frondosa* had no significant effect on the proliferation activity of normal endothelial cells. The three kinds of sea cucumber collagen peptides could significantly alleviate the oxidative damage experience in endothelial cells. Collagen peptides from *Apostichopus japonicus*, *Thelenata anax*, and *Pearsonothuria graeffei* had no significant effect on the MDA content of normal endothelial cells but reduced the content of MDA by 35%–40% in ox-LDL-damaged vascular endothelial cells. The antioxidant capacity of the three kinds of sea cucumber collagen polypeptides was similar. The three kinds of collagen polypeptides significantly promoted NO release from normal vascular endothelial cells but had no significant effect on the activity of NOS. In the model of ox-LDL injury to vascular endothelial cells, cell NOS activity and NO release were significantly increased after pretreatment with the three collagen peptides, and the activities of peptides from different sea cucumbers were similar.

2.1.5 Regulation of Glucose Metabolism

Wang (2021) investigated the antidiabetic effect of sea cucumber hydrolyzate (SCH) from *Holothuria nobilis* through streptozotocin (STZ) combined with a high-fat diet-induced diabetes mellitus type 2 (T2DM) rat model. Compared with the normal group, the levels of TG, TC, LDL-C, and HDL-C in the serum of the model group were increased by about 400%, 150%, 50%, and 475%, respectively. In contrast, metformin and SCH treatment significantly decreased the abnormal increase of TC, TG, LDL-C, and HDL-C in the serum in a dose-dependent manner, and, especially in the SCH high-dose group, the levels of TC, TG, and LDL-C in the SCH-H group were almost identical to those in the normal group after 8 weeks. Oral administration of SCH reduced blood lipid levels in STZ-induced diabetic rats, and, especially, high doses of SCH in particular could reduce blood lipids in diabetic rats to almost the level of the normal group. The hypoglycemic and hypolipidemic effects of SCH peptides might be attributed to the presence of a large number of hydrophobic amino acid, aliphatic amino acid peptides, and hypolipidemic peptides in SCH. Potential molecular mechanism studies have shown that SCH activates the PI3K/Akt signaling pathway, which further regulates the expression of GLUTs and p-GSK-3β proteins, thereby promoting glycogen storage and improving insulin sensitivity. In addition, SCH could improve liver oxidative stress in diabetic rats, and its antioxidant activity was related to the peptides containing a large amount of hydrophobic amino acids and peptides with antioxidant activity in the enzymatic hydrolysis products of sea cucumber.

Dong et al. (2013) compared the effects of the enzymatic hydrolysates of *Acaudina molpadioides* and *Apostichopus japonicus* on the organ coefficients and biochemical indicators of STZ-induced type I and type II diabetes model rats. Results showed that, compared with the model group, both the prevention group and the treatment group could significantly reduce fasting blood glucose and serum glycosylated protein; improve glucose tolerance; clearly reduce urea nitrogen, creatinine, urinary protein, and low-density lipoprotein cholesterol; as well as increase high-density lipoprotein cholesterol. Both sea cucumber polypeptides could significantly improve the glucose and lipid metabolism and other indicators of type I and II diabetic rats, eventually preventing and relieving diabetes.

2.1.6 Antihypertensive Effect

Angiotensin-converting enzyme (ACE) is a zinc-containing dipeptide carboxypeptidase, a key enzyme in the renin-angiotensin system and kallikrein-kinin system, and one which plays an important role in the regulation of blood pressure in the body. The enzyme can be inhibited by metal chelators, heavy metal salts, and specific peptides. At present, a variety of ACE inhibitory peptides have been isolated from aquatic collagenase hydrolysates. Zhao et al. (2012) prepared ACE inhibitory peptides from the body wall of *Acaudina molpadioides* and explored the physiological properties of such peptides. The results of the stability study for collagen peptides from *Acaudina molpadioides* by gastrointestinal proteases showed that after the hydrolysis of collagen peptides with pepsin and chymotrypsin, ACE inhibitory activity was enhanced by 3.5 times. This finding suggests that the collagen peptides from *Acaudina molpadioides* belonged to the prodrug-type ACE inhibitor and have the effect of lowering blood pressure. In vivo experiment further proved that the collagen peptides could significantly reduce the systolic and diastolic blood pressures of spontaneously hypertensive model rats by gavage (the dose was 3 $\mu M \cdot kg^{-1}$), and the antihypertensive effect of oral administration for 4 h was significantly better than that of captopril.

Hu (2017) used *Stichopus japonicus* viscera as raw material to prepare sea cucumber bioactive peptides by enzymatic method, and then they applied two different doses of bioactive peptides – 100 mg/kg (low-dose group) and 300 mg/kg (high-dose group) – to treat spontaneously hypertensive (SHR) rats at one time, with the aim of investigating their effects on blood pressure in rats. The results showed that the blood pressure values of the low-dose group and high-dose group decreased by 36 mmHg and 49 mmHg, respectively, at 8 h after administration and that the blood pressure value of the positive control captopril group (10 mg/kg) decreased by 60 mmHg at 8 h after gavage, indicating that peptides from sea cucumber visceral had a remarkable antihypertensive effect. With the increase of peptide dose, the decline extent of blood pressure in SHR rats gradually increased.

2.1.7 Regulation of Purine Metabolism

Hyperuricemia is a metabolic disease caused by disturbance of purine metabolism in the body, which is the main cause of gout. Hyperuricemia is also closely related to kidney disease, hypertension, diabetes, cardiovascular disease, etc. Wan et al. (2020) compared the ameliorating effects of two hydrolyzate polypeptides (EH-JAP and EH-LEU) of *Apostichopus japonicus* on hyperuricemia. The results showed that both EH-JAP and EH-LEU could alleviate hyperuricemia and renal inflammation in mice with long-term high uric acid-induced hyperuricemia and significantly reduce the content of serum uric acid and creatinine in the model group. Mechanism research results showed that on the one hand, sea cucumber peptides could inhibit the activation of TLR4/MyD88/NF-κB signaling pathway in kidney tissue, and on the other hand, they could regulate the composition and structure of intestinal microflora in mice with hyperuricemia disease, eventually improving the symptoms.

2.1.8 Anti-Fatigue Effect

Exercise-induced fatigue is mainly caused by muscle fatigue and energy exhaustion, which is manifested as a temporary decrease in maximum muscle contraction or maximum output power. The change of energy metabolism mechanism is an important mechanism in the occurrence of exercise-induced fatigue. Mitochondria play a key role in cellular energy metabolism. Mitochondrial dysfunction, structural abnormalities, and energy metabolism disorders are all potential factors in fatigue. Yu et al. (2019) studied the effect of high, medium, and low doses of sea cucumber peptide (300, 150, and 75 mg/kg) administered by gavage for 28 consecutive days on the mitochondrial function of skeletal muscle in exercise-induced fatigue. The results showed that, compared with the normal control group, the activities of

SOD and GSH-PX in skeletal muscle tissue were decreased, and the content of MDA was increased; the opening degree of PTP in skeletal muscle mitochondria was elevated, the activities of respiratory chain enzyme complexes I and III decreased, and the expression of PGC-1α and ERR in mitochondria decreased in the model group. Compared with the model group, the activities of SOD and GSH-PX in the skeletal muscle tissue were increased, the MDA content was decreased, the PTP opening degree of the skeletal muscle mitochondria was weak, the activities of mitochondrial complexes I and III were increased, and the expression of 1α and ERR in skeletal muscle mitochondria increased in the groups treated by high, medium, and low doses of sea cucumber peptides. The results showed that sea cucumber peptide could improve the oxidative stress state of skeletal muscle, reduce the degree of swelling of skeletal muscle mitochondria and mitochondrial membrane permeability, enhance the respiratory function of skeletal muscle mitochondria, and regulate to a higher degree the expression of PGC-1α and ERR in rats experiencing exercise-induced fatigue.

Lu et al. (2009) reported that intragastric administration of sea cucumber peptide to mice could prolong the time of weight-bearing swimming, reduce the levels of urinary blood nitrogen and blood lactic acid, and increase the content of liver glycogen. Guo (2012) administered sea cucumber small molecular peptide at the dose of 2 mg/100 g body weight by gavage to mice and then detected the activities of pyruvate kinase, malate dehydrogenase, and succinate dehydrogenase in skeletal muscle. The results showed that sea cucumber peptide could significantly improve the activity levels of the three enzymes.

Yu (2021) constructed the sports fatigue model by using forelimb grip test, fatigue rod test, and isochronous swimming test in C57BL/6 J male mice. At the same time, the mice were treated with different doses of sea cucumber peptides (0.7 mg/g·d, 1.4 mg/g·d, and 2.8 mg/g·d), respectively. The results showed that, compared with the control group, forelimb grip in the H-Scp group was significantly increased and the fatigue rod rotating time of each sea cucumber peptide group was also significantly improved with a dose-effect relationship. Sea cucumber peptide alleviated the levels of oxidative stress and inflammatory factors and increased the blood glucose level and tissue glycogen content, while the presence of fatigue metabolic markers was reduced in mice after isochronous swimming. Sea cucumber peptide enhanced fatty acid catabolism in liver, heart, and skeletal muscle tissues; increased the expression of PPARα, PGC1α, CPT1, LPL, and PDK4 in the liver and PPARα and CPT1 in the heart; and elevated the expression of PPARδ and CPT1 in the skeletal muscle of mice. Moreover, sea cucumber peptide enhanced gluconeogenesis in the liver and expression of G6Pase and PEPCK and increased the mitochondrial copy number (MtDNA) content in the heart and skeletal muscle in mice. After isochronous swimming, the ATPase activity as well as the COXI, PGC1α, NRF2, NRF1, TFAM, AMPK, and SIRT1 expressions in the heart and skeletal muscle of the sea cucumber peptide group were higher than those of the control group.

2.1.9 Others

Cheng (2021) found that the enzymatic hydrolysate of *Apostichopus japonicus* could reduce the infiltration of inflammatory cells in the wound, improve the level of oxidative stress, and promote angiogenesis, fibroblast proliferation, collagen deposition, and reepithelialization in the wound. In addition, it could activate the TGF-β/Smad signaling pathway and eventually promote collagen synthesis and accelerate wound healing. These comprehensive effects were most prominent in the middle-dose group and the low-molecular-weight group. The in vivo absorption mechanism of hydrolysate in promoting wound healing in rats might be related to the absorption of active peptides such as Ala-Hyp-Gly, Pro-Hyp, Gly-Pro-Hyp, Hyp-Gly, and Ile-Hyp and their various physiological functions. After the rats are treated with hydrolysate, it flows through the intestine, and the di−/tripeptide in hydrolysate stimulates the massive secretion of PepT1 in the small intestinal epithelium, which is then completely absorbed and transported into blood vessels through the PepT1 pathway. It is then transferred to the wound surface through blood circulation and eventually exerts various biological functions.

Jiang (2019) used a scopolamine-induced memory impairment model in mice to explore the auxiliary memory enhancement effect of different doses of protein peptides. The results indicated that the low-dose sea cucumber protein peptide (S-L) group (0.65 mg/(g·d)) had a strong ability to enhance memory in different test periods. Compared with the blank group, the activities of $Na^{+}K^{+}$-ATPase and acetylcholinesterase in the S-L group were decreased slightly, but there was no significant difference, while the content of acetylcholine in the sea cucumber protein peptide and taurine in the S-L group, in the compound low-dose (S + T-L) group of sea cucumber protein peptide and taurine, and in the middle-dose (S + T-M) group of sea cucumber protein peptide and taurine was significantly increased ($P < 0.05$). Compared with the model group, the content of acetylcholine in the S-L group was significantly increased ($P < 0.05$). Compared with the blank group and the model group, the level of lysine acetylation in the S-L group was increased. Morphological results showed that, compared with the blank group, the number of neurons on the hippocampal pyramidal cells of the mice in the model group decreased, which was manifested by the reduction of cell volume, unclear boundary between cell membrane and nuclear membrane, concentration and deep staining of nucleus and cytoplasm, loose arrangement of neurons, obvious loss of neurons, and fewer but unclear levels. The neurons in the hippocampus of mice in the low-dose sea cucumber group were abundant, neatly arranged, and morphologically intact, with compact cells and wide cell bands. Compared with the blank group, the cerebral cortex, liver, colon, and kidney cells of the mice in each test group were regular in shape, neatly arranged, and consistent in size, with clear boundaries and no significant difference. The results showed that the low-dose group of sea cucumber peptide had strong ability to enhance auxiliary memory in both the normal mouse model and memory impaired model, which might be attributed to the low-dose sea cucumber peptide significantly increasing the content of acetylcholine (Ach) and the level of acetylation in hippocampus.

Sea cucumber peptides can effectively inhibit the viability of tumor cells and have obvious inhibitory effect on tumor formation and metastasis. Wang et al. (2007) explored the immunomodulatory and antitumor effects of *Apostichopus japonicus* polypeptides by using the tumor model mice that were treated with *Apostichopus japonicus* homogenate. The results showed that *Apostichopus japonicus* peptides with different doses could significantly inhibit the growth of S180 sarcoma in mice and improve the spleen index and thymus index of mice. It showed that *A. japonicus* could regulate the body's specific and nonspecific immunity function and then affect and kill tumor cells. Zhou et al. (2012) isolated two low-molecular-weight polypeptides of *Stichopus japonicus* (LSCP-1 and LSCP-2) by ultrafiltration membrane separation and analyzed their antitumor activities in vitro. The study showed that although these two fragments did not affect anti-A549 cell line activity, LSCP-1 could significantly inhibit the activity of gastric cancer SGC-7901 cells, and LSCP-2 could significantly inhibit the activity of SGC-7901 and human breast cancer MCF-7 cells. It suggested that small-molecular-weight sea cucumber peptides might be used as raw materials for the treatment of gastric cancer and breast cancer. Sun et al. (2017) studied the antitumor effect of *Cucumaria frondosa* polypeptide on S180 tumor-bearing mice. The results showed that *Cucumaria frondosa* polypeptide could inhibit the growth of tumor cells by reducing the vitality of tumor cells and the growth and reproduction speed of tumor cells.

2.2 Sea Cucumber Vanadium Protein

Vanadium is an essential trace element for humans and other animals. Schecher et al. firstly found that tetravalent VO^{2+} has good insulin-like activity in vitro in 1980 (Shechter and Karlish 1980), attracting attention to vanadium compounds as potential therapeutic reagents. The content of vanadium in ordinary natural substances is generally minimal. Notably, inorganic vanadium compounds have medium to severe toxicity to humans and animals, and their absorption and utilization rates in organisms are also on the low side, which affects its applicability in the field of medicine. By contrast, organic vanadium compounds can improve the absorption rate and bioavailability of vanadium in vivo, and the hypoglycemic effect is significantly better than inorganic vanadium. In particular, they have no obvious toxic side effects. Therefore, it is particularly important to find natural vanadium-rich carriers and effective nontoxic organic vanadium compounds. The sea cucumber is a good carrier of vanadium, and the vanadium in sea cucumbers mainly exists in the form of organic vanadium, with the organic degree higher than 60%. The vanadium concentration of the blood is approximately three times higher than that in the body wall (Liu et al. 2016a, b).

2.2.1 Amelioration of Insulin Resistance

Studies have shown that sea cucumber vanadium protein can effectively improve insulin resistance in mice fed with a high-fat and high-sugar diet (Liu et al. 2016b). A significant increase in fasting blood glucose, in serum insulin, and in the HOMA-IR index was observed in untreated diabetic rats as compared to control groups ($P < 0.01$). However, vanadium-binding protein treatment showed 22.6 and 40.0% reductions in serum insulin levels and HOMA-IR index, respectively. Similar to rosiglitazone treatment, the 12-week treatment regimen with vanadium-binding protein was found to have significantly decreased the high plasma glucose concentration in mice fed with high-fat and high-sugar diet. The results implied that vanadium-containing proteins could remarkably improve insulin resistance.

2.2.2 Inhibition Adipocyte Differentiation

Adipocyte differentiation, known as adipogenesis, is controlled by the balance of a series of negative and positive regulators (Liu et al. 2015a, b). Vanadium-binding protein (VBP) could significantly inhibit adipocyte differentiation through activating the WNT/β-catenin pathway. The number of lipid droplets in VBP-treated groups was significantly decreased compared to the control group. The renal, epididymal, and hypodermic fat weights were significantly lower after 12 weeks in high-fat and high-sugar diet (HFSD)-fed mice treated with vanabins compared to HFSD-fed mice.

2.3 *Digestion and Absorption of Sea Cucumber Oligopeptides*

Sun et al. (2017) screened out a calcium-binding peptide by using sea cucumber ova as raw materials based on amino acid composition and molecular weight distribution through protease controlled enzymatic hydrolysis, peptidomics, and other technologies. The sea cucumber ovum hydrolysates (SCOH) could spontaneously combine with a calcium ion via the carboxyl and amino groups of aspartic acid and glutamate to form peptide calcium chelate, which shows a folded nano network structure. The structure and particle size distribution of sea cucumber ovum peptide calcium chelate during simulated digestion were further studied. The results showed that, under the action of pH, the structure of the peptide calcium chelate expanded, depolymerized, and released calcium ions during the digestion stage of gastric juice; the chelates experienced a sequential structural change of self-assembly, depolymerization, and reassembly in the digestive stage of intestinal juice after gastric juice digestion. In addition, the results of Caco-2 cell model (338.3 nmol/L versus 269.6 nmol/L) and HT-29 cell model (373.9 nmol/L versus 271.7 nmol/L) showed that SCOH had a dose-dependent effect on promoting calcium absorption.

Wang et al. (2022) first synthesized a zinc-chelating peptide from *Apostichopus japonicus* and then found that the zinc-chelating peptide had good stability by simulating the gastrointestinal digestive system in vitro. It was found that the zinc-chelating peptide could increase the content of zinc in rat intestinal sac and Caco-2 cells by fluorescence staining. In an in vitro study, the SCSP-Zn complex could successfully be transported through the intestinal membrane in the model of everted rat gut sacs (nearly 7.5 $M \cdot cm^{-2}$) as well as Caco-2 cells, where the zinc transport reached 0.0014 mg/mL carried by SCSP. In vitro and ex vivo experiments showed that the zinc-chelating peptide had the effect of promoting zinc absorption. It could be used as an effective carrier of zinc to improve the biological activity of zinc.

Cheng et al. (2021) studied the difference of absorption for two kinds of sea cucumber hydrolysates (SCH) in vivo. The changes of free HYP (F-HYP), bound Hyp-Hyp, F-Pro and P-Pro, and Hyp bound small peptides in plasma and skin tissue at multiple time points after intragastric administration of peptides were detected. The gastrointestinal absorption results showed that the gastric emptying rate and intestinal absorption rate of L-Mw-S were faster, there was a higher level of P-Hvp in the L-Mw-S group, and the target P-Hyp might not be obtained by intestinal protease digestion. The changes in the contents of P-Pro and F-Pro showed that SCH could generate P-Pro and F-Pro by gastrointestinal protease digestion. Moreover, L-Mw-S contained more specific di/tripeptides, and its absorption rate was faster than that of any high-molecular-weight group. The results of related mechanism research showed that the reason for the rapid absorption of L-Mw-S might be related to its containing of more di–/tripeptides. Therefore, it could induce the transcription of duodenal PepT1 gene faster and stimulate the persistent expression of PepT1 mRNA in jejunum.

3 Sea Cucumber Polysaccharides and their Oligosaccharides

Sea cucumber polysaccharides are one of the most important bioactive substances in various sea cucumbers, mainly including sea cucumber chondroitin sulfate (SC-CHS) (also called glycosaminoglycan or acid mucopolysaccharide) and sea cucumber fucoidan (SC-FUC). SC-CHS is a kind of chondroitin sulfate with fucose sulfate branched chains including D-N-acetylgalactosamine, D-glucuronic acid and L-fucose branched chain. SC-FUC is a kind of sulfated polysaccharide mainly composed of fucose. Studies have shown that sea cucumber polysaccharides have a wide range of physiological functions, such as anticoagulation, antithrombus, hypolipidemia, antitumor, immune regulation, and metabolism regulation. The branch structure of fucose sulfate is closely related to the biological activities of sea cucumber polysaccharides, which indicates significant structure-activity relationships.

3.1 Sea Cucumber Chondroitin Sulfate

3.1.1 Anticoagulant and Antithrombotic Effects

In 1985, the anticoagulant activity of SC-CHS from *Stichopus japonicus* was identified (Li et al. 1985). Subsequently, the anticoagulant activity of SC-CHS from various sea cucumbers has been continuously studied, and the mechanisms involved have been clarified. Chen et al. reported that the activated partial thromboplastin time (APTT) and prothrombin time (TT), rather than plasma prothrombin time (PT), were significantly prolonged by SC-CHS from *Isostichopus badionotus* (CHS-*Ib*) with different fucose branching structures in vitro (Chen et al. 2011). Subsequently, the anticoagulant activity of SC-CHS from *Pearsonothuria graeffei* (CHS-*Pg*), *Holothuria vagabunda* (CHS-*Hv*), *Stichopus tremulus* (CHS-*St*), and *Isostichopus badionotus* (CHS-*Ib*) were comparatively studied in vitro (Chen et al. 2011). The fucose branched chains were sulfated mainly by 4-SO_4 in CHS-*Pg* and CHS-*Hv*, while those of CHS-*Ib* were sulfated mainly by 2,4-SO_4. The fucose branched chains in CHS-*St* were also sulfated mainly by 2,4-SO_4, while some were also sulfated by 4-SO_4 and 3,4-SO_4. The study in vivo showed that the four SC-CHS had significant anticoagulant activities. Among them, CHS-*Ib* had the strongest activity, and its APTT and TT activities were 183 IU/mg and 157 IU/mg, respectively, which were both better than those of undifferentiated heparin (UFH). The APTT and TT activities of CHS-*St* were 135 IU/mg and 132 IU/mg, respectively, which were similar to those of undifferentiated heparin (UFH). The above results showed that the anticoagulant effect of SC-CHS was closely related to the branched fucose sulfate group. The number of branched fucose sulfate groups was positively correlated with their anticoagulant activity, and the fucose branched chains with 2,4-SO_4 exhibited the best anticoagulant effects (Chen et al. 2011). The results of antithrombotic evaluation in vitro showed that SC-CHS significantly reduced the weight of thrombus compared with the control. CHS-*Ib*, of which fucose branched chains were sulfated mainly by 2,4-O-disulfate, showed superior results to low-molecular-weight heparin (LMWH) and CHS-*Pg* in terms of reducing thrombus weight, and its fucose branched chains were sulfated mainly by 4-O-sulfate. SC-CHS mainly sulfated by 2,4-O-disulfate in the fucose branched chain had better anticoagulant and antithrombotic effect. The results further confirmed that the fucose sulfate group with branched chain structure had better anticoagulant and antithrombotic effect than that with straight-chain structure.

Antithrombin III (AT-III) is a serine protease inhibitor in plasma, which inhibits a variety of activated coagulation factors with serine as the activity center in the blood. Heparin cofactor II (HC-II) is another serine protease inhibitor, which specifically inhibits thrombin. Study on the anticoagulant mechanism of SC-CHS showed that CHS-*Ib* and CHS-*Pg* could inhibit antithrombin III (AT-III)-mediated FIIa activity when only AT-III was present in the reaction system, and the effect of CHS-*Ib* was significantly better than that of CHS-*Pg*, but the inhibitory effect was lower than that of heparin. However, it was significantly better than heparin for the effect of CHS-*Ib* and CHS-*Pg* on inhibiting AT-III-mediated FIIa activity when only heparin cofactor II (HC-II) was present in the reaction system. The results suggested that there was an HC-II binding site in SC-CHS, which was the reason for inhibiting the activity of thrombin. The effects of CHS-*Ib* and CHS-*Pg* on inhibiting AT-III-mediated FXa activity were lower than those of heparin, indicating that the anticoagulant mechanism of SC-CHS was mainly HC-II dependent. In addition, the activity of CHS-*Ib* was stronger than that of CHS-*Pg* in inhibiting AT-III-mediated FXa activity, indicating the activity of SC-CHS with 2,4-O-disulfate substituted fucose branched chains was stronger than that of 4-O-sulfate sulfate substituted fucose branched chain structure.

Further, the anticoagulant effects of CHS-*Ib* oligosaccharides with different structures were studied, and the mechanism involved was explored. The results showed that CHS-*Ib* oligosaccharides had anticoagulant effects mainly by inhibiting the HC-II-dependent endogenous pathway, which was similar to CHS-*Ib*. The minimum active structural fragment of CHS-*Ib* oligosaccharides for FIIa inhibition is heptosaccharide, and the structure is GalNAcOOHβ(4,6S)1-4GlcAβ(Fucα(2,4S) 1–3) 1-3GalNAcβ(4,6S). In addition, the HC-II-dependent pathway was also proven in SC-CHS from *Stichopus japonicus* and *Holothuria leucospilota* (Nagase et al. 1997). In addition, some studies have shown that AT and heparin cofactor II (HCII)-independent anti-FXase activity and AT-dependent FIIa or FXa inhibition activity were also involved in the antithrombotic activity of SC-CHS, which might be the reason for SC-CHS causing less side effects than low-molecular-weight heparin.

SC-CHS and its oligosaccharide showed a structure-activity relationship in anticoagulant activity. Firstly, the coagulation activity was affected by the molecular weight of oligosaccharide and the degree of depolymerization of SC-CHS. The strong anti-FXase activity was shown by nonasaccharide and dodecasaccharide (IC_{50} is 50–100 ng/mL) rather than trisaccharide and hexasaccharide ($IC_{50} > 2000$ ng/mL), which suggested that the degree of polymerization ≤ 6 had no obvious effects (Li et al. 2021). In addition, AT-dependent anti-FIIa and anti-FXa activities decreased with a reduction in molecular weight. SC-CHS with a molecular weight range of 5–12 kDa might be an antithrombotic candidate compound with a selective anti-FXase activity (Wu et al. 2015; Zhou et al. 2020). The activated partial thromboplastin time (APTT) prolonging activity, anti-FXase activity, and FIXa-binding affinity of oligosaccharides of the SC-CHS from *S. variegatus* were decreased, with the decrease in the dp. Nonasaccharide was the minimum structural fragment for anti-FXase activity and bound FIXa with high affinity.

In addition, the anticoagulant and antithrombotic activities of SC-CHS depend on sulfate groups, fucose sulfate (FucS) side chains, and carboxyl reduction (Chen et al. 2011; Gao et al. 2012; Wu et al. 2015; Xiao et al. 2016). The studies on oligosaccharide suggested that the oligosaccharide with Fuc_{2S4S} branches had superior anticoagulant to those with Fuc_{3S4S} branches (Cai et al. 2019; Li et al. 2017a, b). The studies suggested that the anticoagulant and anti-FXase activities of SC-CHS nonasaccharide derivatives were significantly reduced by the reduction of carboxyl groups, whereas carboxymethylation and deacetylation exhibited smaller effects.

3.1.2 Immunomodulatory Effects

Sea cucumber polysaccharide can significantly improve both the innate (nonspecific) and adaptive (specific) immune systems. The immunomodulatory effects of SC-CHS from *Apostichopus japonicus* (CHS-*Aj*) were studied in cyclophosphamide (CPh)-induced immunosuppressed mice. The results suggested that the innate immune system of CPh-treated mice was restored by CHS-*Aj* with improved phagocytic function of macrophage and increased cytotoxicity of splenic natural killer (NK) cells (Wang et al. 2019b). In addition, CHS-*Aj* regulated the adaptive immune system in CPh-treated mice with increased contents of immunoglobulins and cytokines in serum and increased ratio of spleen lymphocyte subsets (Wang et al. 2019b). Mechanism studies suggested that the intracellular Ca^{2+} level of splenic lymphocytes was elevated by CHS-Aj, which further enhanced the proliferation and activation of splenic lymphocytes in CPh-induced mice (Wang et al. 2017a). Furthermore, SC-CHS from *Massinium magnum* (CHS-*Mm*) promoted hematopoiesis in CPh-induced mice (Anisimova et al. 2017).

In addition, the immunomodulatory function of SC-CHS from *Holothuria poli* and *Holothuria tubulosa* was comparatively studied in CPh-induced immunosuppressed mouse models. The results suggested that the two SC-CHS improved the hematopoietic function recovery with suppressed overproductions of inflammatory cytokines in CPh-induced bone marrow-suppressed mice. Notably, α-L-Fuc2,4S was more important to the activity than α-L-Fuc3,4S, which might be attributed to their sulfation patterns, since their molecular weights were similar (Niu et al. 2020).

The phagocytosis test and delayed-type allergy test of mouse peritoneal macrophages showed that the phagocytosis rate of mouse peritoneal macrophages was significantly increased by CHS-*Aj* via intragastric or intraperitoneal injection. The delayed-type hypersensitivity of normal mice, MA-737-bearing breast cancer mice, and CPh-treated mice was significantly increased by CHS-*Aj*. In addition, SC-CHS from *Holothuria leucospilota* (CHS-*Hl*) significantly increased the expression of the macrophage immunoglobulin Fc receptor, promoted the activity of ADCC effector cells and the secretion of spleen cell antibody, and enhanced the immune function accordingly (Chen 1987).

The immunomodulatory effects of SC-CHS from *Stichopus chloronotus* (CHS-*Sc*) were revealed in RAW 264.7 cells. The results suggested that the proliferation and pinocytic activity of RAW 264.7 cells was significantly promoted by CHS-Sc with the production of NO, TNF-α, IL-1β, and IL-6. Mechanism studies further indicated that CHS-Sc could bind to macrophages. Moreover, toll-like receptor 4 (TLR4) and TLR2 were involved in the recognition of CHS-*Sc* (Jiang et al. 2021).

Mechanism research showed that SC-CHS could interfere in the binding of P- and L-selectins on the surface of endothelial cells to sialyl-Lewis X, a tetrasaccharide on the surface of leukocytes (Borsig et al. 2007), which affected leukocyte tethering and rolling and the eventual migration through the endothelial wall (Ley et al. 2007). In addition, SC-CHS inhibited neutrophil recruitment in mice with peritonitis and lung inflammation (Borsig et al. 2007). The repeating trisaccharide unit of SC-CHS from *Holothuria forskali* could adopt a conformation similar to that of sialyl-Lewis X, which might be the reason for the affinity of SC-CHS to P- and L-selectins.

3.1.3 Antitumor Activity

A large number of studies have proven that the SC-CHS had significant inhibitory effects on transplanted tumors in a variety of experimental subjects, in samples including MA-737 breast cancer cells, HepA22 liver cancer cells, U14 cervical cancer cells, S180 sarcoma cells, Lewis lung cancer cells, L795 lung adenocarcinoma cells, gastric fibrosarcoma cells, S37 sarcoma cells, Lio-1 lymphosarcoma cells, B16 melanoma cells, Bcap-37 human breast cancer cells, 95D human high-metastatic lung cancer cells, HO-8910 PM human high-metastatic ovarian cancer cells, and others.

The antitumor effects of CHS-*Ib* were evaluated in a variety of cancer cell lines, and the results showed that CHS-*Ib* had the best inhibitory effect on 95D cells. Its IC_{50} values at 48 h, 72 h, and 96 h were 369.8, 279.2, and 216.3 μg/mL, respectively. Mechanism studies showed that CHS-*Ib* could significantly inhibit proliferation, induce apoptosis, and change cycles, leading to G2/M phase arrest of tumor cell. G1/S and G2/M are two key points in the cell cycle. Cyclin B1 and Cdc2 play key roles in the G2/M phase transition and are closely related to the proliferation, differentiation, metastasis, malignance, and recurrence of tumors, which were overexpressed in a variety of solid tumors. As a tumor suppressor gene, P53 is one of the genes most related to human tumors, and it induces cell growth arrest, apoptosis, differentiation, and DNA repairs. P21 is an inhibitor of broad-spectrum cyclin kinase. P53 can inhibit cells from entering the M phase by inhibiting the activity of cyclin B1/Cdc2 complex. CHS-*Ib* significantly downregulates the expressions of cycling B1 and Cdc2 and increases the levels of P53 and P21, which suggests the effects on inducing G2/M arrest of 95D cell cycle.

B cell lymphoma-extra-large (Bcl-xL) is a transmembrane protein belonging to the B cell lymphoma-2 (Bcl-2) family and located in mitochondria, which inhibits the opening of mitochondrial membrane permeability pore and prevents the release of cytochrome c and other apoptosis-inducing factors. After caspase-3 is activated, it can specifically cleave related substrates and perform apoptosis. CHS-*Ib* significantly reduces the expression of Bcl-xL and upregulates the expression of caspase-3, which suggests that it promotes tumor cell apoptosis (He et al. 2014).

Tumor metastasis, including tumor cell metastasis and angiogenesis, is the key reason for malignant tumor lethality. Two necessary conditions for tumor metastasis are that tumor cells cross the barrier composed of the extracellular matrix and the basement membrane. SC-CHS has also been revealed as inhibitor of metastasis in animal models. The antitumor metastasis effect and molecular mechanism of CHS-*Ib* were systematically studied through the simulation of tumor cell invasion, metastasis, and angiogenesis in vitro and the establishment of mouse tumor metastasis model in vivo. In vitro experiments showed that CHS-*Ib* significantly reduced the adhesion between 95D cells and the matrix and human umbilical vein endothelial cells (HUVEC). Compared with the control group, it decreased by 56.0% ($P < 0.01$) and 22.0% ($P < 0.05$), respectively. In addition, CHS-*Ib* significantly inhibited the migration and invasion of 95D cells, and the migration rate and invasion rate decreased by 40.5% and 72.1% ($P < 0.01$), respectively, compared with the control group. The formation of vascular endothelial cells and angiogenesis in chicken embryo allantoic membrane were significantly inhibited by CHS-*Ib*, suggesting significant antitumor cell metastasis effects. The results of mechanism study showed that CHS-*Ib* significantly downregulated the expression of heparanase (HPA), matrix metalloproteinase MMP-2/−9, and vascular endothelial growth factor (VEGF) and upregulated the expression of TIMP-1/−2, a specific inhibitor of matrix metalloproteinase (He et al. 2014). In addition, the mechanism was involved in CHS-Cf inhibiting the migration of LLC cells that downregulated expressions of MMPs by suppressing ERK1/2/p38 MAPK/NF-κB pathway (Liu et al. 2016c).

The antitumor metastasis effect of CHS-*Ib* in vivo was demonstrated in the Lewis lung cancer spontaneous lung metastasis model and the B16 artificial lung metastasis model. The results of the spontaneous lung metastasis model of Lewis lung cancer in mice showed that CHS-*Ib* significantly inhibited the growth of carcinoma in situ and the number of bilateral lung metastatic nodules, which decreased by 31.5% and 42.3%, respectively. In addition, CHS-*Ib* significantly reduced the number of B16 artificial lung metastases, the content of sialic acid and the activity of γ-glutamyl aminotransferase in serum and the contents of hydroxyproline, aminohexose, and uronic acid in lung tissue, which suggested inhibiting the metastasis of tumor cells by blood circulation (Liu et al. 2016c).

SC-CHS, as the ligand antagonist of P- and L-selectins, disrupted the binding of selectins to tumor cell surface receptors and the further adhesion to platelets, which contributed to inhibiting tumor cell migration (Borsig et al. 2007). Moreover, the antimetastatic properties of SC-CHS were also attributed to its anticoagulant activity, in which it targeted tissue factor and initiator of coagulation and reduced platelet attachment to tumors (Zhao et al. 2013). It has been reported that thrombin-induced platelet activation and aggregation were inhibited by SC-CHS, which led to attenuated cell migration by decreasing the adhesion between platelets and breast cancer cells and the adhesion of platelets/cancer cells to fibrinogen (Qian et al. 2015).

The fucose sulfate branched chain structure is the key to the inhibitory properties of SC-CHS. The interaction between P, L-selectin, and sLex was inhibited by the SC-CHS from *Ludwigothurea grisea*, which is a SC-CHS with 2,4-bisulfate on C-3 position of β-D-uronic acid. The fucose sulfate branched chain was removed by dilute acid, and its anti-P-selectin activity completely disappeared (Borsig et al. 2007).

3.1.4 Regulation of Lipid Metabolism

A growing number of studies have reported the regulating lipid metabolism functions of SC-CHS. Clinical studies have shown that oral administration of SC-CHS from *Holothuria leucospilota* (CHS-*Hl*) for 8 weeks (120 mg/day) significantly reduced the contents of serum total cholesterol (TC), serum triglyceride (TG), and apolipoprotein B (apoB) and increased the level of ApoA I in normal middle-aged and elderly people, as well as convalescent patients with cerebral embolism and patients with ischemic heart disease. ApoB reflects the level of low-density lipoprotein (LDL), which binds to the LDL receptor on vascular endothelial cells to deposit lipids in circulating blood on the vascular wall and forms atherosclerosis as a result. ApoA I is the main component protein of high-density lipoprotein (HDL) and is the activator of lecithin cholesterol acyltransferase. It promotes plasma HDL to receive free cholesterol from peripheral tissues and esterificate it, participating in the reverse transport of cholesterol and thus preventing atherosclerosis. Therefore, the above results suggested that CHS-*Hl* regulated lipid metabolism and protected cardiovascular and cerebrovascular functions (Shen 1997).

Studies indicated that the SC-CHS from *Acaudina molpadioides* (CHS-*Amo*) inhibited adipogenesis both in high-fat high-sucrose diet mice and in 3 T3-L1 adipocytes differentiation. Further studies indicated that the anti-adipogenesis activity of CHS-*Amo* involved the activated Wnt/β-catenin pathway to suppress the expression of peroxisome proliferators-activated receptor γ (PPARγ), CCAAT enhancer-binding proteins (C/EBPα), and sterol regulatory element-binding protein-1 (SREBP-1) (Xu et al. 2015). In addition, CHS-*Aj* decreased the levels of TC, TG, and LDL-C and increased the level of HDL-C in the serum of high-fat diet rats (Liu et al. 2012). Similarly, the SC-CHS from *Isostichopus badionotus* (CHS-*Ib*) and *Pearsonothuria graeffei* (CHS-*Pg*) were also found to have improved the serum lipid profile in high-fat high-fructose diet-fed mice (Wu et al. 2016).

The hyperlipidemic rat model induced by a diet supplemented with 1% cholesterol was fed with SC-CHS from *Metriatyla scabra* (CHS-*Ms*), which was mainly composed of hexuronic acid, aminohexose, and fucose for 6 weeks. The results showed that CHS-*Ms* significantly reduced the levels of TC, TG, and LDL-C and the atherosclerosis index in serum; increased the levels of high-density lipoprotein cholesterol (HDL-C) in serum; and reduced the levels of TC, TG, and phospholipids in liver. The decrease of cholesterol concentration might be related to the inhibition of HMG-CoA reductase activity by CHS-Ms or to the combination with oxidized LDL. In addition, CHS-*Ms* significantly improved the activity of lipoprotein lipase, promoted the decomposition of TG into fatty acids, glycerol and water, and then accelerated the metabolism of cholesterol (Liu et al. 2002). Studies have shown that SC-CHS can directly bind to plasma LDL, thus reducing blood lipid. Compared to SC-CHS with the straight-chain structure, that with fucose branched chain has higher affinity with plasma LDL. The fucose branched chain structure is the key to the interaction between SC-CHS and LDL, since its removal leads to this interaction disappearing (Tovar and Mourão 1996). Notably, depolymerized SC-CHS might retain the antihyperlipidemic activity of its native polysaccharide. For example, CHS-*Ib* with CS-E backbone (4.3 kDa) inhibited body weight gain, hyperlipidemia, and inflammation by inhibiting the lipid synthesis, and enhancing the lipidolysis-related genes in high-fat diet-fed mice (Li et al. 2018).

The injury and dysfunction of vascular endothelial cells mark the occurrence and development of atherosclerosis. The effects of SC-CHS from *Apostichopus japonicus*, *Thelenata anax*, and *Pearsonothria graeffei* on vascular endothelial cells were analyzed and compared in the Ox-LDL-induced injury model of human umbilical vein endothelial cell line (ECV304) in vitro. The results showed that different kinds of SC-CHS significantly inhibited the damage of vascular endothelial cells induced by ox-LDL by inhibiting the apoptosis and promoting the proliferation of vascular endothelial cells.

3.1.5 Regulation of Glucose Metabolism

Insulin resistance is an important characteristic of abnormal glucose metabolism. Studies have shown that insulin resistance induced by high-fat high-sucrose diet and/or high-fat diet is inhibited by various SC-CHS, such as CHS-*Amo*, CHS-Cf, and CHS-*Aj* (Hu et al. 2012, 2013a, 2014a, b; Li et al. 2015; Xu et al. 2022).

The effects and mechanisms of CHS-*Amo* and CHS-Cf on improving insulin resistance were compared in TNF-α-induced 3 T3-L1 insulin-resistant adipocyte model in vitro. The results showed that CHS-*Amo* and CHS-Cf significantly enhanced the glucose transport of model cells in a basal state and an insulin stimulated state, in which CHS-*Amo* had better effects than CHS-Cf did, and synergized with rosiglitazone. Mechanism studies showed that CHS-*Amo* and CHS-Cf significantly promoted the phosphorylation level of the p85 regulatory subunit of key target PI3K and the ser473 and thr308 sites of PKB in an insulin stimulated state and increased the level of GLUT4 protein in the extracellular membrane. In addition, CHS-*Amo* and CHS-Cf significantly promoted the phosphorylation level of ERK1/2. The results showed that CHS-*Amo* and CHS-Cf promoted glucose transport and improved insulin resistance by upregulating the PI3K/PKB insulin signal pathway and activating ERK.

In addition, the study in vitro showed that CHS-*Amo* and CHS-Cf significantly reversed the inhibitory effects on glucose transport of PI3K inhibitor wortmannin, PKB inhibitor MK2206, and ERK inhibitor U0126, respectively. Mechanism studies showed that CHS-*Amo* and CHS-Cf significantly promoted the expression of p-Ser473-PKB, p-Thr308-PKB, p-ERK1/2, and membrane-GLUT4 rather than the expression of p-p85PI3K under the intervention of wortmannin, suggesting that SC-CHS promoted the transmembrane transport of glucose of GLUT4 by activating PKB and ERK other than PI3K. Under the intervention of MK2206, CHS-*Amo* and CHS-Cf significantly increased the expression of p-Ser473-PKB, p-Thr308-PKB, and membrane-GLUT4, suggesting that SC-CHS could promote the transmembrane transport of glucose by upregulating the PI3K/PKB signaling pathway. Under the intervention of U0126, CHS-*Amo* and CHS-Cf significantly promoted the expression of p-ERK1/2 and cytomembrane-GLUT4, suggesting that SC-CHS could promote the transmembrane glucose transport of GLUT4 by activating ERK. The above results showed that both CHS-*Amo* and CHS-Cf improved insulin resistance, by upregulating the PI3K/PKB insulin signaling pathway and activating ERK to promote transmembrane glucose transport of GLUT4, and that the action targets of SC-CHS might be PKB and ERK (Hu et al. 2012, 2014a, b).

The effects of CHS-*Amo* and CHS-Cf on insulin resistance were verified in mice with insulin resistance induced by a high-fat high-sucrose diet. The results showed that CHS-*Amo* and CHS-Cf could significantly reduce the fasting blood glucose level, improve glucose tolerance, reduce insulin level and HOMA-IR value, increase quantitative insulin sensitivity check index (QUICKI) value, reduce body weight and fat weight, increase the level of serum adiponectin, and reduce serum resistin, leptin, and TNF-α, in which CHS-*Amo* was better than CHS-Cf. The above results suggested that CHS-*Amo* and CHS-Cf could effectively improve insulin resistance in vivo (Hu et al. 2013a, b, 2014a, b). Similarly, it has been reported that CHS-*Aj* significantly downregulated the expression of G6Pase and phosphoenolpyruvate carboxykinase (PEPCK), the gluconeogenesis rate-limiting enzyme, in high-fat diet-fed mice and primary hepatocytes through the inhibition of FoxO1, by activating the IRS1/PKB pathway, triggering AMP-activated protein kinase (AMPK), and suppressing cAMP-response element-binding protein (CREB) via the inhibited cAMP-dependent protein kinase (PKA).

The relative deficiency of insulin secreted by islet cells is the main feature of insulin resistance, and the apoptosis of islet cells induced by high glucose is the most immediate cause for the relative deficiency of insulin. In order to explore the inhibitory effect of SC-CHS on islet cell apoptosis in insulin-resistant mice, the microstructure changes of islet tissue in model mice were observed with optical microscope. The results showed that CHS-*Amo* and CHS-Cf significantly reduced the decline in islets in high-fat high-sucrose diet-fed mice, that the morphology of islets in mice recovered to close to normal, and that the surrounding basement membranes and connective tissues were clear and complete, indicating that CHS-*Amo* and CHS-Cf could effectively inhibit the apoptosis of islet cells. The molecular mechanism of SC-CHS inhibiting islet cell apoptosis was further studied. The results showed that CHS-*Amo* and CHS-Cf significantly downregulated the mRNA expression of Bid, Bax, cytochrome c, caspase-9, and caspase-3 apoptosis-promoting genes and upregulated the mRNA expression of Bcl-2 and Bcl-xL. Western blotting results showed that CHS-*Amo* and CHS-Cf significantly increased the protein levels of t-Bid, Bax, caspase-9, and cleaved-caspase-3 genes; promoted the release of cytochrome c from mitochondria; and inhibited the protein levels of Bcl-2 and Bcl-xL. The effects were even more significant when the use was combined with rosiglitazone. It was suggested that CHS-*Amo* and CHS-Cf inhibited the apoptosis of islet cells in model mice by blocking mitochondrial pathway, so as to improve insulin resistance (Hu et al. 2013a, b, 2014a, b). In addition, CHS-Cf also enhanced the value of HOMA-β (homoeostasis model assessment of pancreatic β-cell function), an index reflecting the normal function of β-cell in insulin secretion (Zhu et al. 2020).

Muscle and adipose tissues are the main target organs of glucose utilization, of which more than 75% of glucose treatment occurs in muscle. The PI3K/PKB insulin signaling pathway is the main pathway of glucose transport into cells. In order to explore the mechanism of SC-CHS intervening in insulin resistance and promoting glucose transport in vivo, the effects of SC-CHS on the mRNA and protein expression of key genes in insulin-mediated PI3K/PKB signaling pathway in skeletal muscle and adipose tissue were further studied. The results showed that CHS-*Amo* and CHS-Cf could significantly promote the mRNA expression of IR, IRS-1, PI3K, PKB, and GLUT4 genes in skeletal muscle of mice. Western blotting results showed that CHS-*Amo* and CHS-Cf had no significant effects on the total protein expression of key genes in the PI3K/PKB pathway but significantly promoted p-Tyr-IR-β, p-Tyr612-IRS-1, p-p85-PI3K, p-Ser473-PKB, p-Thr308-PKB, and membrane-GLUT4 in the skeletal muscle of mice. The expression of key genes in the PI3K/PKB signaling pathway for the adipose tissue of model mice was consistent with that in the skeletal muscle, and CHS-*Amo* and CHS-Cf significantly inhibited the level of p-Ser308-IRS-1. It was suggested that both CHS-*Amo* and CHS-Cf promoted the transmembrane transport of glucose by GLUT4, reduced blood glucose, and improved insulin resistance by activating the PI3K/PKB insulin signal pathway in the muscle and adipose tissue of mice (Hu et al. 2013a, b, 2014a, b).

Glucose metabolism is directly related to glucose homeostasis, and glycogen synthesis is another important approach for insulin to process glucose. In order to investigate the effect of SC-CHS on glucose metabolism and glycogen synthesis in vivo, the glycogen content of the liver as well as the activities of glycogen synthesis-related enzymes hexokinase (HK) and pyruvate kinase (PK), glucogen phosphorylase (GP), and glucose-6-phosphatase (G6Pase) were further measured in insulin-resistant mice. The results showed that CHS-*Amo* and CHS-Cf could significantly increase the content of glycogen, enhance the activity of HK and PK related to glycogen synthesis, and reduce the activity of gluconeogenic PK and G6Pase in the livers of insulin-resistant mice. Further, the mechanism of SC-CHS affecting glycogen synthesis was explored in mouse liver, and the results showed that CHS-*Amo* and CHS-Cf significantly promoted the mRNA expression of IR, IRS-2, PI3K, PKB, and GS genes and inhibited the mRNA expression of GSK-3β, the negative regulation gene of glycogen synthesis, in the liver of model mice. It was suggested that both CHS-*Amo* and CHS-Cf could reduce blood glucose and improve insulin resistance by regulating the activities of key enzymes related to glucose metabolism, activating the PKB/GSK-3β insulin signaling pathway, and promoting glucose metabolism and glycogen synthesis in the liver (Hu et al. 2013a, b, 2014a, b).

In summary, SC-CHS improved the insulin resistance of skeletal muscle, liver, and adipose tissues mainly by modulating insulin signaling to regulate the activation of its downstream kinases, such as PI3K, PKB, GSK-3β, and FoxO1, thereby controlling glucose uptake and metabolism.

3.1.6 Antiviral Effect

The effects of CHS from *Stichopus japonicus* (CHS-*Sj*) on herpes simplex virus type I (HSV-I) were studied on Vero cells as the target cell. The results showed that CHS-*Sj* significantly inhibited the proliferation of HSV-I in the concentration range of 0.2–0.25 mg/mL and showed a significant dose-response relationship. Further, CHS-*Sj* significantly prolonged the survival time and reduced the number of deaths for mice with intracranial infection through in vivo experiment (Luo et al. 2008).

The anti-SARS-CoV-2 activity of SC-CHS from *Pentacta pygmaea* (CHS-*Pp*) and (CHS-*Ib*) was ~12 times more efficient than heparin with no cytotoxic effects, which was suggested by IC_{50} values. The dissociation constant (KD) values obtained by surface plasmon resonance of the wild-type SARS-CoV-2 spike (S)-protein receptor-binding domain (RBD) and N501Y mutant RBD in interactions with the heparin-immobilized sensor chip were 94 and 1.8×103 nM, respectively. Competitive surface plasmon resonance inhibition analysis of CHS-*Pp* and CHS-*Ib* against heparin binding to wild-type S-protein showed IC_{50} values (in the nanomolar range) 6 and 25 times more efficient than heparin, respectively (Dwivedi et al. 2021).

In addition, SC-CHS from *Thelenota ananas* (CHS-*Ta*) significantly inhibited the replication and virus entry of human immunodeficiency virus (HIV), especially different strains of HIV-1 in vitro, including HIV-1IIIB, HIV-IRF, HIV-1KM018, HIV-1 TC-2, and T-20-resistant strains. Mechanism research showed that CHS-*Ta* inhibited the attachment and entry of HIV-1 by binding to the recombinant HIV-1 gp120 protein (a factor that promotes HIV entry into target cells) interfering with its normal function (Huang et al. 2013). Importantly, the antiviral activity of SC-CHS was related to its molecular weight and carboxylation, especially sulfated fucose branches (Lian et al. 1830).

Various viruses are attached to cell surfaces for initiating cell entry, which relies on charge-dependent nonspecific interactions with heparan sulfate, a highly sulfated glycosaminoglycan. Therefore, the inhibition of virion-heparan sulfate interactions is a broad-spectrum antiviral strategy. Studies have suggested that enveloped (herpesvirus human cytomegalovirus) and non-enveloped (adenovirus) DNA viruses are inhibited by SC-CHS. The inhibited viral attachment and entry activities by SC-CHS from *Isostichopus badionotus* (CHS-*Ib*) might result from the interactions with virions (Zoepfl et al. 2021).

3.2 Sea Cucumber Fucoidan (SC-FUC)

3.2.1 Improvement of Liver Function

A growing number of studies have shown that sea cucumber fucoidan (SC-FUC) has an alleviating effect on liver injuries. The preventive and therapeutic effects of SC-FUC from *Acaudina molpadioides* on liver injury were studied in mice affected by acute alcoholism, and the results showed that SC-FUC significantly decreased the activities of ALT and AST in mice. In addition, the activities of ADH and ALD, the key enzymes of ethanol metabolism, were significantly increased by SC-FUC, and the activities of CAT and glutathione peroxidase (GSH-Px) related to oxidative stress in the liver were significantly increased by SC-FUC. The above results showed that SC-FUC significantly promoted the metabolism of ethanol and the elimination of a large amount of reactive oxygen species produced in the process of ethanol metabolism, thereby protecting the liver (Teng et al. 2011).

In addition, the preventive and therapeutic effects of SC-FUC on alcoholic liver disease were further explored in a mouse model of alcoholic liver injury caused by long-term excessive intake of alcohol. The results showed that SC-FUC significantly decreased the levels of ALT and AST, increased the activity of GSH-Px and the content of glutathione (GSH), and decreased the content of malondialdehyde (MDA) in serum, suggesting that SC-FUC significantly increased the antioxidant level and inhibited the damage of reactive oxygen species produced by ethanol metabolism to the liver. In addition, SC-FUC significantly decreased the content of serum lipids and liver lipids and increased the activity of ADH and ALDH in the liver. The decrease of liver lipids was related to SC-FUC improving the intracellular environment, restoring the tricarboxylic acid cycle, promoting gluconeogenesis, reducing fatty acid synthesis, and promoting fatty acid extracellular transport. The effect of SC-FUC on the expression of cytochrome P450 (CYP450) in the livers of mice subjected to long-term alcohol consumption was studied. The results showed that SC-FUC significantly inhibited the increase of CYP2e1 and CYP3a mRNA expression induced by ethanol, which is an important physiological substrate of CYP2e1 and often accumulates in large quantities. MEOS was activated and the activity of CYP2e1 was increased. The damage of the liver in this process was mainly manifested in two aspects: Firstly, a large amount of ethanol was oxidized through MEOS in hepatocytes, producing a large number of oxidative stress products, such as OH^-, O_2^-, and H_2O_2, which could activate phospholipase and lipid peroxidation, so as to change the permeability and fluidity of cell membrane. This process further changed the microenvironment of receptors, enzymes, and ion channels combined with it, causing damage to the liver, especially to mitochondria. Secondly, CYP2e1 could also increase the toxicity of some substances, which eventually leads to inflammation and fibrosis and potentially even to cirrhosis and cancer of the liver. Therefore, the balance of CYP2e1 activity is key to the treatment of fatty liver disease. SC-FUC significantly inhibited the increase of CYP2E1 and CYP3A mRNA expression induced by ethanol. The above results suggested that SC-FUC protected the liver by reducing lipid accumulation and promoting ethanol metabolism (Zhu et al. 2012). In addition, it has been reported that SC-FUC from *Acaudina molpadioidea* significantly alleviated inflammatory response via the inhibition of JNK and IKKβ/NF-κB pathways in the liver of high-fat high-sucrose diet mice.

3.2.2 Regulation of Glucose Metabolism

The regulatory effects on glucose metabolism of SC-FUC derived from *Cucumaria frondose*, *Thelenata anax*, and *Acaudina molpadioides* were studied on model mice affected by type II diabetes induced by high-fat high-sucrose diet. The results showed that the three forms of SC-FUC could significantly reduce fasting blood glucose, improve oral glucose tolerance, reduce serum insulin level, and increase glycogen content in the liver and muscle of diabetes mice, indicating that SC-FUC significantly reduced blood glucose and improved insulin resistance. Mechanism study showed that SC-FUC regulated the PI3K/Akt signaling pathway to promote glycogen synthesis mainly by reducing the expression of GSK3β and increasing the expression of IR, PI3K, and AKT2 mRNAs in the liver and muscle. In addition, the effects of SC-FUC on adipocytokines adiponectin, leptin, and TNF-α were detected, and the results showed that the three forms of SC-FUC significantly improved insulin resistance by increasing the ability of adipose tissue to secrete adiponectin and by reducing serum leptin and TNF-α levels (Shi et al. 2013). Consistently, it has also been reported that SC-FUC from *Thelenota ananas* ameliorated hyperglycemia, restored hypertriglyceridemia and hypercholesterolemia, decreased inflammatory status and oxidative stress, and protected against liver injury. In addition, it improved insulin resistance and promoted accumulation of hepatic glycogen by activating IRS/PI3K/AKT signaling and by regulating GSK-3β gene expression in rat models with diabetes mellitus type 2 (T2DM) induced by high-fat diet and rat models with streptozotocin-induced T2DM (Zhu et al. 2020).

3.2.3 Antitumor Effects

The antitumor activities of different components of sea cucumber SC-FUC were studied through analysis of HepG2, HeLa, 95D, HGC-27, and other tumor cell lines. The results showed that 430kD SC-FUC significantly inhibited the growth of HGC-27, and its IC_{50} values were 1158, 400, and 127 μg/mL at 48, 72, and 96 h, respectively. The anti-metastasis activity and mechanism of SC-FUC under hypoxia were studied in 25 μM CoCl2-induced human gastric cancer cell line HGC-27 model for chemical hypoxia in vitro. The results showed that SC-FUC significantly inhibited the growth of HGC-27 cells with time- and dose-dependent relationships, and its IC_{50} values were 169.9 and 100.0 μg/mL at 72 and 96 h, respectively (Yang et al. 2013). In addition, SC-FUC significantly reduced the adhesion between HGC-27 cells and matrix and HUVEC cells and showed a significant time-dependent and dose-dependent relationship (Yang et al. 2013). Migration and invasion are the key steps of tumor cell metastasis. Transwell cell migration and invasion model experiments showed that the migration and invasion ability of HGC-27 cells was decreased significantly by SC-FUC treatment, by 67.96% and 59.73%, respectively.

An important link in the process of angiogenesis is the process by which endothelial cells migrate, fuse with each other, and form a tubular structure. The results of tubule formation experiment in vitro showed that long tubular cell clusters formed by HUVEC cells took shape, following the induction of Matrigel for 24 h, and were cross-linked together to form a network structure. After SC-FUC treatment, the ability of endothelial cells to form the tubular structure was significantly reduced, and endothelial cells could only aggregate into cell clusters, losing the ability of outward cross-linking, with a significant dose-response relationship. This showed that SC-FUC significantly inhibited the formation of vascular endothelial cells. In addition, the effect of SC-FUC on angiogenesis was studied by the chicken embryo allantoic membrane (CAM) method. The results showed that SC-FUC significantly decreased the number and length of branches of blood vessels, suggesting SC-FUC significantly inhibited angiogenesis.

Mechanism studies showed that SC-FUC inhibited the expressions of hypoxia-inducible factor (HIF-1α), heparinase (HPA), matrix metalloproteinase-2 (MMP-2), MMP-9, and vascular endothelial growth factor (VEGF) and increased the expressions of TIMP-1 and TIMP-2, thereby inhibiting tumor metastasis. As a nuclear transcription factor regulating oxygen balance, HIF-1 plays a key role in the process of tumorigenesis and development. The overexpression of HIF-1α can regulate the transcription of many target genes related to tumor cell apoptosis, tumor angiogenesis, and tumor cell energy metabolism, promoting tumor growth, invasion and metastasis. HPA can specifically degrade heparan sulfate proteoglycan in extracellular matrix (ECM). On the one hand, HPA can destroy the integrity of ECM; on the other hand, HPA can release basic fibroblast cytokines and VEGF bound in ECM to promote tumor angiogenesis, tumor invasion, and metastasis. Matrix metalloproteinases (MMP)-2 and MMP-9 are the main enzymes that degrade type IV collagen in ECM. Their overexpression is closely related to tumor angiogenesis, promoting local tumor infiltration and lymph node metastasis. TIMPs are specific inhibitors of metalloproteinases and play an important role in regulating the activity of MMPs. TIMP-1 and TIMP-2 can combine with activated MMP-9 and MMP-9 to form a 1:1 complex that inhibits enzyme activity or combine with zymogen to form a stable complex that prevents zymogen activation. It is suggested that SC-FUC can maintain the balance between MMPs and TIMPs by reducing the ratio of MMPs/TIMPs, so as to protect the integrity of ECM barrier and inhibit tumor cell metastasis. VEGF is a key regulator of tumor-induced angiogenesis, which is regulated by HIF-1α (Wang et al. 2012a; Yang et al. 2013).

Further, the antitumor metastasis activity and mechanism of SC-FUC in vivo were verified in the spontaneous lung metastasis model of Lewis lung cancer in mice. The results showed that SC-FUC significantly inhibited the growth of tumors in mice with an inhibition rate of 45.5% and significantly reduced the number of metastatic nodules on the lung surface with an inhibition rate of 66.0%. Similarly, SC-FUC significantly reduced the expressions of HIF-1α, VEGF, HPA and MMP-2/−9 in tumor tissue of mice (Zhang et al. 2011). In addition, SC-FUC isolated from sea cucumber *Stichopus variegatus* inhibited colony formation of human breast cancer T-47D and MDA-MB-231 cell lines and acted against migration of MDA-MB-231 cells in vitro (Thinh et al. 2018). In conclusion, SC-FUC showed significant antitumor cell metastasis activity, and the mechanism was related to the downregulation of HIF-1α and MMP-9/MMP-2 signaling pathway and the further downregulation of VEGF expression.

3.2.4 Prevention and Treatment of Gastric Ulcer

Gastric ulcers are a common disease of the digestive system and are one of the consequences of ethanol consumption. The effects of SC-FUC from *Acaudina molpadioides* on gastric ulcer were studied in a rat model with gastric ulcer induced by ethanol. The results showed that SC-FUC significantly reduced the ulcer index, and the ulcer inhibition rates reached 66.53%. Histological analysis showed that SC-FUC protected gastric mucosal injury induced by ethanol and the integrity of mucosal cells and reduced the invasion of inflammatory cells.

Recent studies have shown that the hydrolysis and allosteric effect of matrix metalloproteinases (MMPs) on peripheral matrix (ECM) are involved in the pathogenesis and cure of gastric ulcer. MMP-2 is a gelatinase, which can directly destroy gastric mucosa; MMP-9 is an inflammatory protein that participates in inflammatory response and is regulated by NF-κB-mediated regulation of transcriptional control. There is a certain amount of MMP-2 and MMP-9 in normal gastric tissue, and under the induction of ethanol, the expression of MMP-2 is upregulated. However, the expression of MMP-2 is significantly decreased by SC-FUC. Consistent with the appearance of a large number of inflammatory cells, the expression of MMP-9 increases sharply when induced by ethanol, showing a serious inflammatory response. The intake of SC-FUC effectively reduces the expression of MMP-9 and the number of inflammatory cells, indicating that the anti-inflammatory effect of SC-FUC is also involved in inhibiting gastric ulcer. The above studies showed that SC-FUC significantly inhibited the secretion of MMP-2, thereby reducing the degradation of gastric mucosal matrix. In addition, inhibited activation of JNK and ERK signaling pathway was also involved in the inhibition of SC-FUC against ethanol-induced gastric ulcer (Wang et al. 2012b).

Further, the preventive effects of SC-FUC with different molecular weights on ethanol gastric ulcer were compared and studied, and the results showed that the anti-ulcer activity of SC-FUC was closely related to its molecular weight. SC-FUC hydrolyzed to 508 kDa–54 kDa had significantly better effects (about 5 times) on inhibiting ethanol-induced gastric ulcer than those of the original SC-FUC, indicating the existence of the smallest fragment to achieve this activity.

3.3 Characteristics of Digestion and Absorption

The activities of SC-CHS and SC-FUC are closely related to molecular weight. Representative activities of SC-CHS and SC-FUC with different physical and chemical properties are summarized in Table 1. Studies have shown that the characteristics of high molecular weight and high charge may prevent SC-CHS and SC-FUC from entering circulation through the gastrointestinal tract. However, many studies have shown that about 5%–15% of polysaccharides can be absorbed into the body. Absorption efficiency is affected by factors such as polysaccharide molecular weight and charge density, while the biological activity of oral polysaccharides may be related to digestibility. The digestion and absorption of partially degraded SC-FUC (12kD) were studied in rats. The results suggested that the absorption peak was reached at the fifth hour after oral administration of SC-FUC (50 mg/kg) in rats with the serum fucose content of 90 μg/mL, but the recovery rate of fucose in urine was only 0.1%. After injection of SC-FUC (1 mg/kg), the recovery rate of fucose in urine was higher than 5%. The results showed that at least 2% of SC-FUC was absorbed into circulation through the gastrointestinal tract (Imanari et al. 1999).

The digestion and absorption characteristics of SC-FUC with different molecular weights were studied, and the results suggested that serum fucose contents arose to varying degrees after oral administration of SC-FUC with different molecular weights (MW: > 700 kDa, 437.8 kDa, 52.7 kDa, and 20.5 kDa); however, the absorption efficiency decreased with the increase of molecular weight. Similarly, studies in humans, horses, and dogs found that the absorption process of SC-FUC was strongly affected by physical and chemical properties, such as structure and molecular weight (Pescador et al. 1980; Ronca and Conte 1993; Volpi 2002).

It has been reported that the proportion of polysaccharides passing through the intestinal wall is inversely proportional to molecular weight (Adebowale et al. 2002; Cho et al. 2004). Degraded SC-FUC started to enter circulation rapidly from 0.5 h, indicating that the stomach and small intestine play an important role in the digestion and absorption of low-molecular-weight SC-FUC, while high-molecular-weight SC-FUC is mainly absorbed after degradation by colon or cecum microorganisms. The results of 14C-labeled polysaccharides tracer analysis showed that a small amount of polysaccharides can pass through small intestinal epithelial cells completely while the macromolecular polysaccharides are degraded into small molecules by anaerobic bacteria in the colon. In addition, Ronca et al. reported that the main means of gastrointestinal absorption of polysaccharides was endocytosis, but the polysaccharides with low molecular weight and low charge density were digested and absorbed preferentially due to selective absorption by the digestive tract or the desulfurization or depolymerization during endocytosis (Ronca and Conte 1993).

4 Sea Cucumber Lipid

The lipid contents in sea cucumber are relatively low in most cases, but in some sea cucumbers, they are greater than 15% (dry weight). The structure of sea cucumber lipid is complex, mainly including phospholipids, sphingolipids, sterols, and pigments. The saturated fatty acids in sea cucumber belong to C12 ~ C24 even carbon

Table 1 Representative activities of SC-CHS and SC-FUC with different physical and chemical properties

Sea cucumber species	Molecular weight (kDa)	Chemical composition (molar ratios) (GlcA: GalNAc:Fuc:-SO3−)	Functions	References
Sea cucumber chondroitin sulfate				
Apostichopus japonicus	76	~1.0:0.9:1.1:3.6	Anticoagulation	Guan et al. (2019)
Bohadschia argus	70	~1.0:1.0:1.0:3.9	Anticoagulation	Yin et al. (2018)
Cucumaria frondosa	58	~1.0:1.0:1.2:3.4	Anticoagulation	Liu et al. (2016b)
Hemioedema spectabilis	44	~1.2:1.0:1.1:3.9	Anticoagulation	Ustyuzhanina et al. (2018)
Holothuria albiventer	47	~1.0:1.0:1.0:3.2	Anticoagulation	Cai et al. (2019)
Holothuria coluber	55	~1.0:1.0:1.0:3.8	Anticoagulation	Yang et al. (2018b, 2020)
Holothuria mexicana	98	~1.0:1.0:1.3:2.4	Anticoagulation	Li et al. (2018)
Holothuria mexicana	100	~1.0:1.1:1.5:3.5	Anticoagulation	Mou et al. (2017)
Holothuria nobilis	55	~1.0:1.0:0.8:3.0	Anticoagulation	Wu et al. (2015)
Holothuria poli	46	~1.0:1.0:2.0:4.9	Anticoagulation	Ben Mansour et al. (2017)
Holothuria scabra	61	~1.0:0.8:1.0:3.8	Anticoagulation	Wu et al. (2015)
Holothuria scabra	69	~1.0:1.7:2.3:3.3	Anticoagulation	Yang et al. (2018a)
Isostichopus badionotus	109	~1.0:0.7:0.9:3.1	Anticoagulation	Shi et al. (2019)
Massinium magnum	27	~1.0:0.9:0.9:3.9	Anticoagulation	Ustyuzhanina et al. (2017)
Pattalus mollis	60	~1.0:1.0:1.0:3.2	Anticoagulation	Zheng et al. (2019)
Pearsonothuria graeffei	73	~1.0:0.8:1.5:2.6	Anticoagulation	Chen et al. (2013)
Stichopus chloronotus	25	~1.0:0.9:2.0:6.8	Anticoagulation	Ustyuzhanina et al. (2018)
Stichopus herrmanni	64	~1.0:1.0:1.0:3.8	Anticoagulation	Li et al. (2017b)
Stichopus horrens	46	~1.0:1.0:1.0:4.0	Anticoagulation	Shang et al. (2018)
Stichopus tremulus	100	~1.0:0.8:1.2:3.0	Anticoagulation	Chen et al. (2011)

(continued)

Table 1 (continued)

Sea cucumber species	Molecular weight (kDa)	Chemical composition (molar ratios) (GlcA: GalNAc:Fuc:-SO3−)	Functions	References
Stichopus variegatus	70	~1.0:1.0:1.0:3.9	Anticoagulation	Zhao et al. (2015)
Thelenota ananas	66	~1.0:1.0:1.1:3.6	Anticoagulation	Wu et al. (2012)
Apostichopus japonicus	98	~1.0:1.0:1.1:3.9	Immune regulation	Wang et al. (2019a, 2017b)
Assinium magnum	–	–	Immune regulation	Anisimova et al. (2017)
Holothuria poli	47	~1.0:1.0:1.4:4.7	Immune regulation	Niu et al. (2020)
Holothuria tubulosa,	42	~1.0:1.0:1.3:4.7	Immune regulation	Niu et al. (2020)
Stichopus chloronotus	111	~0.8:1.0:0.9:3.0	Immune regulation	Jiang et al. (2021)
Isostichopus badionotus	109	–	Antitumor metastasis	He et al. (2014)
Cucumaria frondosa	12	~1.0:1.0:1.1:3.2	Anticancer	Liu et al. (2016c)
Ludwigothurea grisea	40 or 8	~1.0:0.9:1.2:0.7 or ~1.0:0.9:0.3:0.6	Antitumor metastasis	Borsig et al. (2007)
Stichopus japonicus	–	–	Antitumor metastasis	Zhao et al. (2013)
Holothuria leucospilota	115	–	Antitumor metastasis	Qian et al. (2015)
Isostichopus badionotus	109	~1.0:0.7:0.9:3.1	Lipid metabolism regulation	Wu et al. (2016)
Pearsonothuria graeffei	73	~1.0:0.8:1.5:2.6	Lipid metabolism regulation	Wu et al. (2016)
Apostichopus japonicus	–	–	Lipid metabolism regulation	Liu et al. (2012)
Metriatyla scabra	–	–	Lipid metabolism regulation	Liu et al. (2002)
Acaudina molpadioides	22	~1.0:1.1:1.6:ND	Lipid metabolism regulation	Xu et al. (2015)
Acaudina molpadioides	22	~1.0:1.1:1.6:ND	Glucose metabolism regulation	Hu et al. (2013b, 2014a)
Cucumaria frondosa	15	~1.0:1.5:1.1:ND	Glucose metabolism regulation	Hu et al. (2014b)
Cucumaria frondosa	22	~1.0:1.1:1.6:ND	Glucose metabolism regulation	Hu et al. (2013a)

(continued)

Table 1 (continued)

Sea cucumber species	Molecular weight (kDa)	Chemical composition (molar ratios) (GlcA: GalNAc:Fuc:-SO3−)	Functions	References
Cucumaria frondosa,	–	~1.8:5.0:71.7:ND	Glucose metabolism regulation	Zhu et al. (2020)
Thelenota ananas	–	~1.3:3.2:80.8:ND	Glucose metabolism regulation	Zhu et al. (2020)
Stichopus japonicus	–	–	Antiviral	Luo et al. (2008)
Thelenota ananas	5–23	–	Antiviral	Huang et al. (2013)
Isostichopus badionotus	75	–	Antiviral	Zoepfl et al. (2021)
Pentacta pygmaea	~10–60	–	Antiviral	Dwivedi et al. (2021)
Sea cucumber fucoidan				
Acaudina molpadioides	–	–	Improving liver function	Teng et al. 2011; Zhu et al. (2012)
Thelenata anax	–	~0.2:0.3:1:ND	Glucose metabolism regulation	Shi et al. (2013)
Thelenota ananas	–	~1.3:3.2:80.8:ND	Glucose metabolism regulation	Zhu et al. (2020)
Isostichopus badionotus	–	–	Antitumor metastasis	Yang et al. (2013); Zhang et al. (2011)
Cucumaria frondosa	–	~0.3:0.1:1.0:ND	Antitumor metastasis	Wang et al. (2012b); Yang et al. (2013)
Stichopus variegatus	–	~ 0.8:1.0:1.0:ND or ~0.9:1.0:1.0:ND	Anticancer	Thinh et al. (2018)
Acaudina molpadioides	1614	–	Anti-gastric damage	Wang et al. (2012b)

chains, as well as C15 ~ C23 odd carbon chains. Docosanoic acid (C23:1, n-9) is unique to the sea cucumber. Polyunsaturated fatty acids of sea cucumber phospholipids are mainly C20:5n-3 (EPA) and C22:6n-3 (DHA). In addition, sphingolipids, cerebrosides, gangliosides, and sterols in sea cucumbers have special structural characteristics and possess beneficial biological functions that can contribute to treatment of Alzheimer's disease, fatty liver, tumor, cancer cachexia, and metabolic syndrome.

4.1 Sea Cucumber Phospholipids

Sea cucumber phospholipids account for about 1/3 of the total lipid content. Phosphatidylcholine (PC) and phosphatidylethanolamine (PE) are the main components of sea cucumber phospholipids, accounting for more than 80% of the total phospholipids of sea cucumber. The main fatty acids of PC are C20:5n-3 and C20:4n-6, while the main fatty acids of PE are C20:4n-6, C20:5n-3, and C18:0. In addition, PE is rich in plasmalogen (pPE), of which the content almost accounts for 50%, while the content of plasmalogen in PC is less than 10% (Lou QM 2011).

4.1.1 Brain Function

1. Improvement of Brain Development

DHA accounts for 10%–15% of the fatty acids in the cerebral cortex and participates in a variety of physiological processes in the body, including the construction of the lipid bilayer, signal transduction, neurotransmitter transmission, synapse formation, and the proliferation and differentiation of the neurons, to maintain normal brain development and function. Fu et al. (2022) found that EPA-PC and EPA-pPE supplementation from sea cucumber phospholipids could increase the content of DHA in the brain of congenital n-3 deletion mice. A lack of n-3 polyunsaturated fatty acids (PUFAs) in a mother's diet significantly reduced the amount of DHA in the brain of the offspring, which might affect offspring's brain function. Dietary supplementation with EPA-pPE (0.05%, w/w) and EPA-PC (0.05%, w/w) for 2 weeks could effectively increase DHA levels in the brain (Table 2).

2. Neurodegenerative Diseases

Alzheimer's disease (AD) is a typical disease in this context. AD is a progressively developing neurodegenerative diseases characterized by memory and cognitive impairment with various neuropsychiatric symptoms and behavioral disorders. There are numerous hypotheses about the pathogenesis of AD, such as β-amyloid (Aβ) deposition, the tau protein theory, the gene mutation theory, the cholinergic injury theory, the chronic inflammation theory, oxidative stress, and free radical damage. However, the pathogenesis of AD has not yet been fully elucidated. Wu et al. (2014) studied the effect of EPA-PL from the sea cucumber on learning and memory functions in senescence-accelerated prone mouse strain 8 (SAMP8) and scopolamine-induced mice. SAMP8 mice showed significant learning and memory impairment. After administration of EPA-PL extracted from sea cucumber, the

Table 2 Relative abundance of DHA attached glycerophospholipid of cerebral cortex

DHA%	n-3 Def	EPA-PC	EPA-pPE
PC	7.2 ± 0.2	10 ± 0.3	12 ± 0.5
PE	17 ± 0.4	16 ± 0.4	15 ± 0.4
PS	25 ± 1	28 ± 0.7	28 ± 0.7

learning and memory ability of mice was significantly improved. The Barnes maze experiment showed that sea cucumber phospholipids could shorten the period required by SAMP8 mice to find the target box and could reduce the number of errors; the same experimental results of using the Morris water maze showed that sea cucumber phospholipids could reduce the period required by SAMP8 mice to find the hidden platform, increase the number of times the mice were able to cross the platform, and prolong the time they remained in the target quadrant. The mechanism study showed that the effect of sea cucumber phospholipids on learning and memory impairment was related to decreased lipid accumulation in brain, brain DNA and RNA peroxidation, regulation of oxidative stress in the brain, inhibition of NO cytotoxicity, and reduction of neuronal apoptosis. Zhou et al. (2016) studied the effect of EPA-PL from sea cucumber on learning and memory in the dementia mice model induced by scopolamine injection. The results showed that in behavioral tests, the time required by EPA-PL groups was significantly reduced, while the number of times the test subjects were able to cross the platform and the time spent in the target quadrant were increased. Dietary EPA-PL increased the AChE activity and the SOD activity increased in hippocampus, cortex, and white matter. Dietary EPA-PL also decreased the MAO activity in white matter. Meanwhile, the experiment also compared the phospholipids of different fatty acid compositions with sea cucumber phospholipids. It found that different fatty acid compositions of PC could all diminish cognitive decline and biological damage and protect the brain. EPA-PL from sea cucumber was partly enhanced to the advantageous effects. Wu et al. (2014) studied the protective effect of EPA-PL from sea cucumber on oxidative stress of PC12 cells. The PC12 cells were treated with various concentrations of EPA- PL liposome for 24, 48, and 72 h, and then MTT assay was used to determine the cell viability. There was no toxic effect of EPA-PL on PC12 cells within the tested concentration range. PC12 cells were pretreated with EPA-enriched PL for 24 h, and then the cells were exposed to H_2O_2 or t-BHP for 4 h. After that, MTT assay was used to measure the cell viability. The results showed that EPA-PL resulted in an enhancement of survival in H_2O_2 or t-BHP damaged PC12 cells. EPA- PL intervention made the survival rates of H_2O_2 or t-BHP damaged PC12 cells reach 80% and 110%, respectively.

The accumulation of Aβ1–40 peptides in the brain causes Aβ plaque, which is the recognized pathological basis of Alzheimer's disease. Intraventricular injection of Aβ1–40 in rats will lead to oxidative stress, neuroinflammation, neuronal apoptosis, and synaptic dysfunction, resulting in cognitive impairment. It was used to evaluate the effect of drugs on Alzheimer's disease in the middle and late stages. Wen et al. (2016b) investigated the effects of EPA-PLs from the sea cucumber on Aβ-induced cognitive impairment in rats. The results showed that administration of EPA-PLs (150 mg/kg·day, i.e., 27 days) significantly alleviated Aβ-induced cognitive deficiency. The results of positioning navigation experiments showed that compared with the model group, the time and delay for rats to find the platform in EPA-PL group were significantly shortened and that this improvement persisted until the end of the experiment. The numbers of times the rats in the experimental group crossed the platform and stayed in the space exploration quadrant were both significantly less than those in the sham platform group. The times required for dietary EPA-PL rats to cross the platform and the time they remained in the target quadrant were significantly increased compared with the model group. Behavioral experiments showed that EPA-PL improved Aβ1–40-induced cognitive deficiency in a rat model of Alzheimer's disease.

Parkinson's disease (PD) is another common neurodegenerative disease in middle-aged and elderly people, which mainly results from the progressive degeneration of dopaminergic neurons and pathological changes in the formation of Lewy bodies characterized by prominent clinical features of decreasing dopamine in the striatum, biochemical changes in dopamine, and acetylcholine transmitter imbalance, which cause motor symptoms (tremors, muscle rigidity, bradykinesia, and posture balance disorder) and non-motor symptoms, such as decreased smell sense, constipation, abnormal sleep behavior, and depression (Kalia et al. 2015). The development of PD is significantly influenced by the dietary lipid composition. MPTP (1-methyl-4-phenyl-1,2,3,6-tetrahydropyridine) is highly fat soluble and can easily penetrate the blood-brain barrier. MPTP can be transformed into its effective component MPP+ in the brain under the action of glial monoamine oxidase. After MPP+ is absorbed into the mitochondria of dopaminergic neurons by dopamine transporters, it can inhibit the activity of mitochondrial complex I and lead to the degeneration and death of dopaminergic neurons. Therefore, MPTP-induced mouse model is a commonly used neurotoxin-induced model. Neurotoxins can be used to establish animal models similar to human PD and partially reflect the pathological characteristics of PD, which can provide a good approach for PD therapeutic drug screening and mechanism research. Wang et al. (2016a, b) studied the effects of EPA-PL extracted from the sea cucumber on PD mice induced by MPTP. Dietary addition of 2% EPA-PL for 2 weeks could significantly improve the movement disorder induced by MPTP. The mice supplemented with EPA-PL showed extended period of delay on the rotarod, longer total distance traveled in the open field, and higher frequency entering into the center compared with MPTP group in open field maze test. It showed that the intervention of EPA-PL significantly alleviated the MPTP induced Parkinson's disease. EPA-PL alleviated the loss of dopaminergic neurons cell in MPTP-induced brains significantly. It was found that dietary EPA-PL significantly reduced the expressions of Bax, Bcl-xL, caspase-9, caspase-3, PARP, and other genes related to mitochondrial oxidative stress and apoptosis caused by MPTP in mouse striatum. These studies indicated that EPA-PL suppressed MPTP-induced oxidative stress and apoptosis, thereby alleviating the loss of dopaminergic neurons via the mitochondria-mediated pathway and mitogen-activated protein kinase pathway.

Zhu et al. (2021) studied the oxidative damage of sea cucumber phospholipids on primary hippocampal neurons using a nerve cell model. The function of EPA-pPE and EPA-PE on oxidative damage prevention after H_2O_2 and t-BHP challenge in primary hippocampal neurons was investigated. The results showed that neurons pretreated with EPA-pPE and EPA-PE demonstrated the ability to alleviate oxidative damage, which was proven by the increased cell viability. Moreover, the shape and number of neurons were more similar to those of the control group. Antioxidant activity, apoptosis, and the TrkB/ERK/CREB signaling pathway were studied to explore the mechanisms. EPA-PC and EPA-PE in sea cucumber phospholipids inhibited microglia inflammation. Microglia are considered to be the main cells regulating inflammation in the brain. Persistent chronic inflammation in the brain is one of the important causes of PD. Zhu et al. (2021) studied the anti-inflammatory activity of sea cucumber phospholipids using the LPS-induced microglia model. The results showed that LPS-induced microglia cell bodies became larger and the processes became shorter. The microglia preincubated with EPA-PC and EPA-pPE had bright cell bodies and longer processes. It was found that both EPA-PC and EPA-pPE reduced the inflammation of microglia induced by LPS significantly. Meanwhile, EPA-PC and EPA-pPE significantly slowed the increase of TNF-a induced by LPS. In addition, EPA-PC and EPA-pPE reduced the M1 marker gene expression of COX-2, and EPA-PC increased the M2 marker gene expression of Arg-1, indicating that EPA-PC and EPA-pPE promoted the polarization of microglia from M1 to M2, thus reducing microglia inflammation.

4.1.2 Prevention of Metabolic Syndrome

Metabolic syndrome (MS) is a metabolic disorder syndrome based on the disorder of glucose metabolism and lipid metabolism. It is a main risk factor in cardiovascular diseases. Abnormal glucose metabolism based on insulin resistance leads to elevated fasting blood glucose, which increases the risk of type 2 diabetes. Liu et al. (2014) studied the improvement of sea cucumber phospholipids on metabolic syndrome mice induced by a high-fat diet. It was found that dietary sea cucumber phospholipid could significantly improve the oral glucose tolerance of mice with metabolic syndrome. The results showed that EPA-PL intervention decreased blood glucose at 30 min and the area under the OGTT curve declined significantly. In addition, the administration of EPA-PL decreased the serum insulin level significantly compared with the control group. It was suggested that dietary EPA-PL significantly alleviated insulin resistance in high-fat-diet-fed mice. Adiponectin is an adipocytokine with insulin sensitization. The study also found that dietary EPA-PL increased the level of adiponectin in the blood significantly, indicating that dietary EPA-PL affected the reconstruction of adipose tissue, and then affected the ability of adipose tissue to secrete adiponectin, thus alleviating the metabolic syndrome.

Obesity is accompanied by abnormal levels of inflammatory factors and adipokines. Persistent chronic inflammation is considered to be a main factor of metabolic syndrome. Chronic inflammatory reaction in adipose tissue is considered to be an important factor in the decline of insulin sensitivity. Dietary EPA-PL decreased the levels of serum MCP-1 and IL-6 in mice with metabolic syndrome significantly, indicating that EPA-PL alleviated insulin resistance and metabolic syndrome by regulating the level of inflammation. Hyperlipidemia is a typical feature of metabolic syndrome and is considered to be a major cause of insulin resistance. Liu et al. (2014) studied the effect of dietary intake of EPA phospholipids on blood lipids in mice experiencing hyperlipidemia induced by a high-fat diet. The

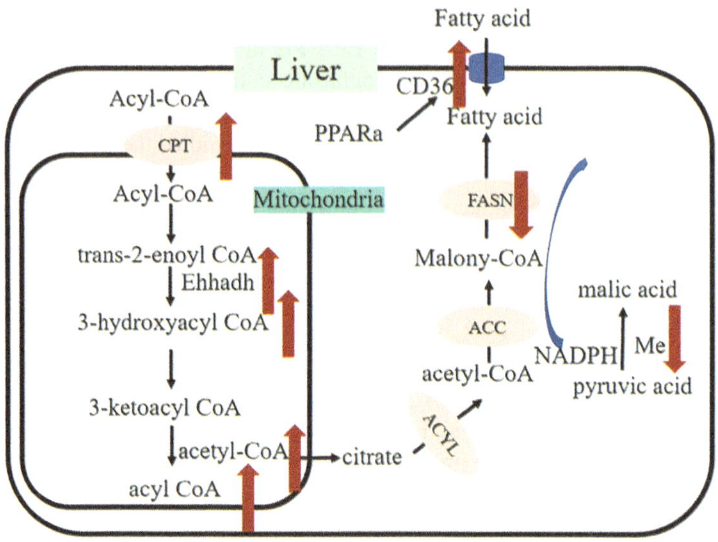

Fig. 1 Effect of EPA-PL on liver lipid metabolism related pathways

results showed that EPA-PL decreased the serum TG level by 50% and decreased the liver TG level by 60% after dietary intervention. In addition, EPA-PL decreased the level of liver cholesterol and increased the level of liver phospholipids significantly. The mechanism study found that dietary EPA-PL from sea cucumber had a significant effect on the lipid metabolism of liver (Fig. 1). Dietary EPA-PL decreased the levels of enzyme activity for FAS and ME genes related to fatty acid synthesis. Dietary EPA-PL increased the enzyme activity related to fatty acid decomposition and the utilization of CPT and peroxisomal β-oxidation significantly in mitochondria. In addition, dietary sea cucumber EPA-PL increased the level of mRNA transcription and the protein level of relevant enzymes for liver CPT1α, acox1, Ehhadh, CD36, and other genes related to fatty acid decomposition and utilization.

EPA-PL improves atherosclerosis by mediating cholesterol metabolism. Atherosclerosis is a lipid metabolism dysfunction-derived inflammatory disease, which is usually considered to be caused by heredity, obesity, and high fat and cholesterol in diet. Ding et al. (Ding et al. 2017) suggested that 1% dietary EPA-PL could significantly reduce atherosclerotic lesions (53.4%) in high -at diet apoE$^{-/-}$ mice. EPA-PL significantly decreased serum and hepatic lipid levels by mediating mRNA and protein levels of genes related to hepatic cholesterol metabolism. In addition, comparative study with EPA ethyl ester and DHA-PL found that the alleviation of atherosclerosis by sea cucumber EPA-PL was related to its special structure (Ding et al. 2020). The sn-1 position of the glycerol skeleton for sea cucumber phospholipids is essentially an acetal bond in the phospholipids structure. This unique fatty acid binding pattern may affect the digestion and absorption of EPA-PL and intestinal microecology, thereby alleviating atherosclerosis.

4.1.3 Others

4.1.3.1 Alleviation of Renal Injury

Acute renal injury is a common clinical syndrome characterized by the rapid reduction of glomerular filtration rate in a short time. It is mainly manifested in the rapid decline of renal function and the accumulation of metabolic waste. Vancomycin is metabolized slowly in the body, and more than 90% of the dose will be discharged from the kidney as a prototype drug. In addition, it will also cause symptoms such as hypoxia and necrosis of renal tubular epithelial cells, which has obvious nephrotoxicity. Shi et al. (Shi et al. 2022) studied the effect of EPA-PL on vancomycin-induced acute renal injury in mice. Intragastric administration of EPA-PL could significantly prolong the survival time of mice with renal injury. Sea cucumber phospholipid supplementation also has a protective effect on mice with cisplatin-induced acute renal injury. The intervention of EPA-PL can significantly alleviate the phenomena of renal tubular necrosis, hyaline tubular type, and renal tubular degeneration caused by cisplatin. Creatinine (SCR), urea nitrogen (BUN), cystatin C (Cys C), and kidney injury molecule 1 (KIM 1) can reflect the glomerular filtration rate function of the kidney. These are the relevant biochemical indexes commonly used to reflect renal function clinically. EPA-PL significantly inhibited the increase of the levels of SCR, BUN, Cys C, and KIM 1 in serum. Sea cucumber phospholipids reduced blood pressure and hypertensive nephrosis in spontaneously hypertensive rats.

Sea cucumber phospholipids alleviated hypertension and hypertensive nephrosis in spontaneously hypertensive rats (SHR). Hypertensive nephropathy is one of the most common chronic complications of hypertension. Treatment with 2% EPA-PL for 3 weeks significantly reduced blood pressure in SHRs. Further study on the mechanism of dietary supplementation with EPA-PL to alleviate renal injury is related to its inhibition of oxidative stress and of the apoptosis associated with the MAPK pathway. Further mechanism studies indicated that dietary EPA-PL remarkably inhibited the activation of TGF-α and Smad 3; elevated the phosphorylation level of PI3K/AKT; suppressed the activation of NF-κB; reduced the expression of pro-inflammatory cytokines, including IL-1 and IL-6; and repressed the oxidative stress and the mitochondria-mediated apoptotic signaling pathway in the kidney. These results indicate that EPA-PL has potential value in the prevention and alleviation of hypertensive nephropathy.

4.1.3.2 Anti-Chronic Fatigue Effect

Wang et al. used the swimming exhaustion experiment to explore the improvement effect of EPA-PL on physical fatigue. The results showed that a single oral administration of EPA-PL to mice could significantly reduce exhaustion due to swimming. The ethyl ester EPA had a certain inhibitory effect on reducing the accumulation of lactic acid but had no significant effect on inhibiting the decrease of blood sugar. Further, the effect of long-term intervention by phospholipid DHA/EPA on anaerobic exercise and aerobic exercise ability was studied by swimming exhaustion and running exhaustion experiments, respectively. The results showed that DHA/EPA phospholipid had a significant effect on the aerobic exercise capacity of mice but had no significant effect on anaerobic exercise. Mechanism studies have shown that phospholipid DHA/EPA is related to improved aerobic exercise endurance by promoting the expression of genes related to muscle lipid synthesis and decomposition and increasing lipid energy supply.

4.1.4 Digestion and Absorption

Dietary phospholipids are primarily hydrolyzed by pancreatic phospholipase to produce lysophospholipids and free fatty acids in the small intestine. The hydrolysates enter the intestinal epithelial cells and are then secreted to the lymphatic system, entering circulation. Wang et al. (2019a) measured the EPA levels in serum within 16 h of administration of EPA-pPE. The proportion of EPA in total fatty acids in serum increased over time, with two peaks, at 5 h and 10 h each. Studies have shown that the digestion and absorption of lipids in vivo are usually completed within 5 h, and the peak is at around 3 h. The digestion and absorption process of EPA-pPE from sea cucumber is significantly later than that of EPA in the form of ethyl esters, triglyceride, and phospholipid, which may be related to EPA-pPE being a unique glycerophospholipid with EPA at the sn-2 position of the glycerol backbone. Generally, there is a lack of lipase that can hydrolyze acetal bond in animal digestive tracts, resulting in slow digestion and absorption processes. In addition, this may be related to the inhibition of liver clearance of lipids in the blood by the products after digestion and absorption of EPA-pPE, but there is still a lack of experimental research support for this hypothesis.

Fu et al. (2022) found that dietary supplementation of EPA could significantly increase the content of EPA and DHA in triglycerides and the effect of EPA-PC was greater than that of EPA-pPE. Dietary EPA changed the fatty acid composition in liver PC, especially when it contained more than 20 carbons at the sn-2 position. The level of DHA increased significantly and the dietary EPA-pPE contributed more to the gain of DHA than did dietary EPA-PC. The DHA content in liver PC after supplementation with EPA-pPE was higher than that in EPA-PC, but the opposite result was shown in triglycerides, indicating that the metabolisms of EPA-pPE and EPA-PC might be different in the liver. Moreover, supplementation with either EPA-pPE or EPA-PC significantly reduced the content of AA and EPA in liver PC.

4.2 *Cerebroside*

Sphingolipids are a class of compounds containing a sphingosine structure and are divided into basic sphingolipids, neutral sphingolipids, and acidic sphingolipids.

4.2.1 Antitumor Effect

Cancer cachexia is a wasting metabolic syndrome, with main pathological features, including progressive decline in fat and skeletal muscle weight, disorder of glucose, lipid and protein metabolism, and systemic inflammation. It is one of the main reasons for the serious decline of quality of life and high mortality of cancer patients. Du et al. studied the effect of sea cucumber cerebroside on cancer cachexia in sarcoma 180 ascitic tumor-bearing mice (Du 2012). The results showed that sea cucumber cerebroside alleviated muscle decomposition and excessive decomposition of adipose tissue in cachexia mice. The final body weight of the model group mice decreased significantly by 19% compared with the initial body weight, indicating that the model group mice showed cancer cachexia symptoms. Dietary intervention with 50 mg/kg sea cucumber cerebroside for 13 days inhibited the weight loss of mice by 17%. The adipose tissue and muscle of mice in the model group were consumed to a greater extent than in normal mice, while sea cucumber cerebroside significantly inhibited the consumption of adipose tissue and muscle, suggesting that sea cucumber cerebroside significantly alleviated the loss of fat and muscle in cancer cachexia mice and prevented organ failure caused by excessive consumption of fat and muscle. At the same time, sea cucumber cerebroside had beneficial effects on the survival time and life prolongation rate of cancer cachexia mice (Fig. 2).

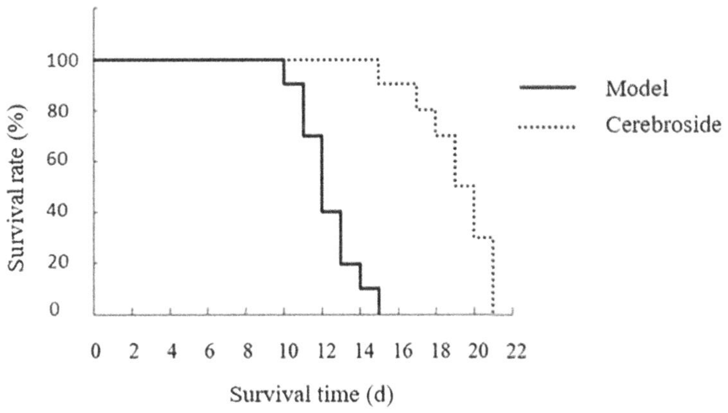

Fig. 2 Life prolongation curve of sea cucumber cerebroside on cancer cachexia-treated mice

Table 3 IC_{50} value of sea cucumber cerebroside and its long-chain base on a variety of tumor cells for 72 h

	HepG2	S180	95D	HGC-27	Caco-2
Cerebroside (μg/mL)	>400	>400	>400	50.31	>400
Long-chain base (μg/mL)	3.33	3.37	1.99	1.63	0.86

Proinflammatory cytokines such as tumor necrosis factor-α (TNF-α), interleukin-1 (IL-1), and interleukin-6 (IL-6) play an important role in the process of cancer cachexia. These cytokines directly affect the appetite of patients and change the metabolic status of the body. A study by Du (2012) indicated that sea cucumber cerebroside inhibited TNF-α, IL-1, and IL-6 levels in the serum of cancer cachexia mice.

After digestion and absorption, 25–60% of cerebrosides are decomposed into ceramide. Ceramide is a kind of amide compound formed by the dehydration of carboxyl group in long-chain fatty acid and long-chain base. To identify the reasons for its effectiveness, Du et al. (2015) prepared long-chain bases from sea cucumber cerebroside. The antitumor effect of sea cucumber cerebroside and its long-chain base was verified in vivo by intragastric administration and intraperitoneal injection in sarcoma 180 ascitic tumor-bearing mice. The results showed that intragastric or intraperitoneal injection of long-chain base significantly prolonged the survival time in sarcoma 180 ascitic tumor-bearing mice and showed good antitumor effect in vivo. Interestingly, only by intragastric administration of sea cucumber cerebroside showed significant antitumor activity, and the antitumor effect was equivalent to that of long-chain base. It was speculated that the antitumor activity of cerebroside was related to its decomposition into long-chain bases in the intestine. The results suggested that the main active unit of sea cucumber cerebroside was its long-chain base.

Du (2012) extracted sea cucumber cerebrosides from *Acaudina molpadioides* and prepared its long-chain base. The inhibitory effects of cerebrosides and long chain base on mouse sarcoma cell line S180, human liver cancer cell line HepG2, human colon cancer cell line Caco-2, human gastric cancer cell line HGC-27, and human highly metastatic lung cancer cell line 95D were compared by MTT assay. The results showed that sea cucumber cerebroside and its long-chain base significantly inhibited the growth of HepG2, S180, 95D, Caco-2, and HGC-27 tumor cells and exhibited different levels of sensitivity to different tumor cell lines, with a certain dose- and time-response relationship. It showed that sea cucumber brain glycosides and their long-chain bases had extensive antitumor cell proliferation activities. Among them, the antitumor cell proliferation activity of long-chain base was much better than that of sea cucumber cerebroside (Table 3). Sugawara et al. (2006) found that the long-chain base hydrolyzed from the cerebroside of *Stichopus variegates* could induce the apoptosis of tumor cells, such as Caco-2, DLD-1, and WiDr, and showed strong antitumor activity.

The antitumor activity of cerebrosides in vivo may be related to their important role as immune stimulators in anti-infection, tumor immune surveillance, autoimmunity, and other function. Du et al. (2015) studied the antitumor activity of sea cucumber cerebroside in sarcoma 180 ascitic tumor-bearing mice to explore the possible antitumor mechanism of sea cucumber cerebroside in vivo from the aspects of immune regulation, antioxidation, and induced tumor cell apoptosis. The results showed that sea cucumber cerebroside significantly reduced the tumor weight and increased the spleen index and thymus index of mice. Further mechanism study found that cerebroside significantly increased the content of glutathione and the activity of glutathione peroxide enzyme and glutathione thiotransferase enzyme in the liver. In addition, sea cucumber cerebroside induced tumor cell apoptosis by regulating the genes related to mitochondrial apoptosis pathway of Bcl-2, Bax, BcL-xL, cytochrome-c, caspase-9, and caspase-3.

4.2.2 Prevention of Metabolic Syndrome

It was found that cerebroside could improve the fatty liver and metabolic syndrome of ob/ob mice by regulating natural killer T cell lymphocytes of the immune system (Margalit et al. 2006). Zhang et al. (2011) studied the effect of sea cucumber cerebroside on the improvement of nonalcoholic fatty liver induced by whey acid. The results showed that the addition of 0.006% sea cucumber cerebroside in the feed reduced the levels of TG and TC in the liver of rats with fatty liver induced by whey acid by 70% and 30%, respectively, and the levels of serum ALT and AST were significantly improved. The above results showed that sea cucumber cerebroside could significantly alleviate the fatty liver in animals. Further studies showed that the effect of sea cucumber cerebroside on fatty liver was related to downregulating the expression of SREBP-1c and its regulated fatty acid synthesis genes (FAS, G6PDH, ME). In addition, the results also showed that after sea cucumber cerebroside was introduced, the fatty acid composition of rat liver changed significantly and the ratio of monounsaturated fatty acid to saturated fatty acid (unsaturated index) decreased significantly. The results suggested that sea cucumber cerebroside could affect the activity of SCD enzyme in the liver and inhibit the synthesis of endogenous fatty acids. Liu et al. (2015b) studied the effect of sea cucumber cerebroside on the abnormal lipid metabolism of high-fat high-fructose-induced C57BL/6 J obese mice. Sea cucumber cerebrosides were prepared from *Acaudina molpadioides* and then administered to high-fat diet-induced obese C57BL/6 J mice at a diet supplement dosage of 0.025% for 5 weeks. The results showed that sea cucumber cerebroside significantly decreased epididymal adipose tissue weights and decreased hepatic triacylglycerol level, indicating that dietary sea cucumber cerebrosides could affect lipid metabolism. Mechanism study found that sea cucumber cerebroside significantly reduced the enzyme activity of lipogenic enzymes, such as FAS and ME in the liver, but had no effect on the mitochondrial CPT of fatty acids β-oxidation and on the enzyme activity of peroxisomal β-oxidation. The mRNA levels of SREBP-1c and FAS were reduced by sea cucumber cerebroside. In addition, sea cucumber cerebroside effectively upregulated the gene expression of SREBP-1c, FAS, ACC, ATGL, and HSL and downregulated the gene expression of

LPL and VLDL-r in the adipose tissue. Since long-chain bases are important products of cerebroside digestion in vivo, the mechanism study also found that the nutritional efficacy of sea cucumber was derived from the long-chain bases and was similar to that of sea cucumber cerebroside; therefore, the nutritional efficacy of sea cucumber cerebroside may be contributed substantially to long-chain bases.

Liu et al. (2015b) also found that sea cucumber cerebroside reduced serum glucose, feeding insulin levels, and HOMA-IR index in metabolic syndrome mice. Yang et al. (2021) systematically compared the effects of sea cucumber cerebroside and its main structural unit ceramide on the improvement of insulin resistance in model rats with the natural course of human insulin resistance induced by high-fructose diet and explored its mechanism of affecting glucose and lipid metabolism. It was found that sea cucumber cerebroside significantly alleviated the glucose tolerance, insulin sensitivity, hyperinsulinemia, and hypertension in high-fructose diet specimens. Endogenous ceramide has long been considered to be closely related to the occurrence of insulin resistance. The increased production of endogenous ceramide will increase the risk of insulin resistance. Interestingly, Yang et al. (2021) found that dietary intake of ceramide from sea cucumber could significantly alleviate insulin resistance in high-fructose diet-fed rats. The beneficial effects of sea cucumber cerebroside and ceramide in terms of effectively alleviating hyperglycemia and hyperinsulinemia and of inhibiting abnormal rise of blood pressure are related to promoting the release of serum nitric oxide. In addition, the study also found that dietary sea cucumber cerebroside improved glucose metabolism in muscle tissue. The results further confirmed that glucosylceramides alleviated insulin resistance in rats by upregulating the IRS/PI3K/AKT signaling pathway in the liver.

$ApoE^{-/-}$ mice were used to evaluate the protective activities of sea cucumber cerebroside on atherosclerosis. In ApoE−/− mice, sea cucumber cerebroside treatment significantly decreased the atherosclerotic lesion formation and attenuated inflammation by decreasing the levels of inflammatory cytokines, such as CRP, TNF-α, and IL-6. Compared to the model group, the sea cucumber cerebroside group showed lower cholesterol levels in serum and liver by mediating the expression of genes related to hepatic LDL uptake and cholesterol excretion.

4.2.3 Brain Function

Che et al. (2017) studied the effect of sea cucumber cerebroside on improving the learning and memory ability of SAMP8 mice. The total distance of SAMP8 mice travelled decreased significantly after dietary sea cucumber cerebroside. This indicated that sea cucumber cerebroside effectively alleviated anxiety in SAMP8 mice. The Barnes maze test was used to determine the spatial memory ability of mice. The results showed that sea cucumber cerebroside intervention decreased the total distance traveled by SAMP8 mice and the time required to find the target box decreased significantly from the third day. The results of behavioral experiments, such as the Morris water maze, showed that the intervention of sea cucumber cerebroside effectively alleviated severe learning and cognitive impairment in

SAMP8 mice. The mechanism study showed that sea cucumber cerebroside reduced the Aβ content in the hippocampus of SAMP8 mice, improved the antioxidant enzyme activity in the brain, inhibited the toxicity of NO, and improved the peroxidation of lipid, DNA, and RNA in the brain, thus improving the learning and memory ability of mice.

Sea cucumber cerebroside improves the learning and memory ability of scopolamine model mice. Fu et al. evaluated the effect of sea cucumber cerebroside on the learning and memory deficits in mice induced by scopolamine through the Morris water maze behavioral experiment (Fu et al. 2016). Studies have shown that sea cucumber phospholipid and sea cucumber cerebroside improved the learning and memory ability of mice. In the positioning navigation experiment, the escape time required by mice fed with sea cucumber phospholipid and sea cucumber cerebroside was shortened, the total distance was reduced, and the proportion of residence time in the target quadrant was increased. In the space exploration experiment, for mice fed with sea cucumber phospholipid and sea cucumber cerebroside, the time spent in the target quadrant and the number of times that specimens were able to cross the platform increased significantly. Sea cucumber cerebroside could also improve the learning and memory ability of mice with dementia induced by scopolamine. In the further mechanism study, it was found that sea cucumber cerebroside reduced the content of MDA, increased the activity of SOD, and effectively inhibited the activity of TChE, so sea cucumber cerebroside had a certain protective effect on learning and memory.

An oxidative damage model of nerve cells in vitro was established using H_2O_2 and t-BHP induced PC12 cells (Che et al. 2017). The survival rate of PC12 cells was increased, and the cell morphology was restored through treatment with sea cucumber cerebroside, indicating that sea cucumber cerebroside had a protective effect on oxidative damage of PC12 cells. The results of mechanism study showed that sea cucumber cerebroside reduced the leakage of lactate dehydrogenase in cells; protected the integrity of cell membrane; improved the activity of total antioxidant capacity and superoxide dismutase in cells; increased the antioxidant capacity of cells; downregulated the proapoptotic genes Bax, caspase-9, and caspase-3; and promoted the expression of antiapoptotic gene Bcl-2, thereby inhibiting neuronal apoptosis. It was suggested that the protective mechanism of sea cucumber cerebroside against oxidative damage might depend on its antioxidant activity and antiapoptotic activity.

4.2.4 Others

A hyperuricemia model for the effects of sea cucumber cerebroside and ceramide on mice was established through feeding 20% yeast extract powder to male Kunming mice to study the effect of dietary sea cucumber cerebroside on hyperuricemia (Dong et al. 2013). The results showed that the cerebroside group could significantly reduce the concentration of serum uric acid in mice. The cerebroside group could significantly inhibit the activities of xanthine oxidase and adenosine deaminase in the liver of model mice.

4.2.5 Digestion and Absorption of Cerebroside

Sugawara et al. (Sugawara et al. 2003) reported that cerebroside was hydrolyzed into ceramide and free sphingosine base in the rat digestive tract and the degradation products 2-hydroxyfatty acid and ceramide entered the lymph. Most dietary sphingosine was converted to palmitic acid in the intestinal mucosa and bound to chyle particles, which was then transported to the lymph (Ishikawa et al. 2009; Sugawara et al. 2010). In the gastrointestinal tract, cerebrosides were hydrolyzed to ceramide, sphingoid bases, and free fatty acids by the effects of glucosylceramidase and ceramidase. Duan et al. (2016) studied the characteristics of in vivo digestion and absorption for cerebroside derived from sea cucumber and found that part of the sea cucumber cerebroside was utilized by microorganisms in the intestine and transformed into acetic acid and propionic acid. In addition, odd chains containing cerebrosides (d17:1-GlcCer; 19:2-GlcCer) from sea cucumbers were better absorbed than d18:2-GlcCer but were still not comparable to d18:1-GlcCer.

4.3 Sea Cucumber Gangliosides

Gangliosides (GLS) are enriched in the central nervous system of animals, and the glycosyl groups carried by GLS and their negatively charged sialic acid moieties are closely related to cellular recognition, connectivity, and information transduction and play a crucial role in maintaining cellular functions (Meisen et al. 2003). Therefore, gangliosides are essential for the development of the nervous system in higher animals. As early as 1991, Higuchi et al. found that marine echinoderm-derived ganglioside (sea star *Astropecten latespinosus*) GP-2 increased the survival of rat fetal cortical cells and was potentially more active than mammalian ganglioside GM1 (Higuchi et al. 1991). Yamada et al. progressively developed a variety of gangliosides from various sea cucumbers, including *Cucumaria echinate* (Yamada et al. 1998), *Holothuria leucospilota* (Yamada et al. 2000), and *Stichopus chloronotus* (Yamada et al. 2003) and were the first to discover tri-sialic acid. Kaneko et al. not only demonstrated that *Stichopus japonicus* gangliosides could promote the axonal growth of PC12 cells in vitro but also found that their activity was superior to that of mammalian gangliosides GM1 (Kaneko et al. 2003). It seemed that the neuroprotective activity of GLS was influenced by its structure, especially the structure of its oligosaccharide chains (Cong 2012). Specifically, the neurodifferentiation and protective activity of GLS were positively correlated with the amount of sialic acid and the number of sulfate groups they contained (Frey and Lee 2013; Wang et al. 2017a).

It has been suggested that continuous injection of GM1 into the lateral ventricles of patients with early Alzheimer's disease could alleviate AD symptoms and improve cognition, reading, and writing abilities (Svennerholm et al. 2002). Marine food-derived GLS could counteract AD symptoms by reducing neurosynaptic loss and blocking mitochondrial apoptosis pathways in mice, with dose-dependent and constitutive effects (Wang et al. 2017b). However, it has also been reported that lateral ventricular injection of GM1 increased Aβ deposition in the brain of late APP/PS1 double transgenic mice, which showed the opposite effect of peripheral administration (Matsuoka et al. 2003); therefore, the role of exogenous gangliosides in the prevention and treatment of Alzheimer's needs to be further explored. Wang et al. found that differentially glycosylated sea urchin GLS reduced neuronal loss and alleviated Aβ1–42 accumulation in AD models (Aβ25–35-induced PC12 cells and SAMP8 mice), a result that holds great promise for the alleviation of AD through the use of sea cucumber GLS (Wang et al. 2017b).

Moreover, GLS plays a critical role in neurological health of the brain because Sandhoff et al. noted that defective endogenous GLS metabolism led to fatal human diseases (Sandhoff et al. 2018). The active functions of exogenous GLS, especially of dietary origin, in immunomodulation, cancer, and metabolic diseases have also been increasingly reported (Wang et al. 2021). However, the question is whether GLS in sea cucumber possesses these aspects of efficacy and whether they are superior to mammal-derived GLS in aspects other than neurological function. These questions have not yet been clearly addressed at present, but future studies may reveal sea cucumber GLS to offer further benefits to human health.

4.4 *Sea Cucumber Sterol*

Compared with phytosterols, sea cucumber sterol exhibited superior biological activity because of the sulfate group at the C-3 position. Sea cucumber sterol is mainly composed of two monomers – cholesterol sulfate (CS) and 24-methylene cholesterol sulfate (24-MCS) – of which cholesterol sulfate is widely distributed in the human body and can be synthesized endogenously with cholesterol.

4.4.1 Modulation of Lipid Metabolism

4.4.1.1 Improving Hyperlipidemia

Zeng et al. established an obese C57BL/6 J mice model with a high-fat, high-fructose diet containing 20% lard and fructose and investigated the effects of sea cucumber sterol (SCS) on lipid metabolism (Zeng et al. 2020). The results showed that the dietary supplementation of sea cucumber sterol significantly inhibited the weight gain of mice without changing the energy intake and in terms of lipid regulation. SCS significantly reduced the serum total cholesterol level and LDL-C level and increased the HDL-C level of hyperlipidemic mice, reducing the arteriosclerosis index significantly by 60.2%. It was suggested that besides inhibiting weight gain and lipid accumulation, sea cucumber sterol might also have beneficial effects on atherosclerosis, which is mainly characterized by hypercholesterolemia.

4.4.1.2 Inhibiting Hepatic Lipogenesis and Promoting Fatty Acid β-Oxidation

Zeng et al. investigated the effect of dietary sea cucumber sterol on hepatic lipid metabolism. The results showed that sea cucumber sterol significantly inhibited excessive accumulation of hepatic TG. A dietary addition of 0.4% sea cucumber sterol could decrease hepatic TG by 25% and similar effects on total hepatic cholesterol were observed (Zeng et al. 2020). The authors also analyzed the expressions of genes involved in lipogenesis and fatty acid β-oxidation in the liver using quantitative RT-PCR and Western blotting. The results showed that dietary sea cucumber sterol significantly decreased the expression levels of genes involved in lipogenesis, such as the nuclear transcription factor SREBP-1c and its downstream ACC, FAS, and SCD1; on the other hand, dietary sea cucumber sterol significantly increased the mRNA expression of PPARα, CPT1, CPT2, and ACOX1, which are key genes of fatty acid β-oxidation, and the protein expression of rate-limiting enzyme CPT1 was also significantly increased. The results suggested that dietary supplementation with sea cucumber sterol could effectively inhibit lipogenesis and promote fatty acid β-oxidation, thereby reducing lipid accumulation in the liver.

4.4.1.3 Reducing Inflammation in Adipose Tissue

The accumulation of triglycerides in white fat leads to an increase in adipocyte volume, further causing adipose tissue hypoxia and inducing inflammation accordingly, which can be reflected by the levels of serum inflammatory factors. The results found that sea cucumber sterol significantly reduced the circulating levels of inflammatory factor MCP-1 and tumor necrosis factor TNF-α and increased the levels of anti-inflammatory peptide ADP in mice with metabolic syndrome, thus reducing inflammation of the body (Zhang et al. 2020a, b).

4.4.1.4 Regulating Cholesterol Metabolism and Reducing Atherosclerosis

By experimenting on palmitic acid and oleic acid-induced disorders of lipid metabolism in HepG2 cells, Ding et al. investigated the effects of low and high doses of sea cucumber sterol on lipid metabolism regulation in vitro (Ding et al. 2021). The results showed that sea cucumber sterol had no significant effect on the TG content of HepG2 cells but could significantly reduce TC levels, suggesting that sea cucumber sterol might have a direct regulatory effect on cholesterol metabolism. The results showed that dietary sea cucumber sterol significantly reduced the area of atherosclerotic lesions.

The effect of sea cucumber sterol on serum lipid composition of ApoE$^{-/-}$ mice was basically consistent with the animal experimental study by Zeng et al. (2020). Dietary sea cucumber sterol significantly reduced the serum total cholesterol level and substantially reduced the LDL-C level in atherosclerotic mice. Unlike the results of the animal experimental study by Zeng et al., sea cucumber sterol had no significant effect on serum HDL-C levels in ApoE$^{-/-}$ mice. The TG and TC contents of lipoproteins in serum were measured by continuous separation according to particle size using FPLC (fast protein liquid chromatography), and the results showed that the TG level in very low-density lipoprotein (VLDL/LDL) particles did not change, while the cholesterol content was significantly reduced. This suggested that the reduction of plasma VLDL/LDL-c might be the main reason for the anti-atherosclerotic effect of sea cucumber sterol.

4.4.1.5 Promoting Dissimilation of Cholesterol

By measuring the content and composition of cholesterol and bile acids, as well as the expressions of related genes in liver, gallbladder, and feces, which were the major sites of cholesterol catabolism, Ding et al. investigated the mechanisms of sea cucumber sterol in regulating cholesterol metabolism. The results showed that sea cucumber sterol significantly reduced the cholesterol content in liver, gallbladder, and serum and significantly increased the fecal cholesterol efflux. Sea cucumber sterol significantly reduced the protein expression of the rate-limiting enzyme of hepatic cholesterol synthesis (HMGCR) and the level of SREBP2, a nuclear regulatory transcription factor in cholesterol synthesis. By contrast, the protein expression of CYP7a1 was significantly increased, and the protein expression of LXR and nuclear FXR, key regulators of bile acid metabolism, were significantly increased. Moreover, the mRNA expressions of genes related to bile acid synthesis, such as CYP7a1, CYP8b1, CYP27a1, CYP7b1, Beat, and slc27a5 significantly increased or exhibited a trend of increase. The results suggested that sea cucumber sterol could inhibit cholesterol synthesis and promote bile acid synthesis and cholesterol efflux in the liver. In addition, changes of bile acid composition in serum, liver, gallbladder, and feces, characterized by a significant increase of TMCA synthesized from CDCA in liver, blood, and gallbladder and a significant increase of secondary bile acids UDCA, HDCA, and LCA in feces, suggested that sea cucumber sterol might affect intestinal microbiota.

4.4.2 Regulation of Glucose Metabolism

Using a high-fat, high-fructose diet containing 20% lard and fructose to establish a model of insulin resistance with C57BL/6 J mice, Zhang et al. investigated the regulatory effects of sea cucumber sterol on glucose metabolism (Zhang et al.

2020a, b). In mice with dietary intervention of sea cucumber sterol, the glucose content in serum during OGTT was lower than that in the model group after 0.5 h, 1 h, 1.5 h, and 2 h of glucose gavage. After 30 min of glucose gavage, serum glucose was significantly reduced by 24.6% in the sea cucumber sterol group, which almost reached the level of the normal group. Meanwhile, dietary supplementation of sea cucumber sterol significantly reduced fasting insulin to normal levels in insulin-resistant mice. In addition, dietary sea cucumber sterol significantly alleviated the decrease of glycogen content in liver and muscle tissues caused by a high-fat, high-fructose diet and improved insulin sensitivity of the tissues. The above results suggested that dietary sea cucumber sterol could significantly alleviate high-fat, high-fructose diet-induced glucose tolerance impairment and increase insulin sensitivity. The mRNA expressions of genes related to insulin signaling pathways, such as InsR, p-IRS1/IRS1, and p-AKT/AKT, and of the genes related to glycogen synthesis and catabolism, such as p-GSK3/GSK3 and G6Pase in skeletal muscle and liver, were investigated mechanistically, and the results showed that the ameliorative effect of sea cucumber sterol on insulin resistance was related to the activation of PI3K/Akt signaling pathway.

4.4.3 Digestive Absorption and Tissue Distribution

Li et al. measured the distribution of two major monomers of sea cucumber sterol (Fig. 3) in various tissues by oral forced feeding and tail vein injection of 0.05 mg/g body weight of sea cucumber sterol. Through this procedure, they investigated the absorption characteristics, pharmacokinetics, and tissue distribution of sea cucumber sterol and provided a theoretical basis for the exploration of functional foods related to sea cucumber sterol (Li et al. 2022).

The digestion and absorption of the two main monomers of sea cucumber sterol were determined through the measurement of their contents in the small intestine contents, the small intestinal wall, the liver, and blood at different times after oral administration of sea cucumber sterol. It was found that sea cucumber sterol disappeared rapidly from the digestive tract after 2 h and entered the small intestinal

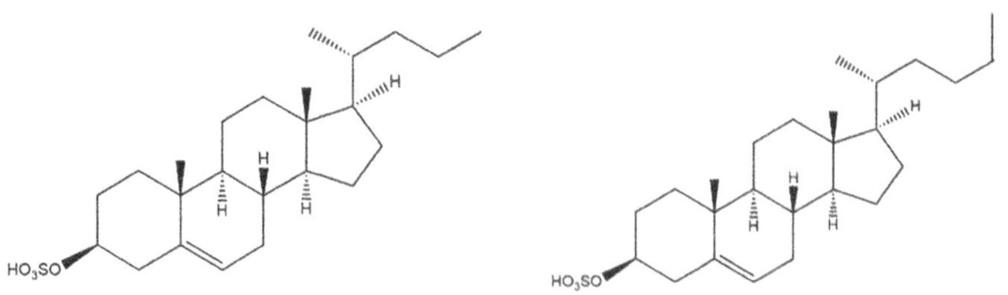

Cholesterol sulfate (CS)

24-methylene cholesterol sulfate (24-MCS)

Fig. 3 Two major monomers of sea cucumber sterol

endothelium and blood. In addition, compared to 24-MCS, CS was able to enter the blood and liver faster and maintain a higher concentration in blood and liver. The pharmacokinetic properties of sea cucumber sterol were investigated by tail vein injection, and the results showed that the C_{max} values of CS and 24-MCS were 47.89 and 34.53 μg/mL, respectively, and that the $t_{1/2}$ values were 47.60 and 22.36 min, respectively, indicating that CS exhibited a relatively longer duration of efficacy than 24-MCS did. The two major monomers appeared in large amounts in urine at 3 h after injection, indicating that sea cucumber sterol could be excreted via the urine through the kidneys.

On that basis, Srihera et al. replaced cholesterol with sea cucumber sterol to develop new multifunctional nanoliposomes stabilized with sea cucumber sterol by thin film hydration to avoid multiple unhealthy risks from cholesterol (Srihera et al. 2022). The astaxanthin was encapsulated in liposomes with a 3:1 mass ratio of egg yolk lecithin to sea cucumber sterol. The liposomes showed significantly higher antioxidant activity in terms of DPPH scavenging activity and reduced capacity compared to the mixture of astaxanthin and blank sea cucumber sterol liposomes. In vivo digestion and absorption results showed that the bioavailability of liposome-encapsulated astaxanthin was significantly higher and sea cucumber sterol had great potential as a multifunctional and efficient carrier in functional food exploration and biomedical applications.

4.5 Other Fatty Acids

Yang et al. used sea cucumber-derived 12-methyltetradecanoic acid (12-MTA) to incubate human prostate cancer cell PC3 and humanized 5-LOX, COX-1, and COX-2 enzymes at the semi-inhibitory amount of 25 ug/mL, with the aim of investigating the antitumor activity of 12-MTA and the possible involved underlying mechanism (Yang et al. 2003). Over time, the cells gradually exhibited apoptotic phenomena, such as rounding, cytoplasmic vacuolization, mitochondrial aggregation, and nuclear sequestration. The kinetic inhibition of LOX and COX enzyme lines suggested that the in vitro pro-apoptotic activity of 12-MTA might be related to the inhibition of lipid oxidase lines.

Nguyen et al. isolated two unsaturated fatty acids, 7(Z)-octadecenoic acid and 7 (Z), 10(Z)-octadecadienoic acid, from *Stichopus japonicus* and determined their IC_{50} values on *B. stearothermophilus* α-glucosidase, *S. cerevisiae* α-glucosidase, and rat-intestinal sucrase and maltase, respectively (Nguyen et al. 2011). The results showed that 7(Z), 10(Z)-octadecadienoic acid had stronger inhibitory activity and that both compounds could maintain stable activity for at least 40 min at a high temperature of 100 °C and acidic conditions at pH = 2. It was suggested that sea cucumber fatty acids might prevent diabetes by inhibiting α-glucosidase to retard the absorption of glucose.

5 Sea Cucumber Saponins

The primary secondary metabolites in sea cucumber are saponins, which is a class of the main active compounds in sea cucumber. As a chemical defense substance, saponin has functional properties of antitumor, anti-hyperlipidemia, anti-fatty liver, fat accumulation inhibition, blood glucose regulation, hyperuricemia improvement, gout prevention, bone marrow hematopoietic function promotion, antibacterial and antiviral activity, reproductive development properties regulation, etc.

5.1 Antitumor Activity

It has been reported that sea cucumber saponins could suppress tumor cell proliferation and angiogenesis, as well as having strong antitumor efficacy in vivo and in vitro. The five triterpene glycoside compounds 24-dehydroechinoside A, echinoside B, echinoside A, holothurin B, and holothurin A isolated from sea cucumber all demonstrated inhibitory activity on the proliferation of human gastric cancer cell MKN-45 and colon cancer cell HCT-116. Sea cucumber saponins isolated from sea cucumber *Mensamria intercedens* Lampert showed certain proliferation inhibitory activity against 10 tumor cell lines in vitro and at the same time had significant inhibitory activity on the growth of Lewis lung cancer and S180 sarcoma in vivo (Zou et al. 2003, 2005). Zhao et al. studied the active compounds philinopside A and philinopside E isolated from *Pentacta quadrangularis* Lesson and showed that saponins could not only significantly inhibit the growth of human gastric adenocarcinoma cell MKN-28, colon cancer cell HCT-116, breast cancer cells MDA-MB-468 and MCF-7, liver cancer cell BEL-7402, lung cancer cell SPC-A4, ovarian cancer cell HO-8910, and other tumor cell lines but also had a strong antitumor angiogenesis effect, with a significant dose- and time-dependent relationship (Zhao et al. 2018). Moreover, another study reported the effect of the saponin-rich component holotoxin A1 extracted from *Apostichopus japonicus* on HL-60, B16F10, MCF-7, and Hep3B tumor cell lines and found that the sea cucumber saponin holotoxin A1 could inhibit the proliferation of those tumor cells (Dai et al. 2020).

Zhao et al. studied the inhibitory effects on the proliferation of various tumor cell lines of sea cucumber saponins holothurin A1 (HA) and 24-dehydroechinoside A (DA) isolated from *Pearsonothuria graeffei*. The results showed that HA and DA had extensive inhibitory effects on the proliferation of a variety of tumor cell lines and exhibited significant time- and dose-dependent effects. Among the tumor cell lines, HepG2 cells were most sensitive to HA and DA intervention, and the anticancer activity of DA was stronger than that of HA (Zhao et al. 2010). A further study evaluated the antitumor activity of sea cucumber saponins EA and its deglycosylated metabolites ds-echinoside A (DSEA) in vitro and found that both EA and DSEA significantly inhibited the proliferation of HepG2, B16, Caco-2, Hela,

Table 4 Antitumor activity of sea cucumber saponins in vitro

Saponins	Experimental model	IC_{50}	Reference
Holothurin A1 (HA)	Human umbilical vein endothelial cells (HUVECs)	4.8 μmol/L	Zhao et al. (2011a)
24-dehydroechinoside A		3.82 μmol/L	
Ds-echinoside A	Hep G2 cells	2.65 μmol/L	Zhao, Liu et al. (2011)
Echinoside A	K562, MCF-7, KB cell lines	5.42, 1.32, 0.95 μmol/L	Li et al. (2010)
Colochiroside A	BEL-7402, HO-8910, MDA-MB-435, MDA-MB-468, A431 cell lines	1.75–3.66 μg/ml	Zhang and Yi (2011)
Frondoside A	A549 cell	0.6 μmol/L	Nguyen et al. (2017)
Arguside A	HCT-116 cell	0.14 μmol/L	Liu et al. (2007b)

P388, and S180 cell lines. The study showed that there were significant time- and dose-dependent effects, indicating that EA and DSEA had a wide range of activities to inhibit tumor cell proliferation. Moreover, HepG2 cells were the most sensitive to the two saponins, and the IC_{50} values at 24 h were 2.49 μM and 2.46 μM, respectively. The antitumor activity of DSEA was higher than that of EA except in B16 cells (Zhao et al. 2011a, b, 2012). The antitumor activities of some sea cucumber saponins in vitro are shown in Table 4.

The results of the above in vitro studies showed that sea cucumber saponins demonstrated remarkable antitumor activity. As for in vivo studies, Zhang et al. (Zhang and Yi 2011) studied the antitumor activity of colochiroside A (CA), a suspected saponin extracted from *Colochirus anceps*. Taking mouse sarcoma S180 and mouse liver cancer H22 cells as transplanted tumor models, the in vivo antitumor activity of CA and its effect on immune organs were investigated. It was found that the inhibition rate of S180 sarcoma in mice was 35–70%, and the inhibition rate of H22 liver cancer in mice was 35–50%. There was no significant effect on the weight of immune organs in tumor-bearing mice.

5.1.1 Antitumor Mechanism of Sea Cucumber Saponins

Studies have shown that sea cucumber saponins can block cell cycle and division. Flow cytometry analysis showed that the cell cycle of HepG2 cells was blocked in the G0/G1 phase after intervention with sea cucumber saponins for 12 h. The percentages of cells in the S phase and G2/M phase in the sea cucumber saponin-treated groups were significantly decreased, and the expression of genes related to inhibiting cell proliferation increased, indicating that sea cucumber saponins could hinder tumor cells in the G0/G1 phase and prevent them from entering the S and M phase, consequently inhibiting cell proliferation (Zhao et al. 2012). Apoptosis is a highly regulated process of cell death by gene expression. After treatment with sea cucumber saponins, cells began to undergo apoptosis, and the level of apoptosis increased with time, indicating that sea cucumber saponins alone could significantly induce tumor cell apoptosis and increase the expression of apoptosis-related genes in tumor cells (Yu et al. 2015).

Sea cucumber saponins can significantly increase the number of antibody-forming cells and serum hemolysin content in immunocompromised mice, promote humoral immune function and delayed-type allergy reactions in mice, and at the same time improve the proliferation capacity of spleen lymphocytes, thereby improving cell immune function in mice. In addition, sea cucumber saponins can significantly improve the phagocytosis rate and phagocytosis index of chicken erythrocytes by mouse peritoneal macrophages and promote the nonspecific immune function of mice. For immunocompromised mice, under the intervention of low-dose saponins, the phagocytic index of macrophages can be increased by 2.5 times, and the phagocytic ability returns to normal levels with the increase of saponin dose, indicating that sea cucumber saponins can effectively regulate the body's immune system.

Wang et al. studied the effects of saponins isolated from *Pearsonothuria graeffei* on humoral and cellular immunity in cyclophosphamide-induced mouse model. Studies have found that saponins can transform some previously unresectable tumors into operable tumors by inhibiting tumor angiogenesis, reducing the staging of solid tumors, and inhibiting the formation of metastatic lesions (Wang et al. 2010b). Previous studies showed that holothurin B, a sea cucumber saponin, could induce apoptosis of vascular endothelial cells (HUEVC) (Yegdaneh et al. 2021). NF-kB is a nuclear transcription factor related to tumor cell metastasis and plays an important role in inhibiting tumor cell apoptosis. Patagonicoside A extracted from *Psolus patagonicus* can induce the activation of NF-KB, thereby inhibiting tumor cell metastasis and promoting tumor cell apoptosis (Careaga et al. 2009). Matrix metalloproteinases (MMPs) family can promote VEGF expression, which is closely related to cancer cell migration and tumor angiogenesis. Tong et al. assessed the effects of philinopside A on tumor angiogenesis and tumor growth in a series of models in vitro and in vivo. The findings revealed that philinopside A was a potential anticancer drug with combined cytotoxic and antiangiogenic properties, possibly due to VEGF receptor inhibition (Tong et al. 2005).

5.2 *Improvement of Metabolic Syndrome*

Metabolic syndrome (MetS) is an aggregation of multiple metabolic-related risk factors, including glucose metabolism, abnormal lipid metabolism, and obesity. Sea cucumber saponins have significant effects in terms of alleviating metabolic syndrome and have the functions of regulating blood glucose level, lowering blood lipids, improving nonalcoholic fatty liver, and inhibiting fat accumulation. The related mechanisms of sea cucumber saponins on metabolic syndrome mainly include the regulation of lipid metabolism, glucose metabolism-related genes, and enzymes.

5.2.1 Regulating Glucose Metabolism

Studies have reported that saponins have a significant effect on blood glucose regulation. Saponins can regulate blood glucose concentration in type 2 diabetes mellitus mice. The antidiabetic effects of sea cucumber may be the result of the regulation of insulin secretion, glucose uptake, anti-oxidative stress, and anti-inflammatory pathways. Sea cucumber saponins have a similar chemical structure to plant saponins and might have significant effects on blood glucose regulation.

Chen et al. established a mouse model with insulin resistance using a high-fat diet and studied the effect of a mixture of sea cucumber saponins EA and HA on glucose metabolism. It was found that an appropriate amount of sea cucumber saponins supplementation in the diet could significantly improve impaired glucose tolerance, decrease fasting insulin level in mice, and improve insulin resistance (Chen et al. 2018). Mechanism studies have shown that sea cucumber saponin liposomes can effectively regulate the p-ERK/cPLA2/COX1 pathway and ultimately reduce the level of PGE2 in adipose tissue, thereby improving insulin resistance and further regulating blood glucose by increasing the utilization of glucose by the liver and the absorption by peripheral tissues.

C57BL/KsJ(db/db) is an animal model of hereditary obesity type 2 diabetes mellitus, which is characteristic of hyperglycemia, hyperinsulinemia, hypertriglyceridemia, and insulin resistance. In this subject, the blood glucose gradually rose and obesity symptoms appeared from 4 to 6 weeks after birth. Wen et al. used this animal model to study the effects of sea cucumber saponins on db/db mice and found that a diet supplemented with 0.07% sea cucumber saponins could significantly inhibit the increase in blood glucose level after oral glucose administration. Mechanism studies suggested that the improvement might be achieved through the inhibition of the activity of α-glucosidase in the brush border small intestine, thereby reducing glucose absorption. Sea cucumber saponins can also improve insulin resistance in spontaneously diabetic mice (C57db/db) simultaneously. Compared with the model group, the fasting insulin level in the sea cucumber saponin-treated group decreased by 20%, and the fasting blood glucose value decreased by 24.6%. It was speculated that sea cucumber saponins might increase the utilization of glucose by improving the sensitivity of insulin to control blood glucose (Wen et al. 2013).

5.2.2 Improving Hypertension

The regulation of blood pressure is a complex process involving multiple systems, and the occurrence and development of hypertension is a result of the joint action of multiple genes and factors and is closely related to the environment. At present, blood pressure-regulating factors mainly include the nervous system, renin-angiotensin system (RAS), adrenergic system, bradykinin-prostaglandin system, endothelial-derived vasoactive substances, and vasopressin.

Zhang et al. (2015) studied the effect of sea cucumber saponins on blood pressure reduction in db/db obese mice, which had significantly higher systolic and diastolic blood pressures than normal mice. The results showed that after the intervention of sea cucumber saponins, the systolic blood pressure of db/db mice was significantly reduced by 20%, which reached the normal pressure level. While the diastolic blood pressure did not change significantly after the intervention. Mechanism studies suggested that sea cucumber saponins might improve hypertension during obesity progression in mice by regulating the renin-angiotensin system.

5.2.3 Improving Hyperlipidemia

Hyperlipidemia is the abnormalities of blood lipids with excessive triglycerides, cholesterol, and low-density lipoprotein due to a variety of factors (Nelson 2013). It has been reported that the sea cucumber could effectively reduce the serum cholesterol, low-density lipoprotein, triglyceride levels, and atherosclerosis index in the model rats with hyperlipidemia. Diets containing sea cucumber (*Isostichopus badionotus*) exhibited hypocholesterolemic activity in the model (Olivera-Castillo et al. 2013). Wen et al. found that sea cucumber saponin significantly reduced serum lipid levels (Wen et al. 2016a). Another study was conducted to evaluate the effect of sea cucumber saponin liposomes in mice on a high-fat diet (Chen et al. 2018). Ding et al. used ApoE knockout mice for intervention with 0.07% sea cucumber saponins for 8 weeks to study the effect of saponins on hyperlipidemia. It was found that sea cucumber saponins could significantly reduce serum total cholesterol (19.5%) and liver free cholesterol (49.2%) levels. Mechanistically, saponins from sea cucumber regulate serum and hepatic lipid levels by modulating the mRNA and protein levels of genes associated with hepatic cholesterol reserve transport. Furthermore, dietary saponins significantly suppressed inflammation by downregulating the expression of pro-inflammatory cytokines in vascular and peritoneal macrophages (Ding et al. 2018).

Meng et al. found that saponins in sea cucumber played an important role in hyperlipidemia and systematically compared the effects of different sources and doses of saponins on lipid metabolism in rats. The study indicated that after the intervention of ginsenosides and sea cucumber saponins in a high-fat diet-induced obesity model for 8 weeks, both of the saponins could reduce serum TC and TG concentrations, of which 0.02% sea cucumber saponins and ginseng saponins reduced blood lipids at an equivalent level. In addition, the serum TG concentration in the 0.02% sea cucumber saponin-treated group was significantly lower than that in the saponin groups with different sources, which was 20% lower than that in the model group. In the 0.08% saponin group, the reducing effect of sea cucumber saponins on blood lipids and liver lipids was significantly stronger than that of the same dose of ginsenosides. Consumption of sea cucumber saponins and ginsenosides had no significant effect on the serum HDL-C content of rats, and the content of TC and TG in liver showed the same change rule as that in serum (Meng et al. 2018). The above results showed that both sea cucumber saponins and ginsenoside could reduce blood lipids and liver lipids and that the high dose of sea cucumber saponins (0.08%) was better than the same dose of ginsenosides.

5.2.4 Improving Nonalcoholic Fatty Liver Disease

Guo et al. (2018) studied the effects of sea cucumber saponins and EPA phospholipids on nonalcoholic fatty liver disease (NAFLD) by comparing 1% EPA phospholipids, 0.028% sea cucumber saponins, and the combination of phospholipids and saponins in different ratios (0.5:0.5, 1:1). It was found that sea cucumber saponins and EPA phospholipids could significantly reduce liver lipids, both individually and in combination. As for liver TG, 0.028% sea cucumber saponins had no significant effect on reducing liver lipids, while 1% EPA phospholipids had an obvious effect on reducing liver TG. Moreover, saponins, EPA phospholipids, and their combination could significantly reduce TC compared with the model group, but there was no significant difference between the groups subjected to these interventions.

Zhang et al. (2020a, b) studied the effect of sea cucumber saponin EA and its metabolites on nonalcoholic fatty liver disease. It was found that both EA and their metabolites could significantly reduce liver TC (by about 50%) and TG (by about 45%), indicating that EA and their metabolites could inhibit lipogenesis to reduce abnormal lipid storage in the liver. Furthermore, the result of enzyme activity in vitro and lipid metabolism-related genes indicated that the EA aglycone of demonstrated obvious advantages in stimulating lipolysis and lipid metabolism and could significantly reduce fatty acid synthase (FAS), glucose 6-phosphate dehydrogenase (G6PDH), and malic enzyme (ME) in the cytosol fraction and gene expression compared with EA.

Han et al. studied the effect of saponins extracted from *Cucumaria frondose* on the alleviation of nonalcoholic fatty liver induced by orotic acid in rats and found that liver TG, cholesterol, and TC levels decreased remarkably. According to a mechanistic investigation, sea cucumber saponins reduced orotic acid-induced nonalcoholic fatty liver by blocking fatty acid synthase activity, downregulating fatty acid synthesis associated gene expression, and enhancing the activity of fatty acid oxidation related enzymes (Han et al. 2014).

5.2.5 Inhibiting Fat Accumulation

Obesity is associated with a multitude of physiological metabolisms and pathological changes in the body as a result of an energy imbalance, which might include excessive or insufficient energy consumption. Hu et al. employed a high-fat diet-induced obesity mouse model to investigate the inhibitory effect of sea cucumber saponins on fat accumulation and found that sea cucumber saponins EA and HA could significantly reduce body fat. The mechanism of weight loss is related to significantly promoting the excretion of fecal lipids and hindering the digestion and absorption of exogenous fats, and sea cucumber saponins can increase the levels of hormones that promote metabolism and inhibit the activity of enzymes related to fatty acid synthesis in the liver (Hu et al. 2012).

5.2.6 Improving Hyperuricemia

Hyperuricemia is significantly positively correlated with obesity, hyperlipidemia, hypertension, diabetes, and atherosclerosis. Widespread changes in dietary structure and living habits have led to increasing incidence of hyperuricemia. Previous studies have reported that hyperuricemia is an important biochemical basis for gout, with 5% ~ 12% of hyperuricemia patients suffer from gout. Hyperuricemia causes acute gout and chronic arthritis, while affecting kidney function and causing chronic interstitial diseases, nephritis, and uric acid nephrolithiasis. Xu et al. studied the effects of sea cucumber saponins echinoside A (EA) and holothurin A (HA) on diet-induced hyperuricemia in mice by feeding them a diet supplemented with 20% yeast extract. It was found that the serum uric acid level of the mice administrated with sea cucumber saponins significantly decreased on the ninth day, indicating that saponins could improve hyperuricemia and prevent gout (Xu et al. 2011).

5.2.7 Regulating Rhythm

The effect of sea cucumber saponin EA on lipid metabolism was studied over a 24-h period in time-course research by Wen et al. At most time points over the 24 h, EA reduced the levels of TC and TG in both serum and liver. EA suppressed and increased the activities of hepatic lipogenic and lipolytic enzymes to various degrees at different periods. Simultaneous fluctuation tendencies in gene expression involved in fatty acid production and beta-oxidation were also detected (Wen et al. 2016a).

Regarding the mechanism of sea cucumber saponins regulating glucose metabolism, a previous study found that sea cucumber saponins mainly regulated glucose metabolism by regulating cytokines and inhibiting α-glucosidase activity. The improvement mechanism of sea cucumber saponins on high-fat diet-induced insulin resistance showed that sea cucumber saponins could regulate glucose metabolism by increasing the content of adiponectin and reducing the content of TNF-α. Different concentrations of sea cucumber saponins have different inhibitory effects on α-glucosidase, resulting in different degrees of improvement of glucose metabolism with different concentrations of sea cucumber saponins.

The sterol regulatory element-binding protein 2 (SREBP-2) is one of the important nuclear transcription factors in lipid metabolism, which is related to cholesterol synthesis and regulation. Inhibition of 3-hydroxy-3-methyl glutaryl coenzyme A (HMG-CoA) reductase could suppress cholesterol synthesis. Wang et al. found that saponins could alleviate orotic acid-induced NAFLD by lowering liver lipid levels, which might be due to SREBP-2 and HMG-CoA reductase inhibition (Wang et al. 2016b).

The digestive enzyme lipase, of which pancreatic lipase is a major component, is responsible for fat digestion. It has been proposed that inhibiting pancreatic lipase activity can decrease fat breakdown and absorption in the small intestine, resulting in a reduction in exogenous fat intake and weight loss. Sea cucumber saponins have been shown to have a significant function in lipid metabolism and help to minimize fat accumulation. The mechanism of sea cucumber saponins regulating lipid metabolism is related to the digestion and absorption of lipids and fat synthesis. Sea cucumber saponins can effectively inhibit the activities of pancreatic lipase and FAS, so as to reduce the digestion and absorption of lipids and fat synthesis, inhibiting the accumulation of fat in the body (Hu et al. 2012).

Moreover, LXRs, such as LXR-α and LXR-β, are important in maintaining cholesterol, triglycerides, fatty acids, and glucose homeostasis. Guo et al. found that sea cucumber saponins extracted from *Pearsonothuria graeffei* exhibited an anti-obesity effect through the inhibition of pancreatic lipase activity and the upregulation of LXR-β signaling in C57BL/6 obese mice induced by high-fat diets (Guo et al. 2016).

Adenosine deaminase (ADA) is a sulfhydryl enzyme, which can catalyze the deamination of adenine nucleoside or deoxyadenosine nucleoside to produce hypoxanthine nucleoside or hypoxanthine deoxynucleoside, producing hypoxanthine by nucleoside phosphorylase catalysis. Xanthine oxidase (XOD) participates in the metabolism of purine, xanthine, and hypoxanthine, with the end product of uric acid. Both ADA and XOD play an important role in the formation of uric acid in vivo (Xu et al. 2011). Further studies showed that sea cucumber saponins might improve hyperuricemia by inhibiting the activity and mRNA expression of liver XOD and ADA, as well as the mRNA expression of renal transporter. The results showed that sea cucumber saponins significantly inhibited the activities of XOD and ADA in the liver of hyperuricemia mice, indicating that the reason why sea cucumber saponins alleviating hyperuricemia was related to the inhibition of XOD and ADA activities. In addition, sea cucumber saponins can significantly downregulate the expression of XOD and ADA mRNA in mouse liver, so as to inhibit the activity of related enzymes and improve hyperuricemia. Compared with hyperuricemia mice, sea cucumber saponin can downregulate the expression of GLUT9 mRNA in mouse kidney, indicating that the improvement of saponins on hyperuricemia was related to inhibiting the expression of GLUT9 mRNA in mouse kidney.

5.3 Promotion of Bone Marrow Hematopoietic Function

The hematopoietic process is an active process of continuous cell proliferation, differentiation, and release. Pluripotent stem cells maintain a constant amount of self-renewal, and proliferate and differentiate from pluripotent hematopoietic stem cells into committed progenitor cells, and then further proliferate and differentiate into mature blood cells, which are released to the peripheral blood circulation. Chen et al. (2003) found that total ginseng saponins could significantly increase the expression of GM-CSF mRNA in bone marrow stromal cells, endothelial cells, monocytes, thymocytes, and splenocytes and directly or indirectly stimulate bone marrow stromal cells in the hematopoietic-inducing microenvironment and lymphocytes to synthesize and secrete GM-CSF-like active substances, thereby promoting the proliferation and differentiation of CFU-GM. This indicated that saponins could synergistically promote the proliferation and differentiation of bone marrow cells by inducing the body to secrete hematopoietic growth factors, so as to achieve the role of hematopoietic function.

Li et al. (2011) studied the effect of total saponins from sea cucumber on peripheral blood cells, bone marrow nucleated cells, and colony formation of hematopoietic stem/progenitor cells in mice. The results showed that saponins from sea cucumber could significantly increase the number of white blood cells (WBC), platelets (PLT), and peripheral blood red blood cells (RBC) in mice, as well as the content of peripheral blood hemoglobin Hb compared with the hypohematopoiesis model. In addition, sea cucumber saponins also significantly increased the number of reticulocytes in mice.

The number of bone marrow nucleated cells in the model group was significantly lower than that in the normal group on the first day after administration, and the high-dose saponin group significantly promoted the proliferation of bone marrow nucleated cells, which was 70% higher than that in the model group. On the third and fifth day, the bone marrow nucleated cells of the mice in the low-dose saponin group recovered significantly, which were 40% and 27% higher than those in the model group, respectively. Li et al. (2011) also studied the effect of sea cucumber saponins on the proliferation and differentiation of hematopoietic stem cells in mice and found that after sea cucumber saponins treatment, the number of spleen colony-forming units in the mice increased significantly and the spleen nodules in mice were large in both size and number. This showed that sea cucumber saponins promoted the proliferation and differentiation of myeloid cells and the recovery of hematopoietic function of bone marrow cells. The effects of sea cucumber saponins on the secretion of colony-stimulating factors that could promote hematopoiesis in mouse lung-conditioned medium (LCM), spleen-conditioned medium (SCM), and peritoneal macrophage-conditioned medium (PMΦCM) were studied by in vitro experiments. The results showed that saponins could significantly promote the synthesis and secretion of colony-stimulating factors, such as GM-CSF/G-CSF.M-CSF, IL-3, EPO, TPO, and IL-6, improving monolineage, erythroid, and megakaryocyte hematopoiesis.

Flow cytometry analysis showed that sea cucumber saponins could inhibit the S phase arrest of bone marrow cells in mice and transform S phase cells to G2/M phase, suggesting that the mechanism of saponins affecting bone marrow hematopoietic function might be that sea cucumber saponins could promote bone marrow cell mitosis, inhibit cell apoptosis, and preserve cell integrity, thereby promoting bone marrow cell proliferation. The result of gene expression further proved that sea cucumber saponins significantly promoted the expression of cyclin B1, Bcl-2, and Bcl-xL, suggesting that sea cucumber saponins inhibited cell cycle arrest and bone marrow cell apoptosis. Moreover, sea cucumber saponins could upregulate the expression of EPOR, GM-CSF, and TPO in bone marrow cells, suggesting that sea cucumber saponins could stimulate the body to synthesize and secrete EPOR, GM-CSF, and TPO, promoting the hematopoietic function.

5.4 Improvement of Immunity

Sea cucumber saponin has progressively gained attention as a valuable marine bioactive component that might boost the body's immunity (Aminin 2016). Macrophages, as major immunocytes, have a wide range of functions in the immune system, which are critical for maintaining physiological equilibrium. By attaching bacterial compounds to receptors on the surface of macrophages, macrophages may engulf a large number of microbes or other cells, making them the most effective phagocytic cells. Chemokines are produced by macrophages, which boost cellular immunity. The spleen contains a variety of lymphocytes, most of which are B cells (approximately 60%) and T cells, with a minor number of natural killer cells. When infections infiltrate the body, the immune cells in the spleen mount an immunological response (Cesta 2006). Bioactive components from *Isostichopus badionotus*, *Isostichopus fuscus*, and *Cucumaria frondosa* have been shown to improve cellular immune and nonspecific immunological function, plantar thickening, and phagocytic capability of peritoneal macrophages in normal mice. It was also discovered that saponins echinoside A and holothurin A increased the number of antibody-forming cells and serum hemolysin levels in immunocompromised mice, promoted humoral immune function, and delayed allergic reactions, splenic lymphocyte proliferation, and the phagocytic rate and index of mouse peritoneal macrophages on chicken red cells, improving nonspecific immune function in an all-round way (Li et al. 2011; Wang et al. 2010b). Aminin et al. (2008) discovered that a low dose of *Cucumaria frondosa* saponin increased antibody production to promote monocyte phagocytosis and the release of cytokines like IL-6, thereby improving humoral immunity and nonspecific immune function in mice, but had no effect on Th1-mediated delayed-type allergy. Cucumarioside A2–2, a sea cucumber saponin, also had a substantial immunomodulatory impact (Pislyagin et al. 2017).

5.5 Antibacterial and Antivirus Effects

Sea cucumber saponins can inhibit the growth of pathogenic bacteria and pathogenic fungi. Kitagawa et al. (1978) found that holotoxins A, B, and C isolated from *Stichopus japonicus* had strong growth inhibitory activity (MIC80 = 3.12 μg/mL) against *Trichophyton sp.*, *Candida sp.*, *Trichomonas sp.*, etc. Wu (2005) studied the antifungal activities of nobilisides A isolated from sea cucumber, and the results showed that nobisliideA had strong inhibitory effects against *Candida albicans*, *Cryptococcus neoformans*, *Candida pseudotropicalis*, *Candida parapsilosis*, *Sporothrix schenckii*, *Aspergillus fumigatus*, and *Trichophyton rubrum*, and their MIC80 values were 1.1 μg/mL, 4.4 μg/mL, 2.2 μg/mL, 1.1 μg/mL, 2.2 μg/mL, 2.2 μg/mL, and 4.4 μg/mL, respectively. Sea cucumber saponins A and B isolated from sea cucumber have been used for clinical treatment of beriberi and trichophyton mentagrophytes infection at present. Zhang et al. (2010) studied the bacteriostatic effect of lecanorosides A and C on *Aspergillus fumigatus*, *Candida albicans*, *Candida tropicalis*, *Cryptococcus neoformans*, *Trichophyton rubrum*, *Microsporum gypseum*, and *Fonsecaea compacta*. The MIC80 values of *Microsporum gypseum* and closely pigmented fungi were all lower than 32 μg/mL. Careaga et al. (2011) also found that two new triterpene glycosides (patagonicosides B and C), patagonicoside A, and their desulfated analogs showed good antifungal activities against phytopathogenic fungus *Cladosporium cladosporioides* in a dose-dependent fashion.

The antibacterial activity of sea cucumber saponins is closely related to the chemical structure. Chludil et al. (2002) studied the antifungal effects of two new sulfated triterpenoid saponins hemoiedemoside A and hemoiedemoside B isolated from sea cucumber and of their semisynthetic derivatives. The results showed that the antifungal activities of hemoiedemoside A and hemoiedemoside B were stronger than their derivatives. Kitagawa et al. (1980) isolated two antifungal components echinosides A and B from the methanol extract of sea cucumber body wall and modified the structure of echinosides A to obtain its desulfurization derivatives. The results showed that the desulfurization derivatives of echinosides A exhibited a stronger inhibitory effect on microorganisms. Further analysis showed that sea cucumber saponins lost their antibacterial activity when the keto group at position 16 and the γ-lactone rings at positions 18 and 20 of the aglycone were opened. When the degree of partial hydrolysis of sugar chain was different, the antibacterial efficacy and antibacterial spectrum were also slightly different. Xylose, quinoa, glucose, and 3-O-methyl glucose must be retained to avoid impact on their antibacterial efficacy.

As it is the most widely utilized receptor by HIV-1 strains and is considered to be critical in viral transmission, chemokine receptor subtype 5 (CCR5) has been a particularly appealing target. CCR5 inhibition as an antiviral treatment has received a lot of interest. Hegde et al. (2002) discovered that the aqueous methanolic extract of a sea cucumber containing two triterpene glycosides had selectivity in inhibiting CCR5.

5.6 *Other Activities*

Sea cucumber has the largest saponin concentration before and after ovulation. Saponins can prevent oocyte maturation and control sea cucumber fertility accordingly. Mats et al. (1990) discovered that a combination of triterpene glycosides from the Far Eastern holothurian *Stichopus japonicus* Selenka (holotoxins A1 and B1) possessed contraceptive function by suppressing ovulation and promoting uterine contractility.

5.7 *Metabolism of Sea Cucumber Saponins*

Sea cucumber saponins are digested and absorbed by small intestinal epithelial cells. Structural changes in the process of digestion and absorption will produce disaccharide metabolites of saponins (Fig. 4). Song et al. (2017b) studied the metabolic changes of sea cucumber saponins from *Pearsonothuria graeffei* and found that HA and EA could be directly absorbed into the blood from the intestine, with peaks in the blood at about 3 h and 7 h.

The absorption rate of sea cucumber saponins is limited, and the oral bioavailability of HA and EA is only 16.17% and 7.56%, respectively (Song et al. 2016). It can be seen that most sea cucumber saponins are not absorbed in the intestinal tract or in the form of prototypes after administration. Studies have reported that ginsenosides undergo metabolic events, such as deglycosylation and dehydrogenation under the influence of intestinal flora, and the metabolites produced can be absorbed by the human body via the gut, with some metabolites exerting higher physiological activity. The bioavailability of the prototype sea cucumber saponins is poor. Except for the saponins that cannot be absorbed by the intestinal tract, metabolites may also be produced in the intestinal tract under the action of intestinal microorganisms. After the co-incubation of sea cucumber saponins HA and EA with intestinal microbes, six kinds of metabolites of EA and HA were detected and determined by HPLC MS/MS. The six metabolites obtained were all products of

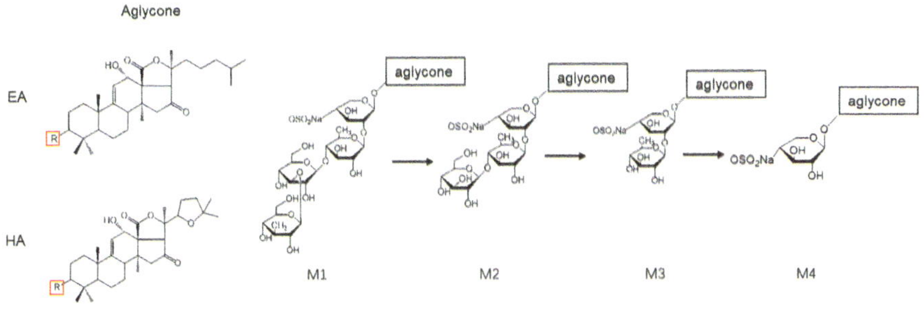

Fig. 4 Chemical structure of sea cucumber saponins and their metabolites

the deglycosylation of HA and EA, and it was clear that the step-by-step deglycosylation reaction was the main metabolic pathway of sea cucumber saponins in the intestinal tract. The metabolic profile of sea cucumber saponins in blood was investigated, and it was discovered that EA and HA were mostly metabolites of two sugars eliminated from the circulation.

Triterpene glycosides can be digested and absorbed by gastrointestinal tract. Joo et al. (2010) measured the ginsenoside level in blood following oral administration to investigate its digestion and absorption feature and discovered two peaks for absorption in the stomach and intestines. The digestion and absorption of two sulfated triterpenoid saponins from *Pearsonothuria graeffei*, echinoside A and holothurin A, were studied in pharmacokinetic research. The blood concentration of echinoside A rapidly reduced 5 min following intravenous treatment, but the serum concentration of holothurin A dropped to its lowest level after 1 h. Further research revealed that echinoside A peaked in serum at 3 and 7 h after oral treatment, whereas holothurin A peaked at 3 and 9 h, indicating that echinoside A and holothurin A might be absorbed directly into the bloodstream through the intestinal tract (Song et al. 2016). Before being taken into the body, most dietary components underwent metabolic changes into different forms and structures by gut microorganisms. The metabolic properties of echinoside A and holothurin A were previously explored by gut bacteria in a prior work. The results showed that the predominant intestinal microflora-mediated metabolic route for sea cucumber saponins was deglycosylation and that the resultant deglycosylated metabolites could be absorbed by the gut (Song et al. 2017b).

6 Other Functional Ingredients

Studies have shown that lectins are a kind of glycoproteins with antigen specificity, which can promote cell agglutination, and play an important role in humoral and cellular immunity (Kuwahara et al. 2002; Gowda et al. 2008a). The lectins in different kinds of sea cucumbers have different biological activities. For example, the lectins isolated from *Cucumaria echinata* have strong and rapid hemolytic activity toward human and rat erythrocytes in the presence of serum (Yoshida et al. 2007), while the lectins obtained from *Holothuria scabra* have antibacterial activity (Gowda et al. 2008b). Sea cucumber also contains pigments and trace substances, such as methionine, taurine, vanadium, selenium, germanium, and vitamin PP. These substances are necessary for human growth and development and have irreplaceable physiological functions (Wang et al. 2009).

7 Conclusion

This chapter systematically introduces the research progresses regarding bioactive food components in sea cucumbers. Although the sea cucumber has a long history of consumption as an important source of both food and medicine in the Asian diet and especially in the health maintenance theory system of Chinese medicine, this chapter provides the scientific basis of modern nutrition for the sea cucumber as both food and medicine. However, there are still many problems in this field that have not yet been solved. For example, there are many kinds of sea cucumbers, but few species are part of actual dietary habits, especially those found in the oceans outside Asia, where many sea cucumbers are not cultivated for consumption; further research is therefore required on how to exploit these species as a means of alleviating health problems faced by modern people. The content and structure of bioactive components vary greatly among different types of sea cucumbers, which, theoretically, have different physiological regulating effects, and studies of the structure-effect relationship need to be further developed. In addition, although there are some related studies on the digestion and absorption process, changes in the digestive tract, the metabolic processes, and the rules in vivo, most of the studies are still in the exploration stage and cannot yet be regarded as comprehensive. Solutions of these basic scientific problems will have an important impact on the development of sea cucumber-related products (food, medicine, and biological agents),and are also critical for the sustainable development of the sea cucumber industry.

References

Adebowale A, Du J, Liang Z, Leslie JL, Eddington ND (2002) The bioavailability and pharmacokinetics of glucosamine hydrochloride and low molecular weight chondroitin sulfate after single and multiple doses to beagle dogs. Biopharm Drug Dispos 23(6):217–225

Aminin D (2016) Immunomodulatory properties of sea cucumber triterpene glycosides. Marine and Freshwater Toxins:381–401

Aminin DL, Agafonova IG, Kalinin VI, Silchenko AS, Avilov SA, Stonik VA et al (2008) Immunomodulatory properties of frondoside a, a major triterpene glycoside from the North Atlantic commercially harvested sea cucumber *cucumaria frondosa*. J Med Food 11(3): 443–453

Anisimova N, Ustyuzhanina N, Bilan M, Donenko F, Usov A, Kiselevskiy M et al (2017) Fucoidan and fucosylated chondroitin sulfate stimulate hematopoiesis in cyclophosphamide-induced mice. Mar Drugs 15(10)

Ben Mansour M, Balti R, Ollivier V, Ben Jannet H, Chaubet F, Maaroufi RM (2017) Characterization and anticoagulant activity of a fucosylated chondroitin sulfate with unusually procoagulant effect from sea cucumber. Carbohydr Polym 174:760–771

Borsig L, Wang L, Cavalcante MC, Cardilo-Reis L, Ferreira PL, Mourąo PAS et al (2007) Selectin blocking activity of a fucosylated chondroitin sulfate glycosaminoglycan from sea cucumber: effect on tumor metastasis and neutrophil recruitment. J Biol Chem 282(20):14984–14991

Cai Y, Yang W, Li X, Zhou L, Wang Z, Lin L et al (2019) Precise structures and anti-intrinsic tenase complex activity of three fucosylated glycosaminoglycans and their fragments. Carbohydr Polym 224:115146

Careaga VP, Bueno C, Muniain C, Alché L, Maier MS (2009) Antiproliferative, cytotoxic and hemolytic activities of a triterpene glycoside from *psolus patagonicus* and its desulfated analog. Chemotherapy 55(1):60–68

Careaga VP, Muniain C, Maier MS (2011) Patagonicosides B and C, two antifungal sulfated triterpene glycosides from the sea cucumber *psolus patagonicus*. Chem Biodivers 8(3):467–475

Cesta MF (2006) Normal structure, function, and histology of the spleen. Toxicol Pathol 34(5): 455–465

Chai YC, Liu Y, Jang XX, Cheng HL, Song SL, Ji AG (2015) Effect of stichopus japonicas polypeptide on NIH/3T3 cells and the comparison with other anti-wrinkle ingredients. Flavour Fragr Cosmet 6:51–55

Che H, Du L, Cong P, Tao S, Ding N, Wu F et al (2017) Cerebrosides from sea cucumber protect against oxidative stress in SAMP8 mice and PC12 cells. J Med Food 20(4):392–402

Chen Z (1987) Effect of acidic polysaccharide from *holothuria leucospilota* on immunologic function of mice. Tianjin Medical Journal

Chen D, Wang SL, Wang YP, Wang L, Jiang R (2003) Experimental study on expressing GM-CSF from lymphocytes by total saponins of Panax ginseng. J Immunol Res 19(6):411–414

Chen S, Xue C, Yin L, Tang Q, Yu G, Chai W (2011) Comparison of structures and anticoagulant activities of fucosylated chondroitin sulfates from different sea cucumbers. Carbohydr Polym 83(2):688–696

Chen S, Li G, Wu N, Guo X, Liao N, Ye X et al (2013) Sulfation pattern of the fucose branch is important for the anticoagulant and antithrombotic activities of fucosylated chondroitin sulfates. Biochim Biophys Acta 1830(4):3054–3066

Chen C, Han X, Dong P, Li Z, Yanagita T, Xue C et al (2018) Sea cucumber saponin liposomes ameliorate obesity-induced inflammation and insulin resistance in high-fat-diet-fed mice. Food Funct 9(2):861–870

Cheng MJ (2021) Effect of sea cucumber hydrolysate on skin wound healing in SD rats. Thesis, Jiangnan University

Cheng MJ, Hu JH, Tang X, Yang TT, Zhou P, Hu W (2021) Effect of sea cucumber enzymatic hydrolysate on skin wound healing in SD rats. Food Ferment Ind 47:94–101

Chludil HD, Muniain CC, Seldes AM, Maier MS (2002) Cytotoxic and antifungal triterpene glycosides from the Patagonian Sea cucumber *hemoiedema* spectabilis. J Nat Prod 65(6): 860–865

Cho SY, Sim JS, Jeong CS, Chang SY, Choi DW, Toida T et al (2004) Effects of low molecular weight chondroitin sulfate on type II collagen-induced arthritis in DBA/1J mice. Biol Pharm Bull 27(1):47–51

Cong PX (2012) Isolation and identification of monosialogangliosides in sea urchins and sea cucumbers. Dissertation, Ocean University of China

Cui FX (2007) Biochemical properties and collagen peptide activity of sea cucumber collagen. Dissertation, Ocean University of China

Dai YL, Kim EA, Luo HM, Jiang YF, Oh JY, Heo SJ et al (2020) Characterization and anti-tumor activity of saponin-rich fractions of south Korean sea cucumbers (*Apostichopus japonicus*). J Food Sci Technol 57(6):2283–2292

Ding L, Zhang L, Wen M, Che H, Du L, Wang J et al (2017) Eicosapentaenoic acid-enriched phospholipids improve atherosclerosis by mediating cholesterol metabolism. J Funct Foods 32: 90–97

Ding L, Zhang TT, Che HX, Zhang LY, Xue CH, Chang YG et al (2018) Saponins of sea cucumber attenuate atherosclerosis in $ApoE^{-/-}$ mice via lipid-lowering and anti-inflammatory properties. J Funct Foods 48:490–497

Ding L, Zhang LY, Shi HH, Wang CC, Jiang XM, Xue CH et al (2020) Eicosapentaenoic acid-enriched *phosphoethanolamine* plasmalogens alleviated atherosclerosis by remodeling gut microbiota to regulate bile acid metabolism in LDLR(−/−) mice. J Agric Food Chem 68(19): 5339–5348

Ding L, Xu ZJ, Shi HH, Xue CH, Huang QR, Yanagita T et al (2021) Sterol sulfate alleviates atherosclerosis via mediating hepatic cholesterol metabolism in ApoE(−/−) mice. Food Funct 12(11):4887–4896

Dong Z, Ding N, Cui J, Zhang H, Xue CH, Wang YM et al (2013) Effects of holothurian cerebroside and ceramide on hyperuricemia in mice. Chinese J Marine Drugs 32(6):65–70

Du L (2012) Study on anti-tumor activity and ameliorating cancer-associated cachexia effect of cerebrosides derived from *Acaudina molpadioides* and *Asterias amurensis*. Thesis, Ocean University of China

Du L, Xu J, Xue Y, Takahashi K, Xue CH, Wang JF et al (2015) Cerebrosides from sea cucumber ameliorates cancer-associated cachexia in mice by attenuating adipose atrophy. J Funct Foods 17:352–363

Duan J, Ishida M, Aida K, Tsuduki T, Zhang J, Manabe Y et al (2016) Dietary cerebroside from sea cucumber (*Stichopus japonicus*): absorption and effects on skin barrier and cecal short-chain fatty acids. J Agric Food Chem 64(37):7014–7021

Dwivedi R, Samanta P, Sharma P, Zhang F, Mishra SK, Kucheryavy P et al (2021) Structural and kinetic analyses of holothurian sulfated glycans suggest potential treatment for SARS-CoV-2 infection. J Biol Chem 297(4):101207

Frey SL, Lee KY (2013) Number of sialic acid residues in ganglioside headgroup affects interactions with neighboring lipids. Biophys J 105(6):1421–1431

Fu XY, Li FH, Zhang LY, Xu J, Xue Y, Xue CH et al (2016) Protective effects of sea cucumber polar lipids on dementia induced by scopolamine in mice. J Chin Inst Food Sci 16(9):7

Fu SS, Wen M, Zhao YC, Shi HH, Wang YM, Xue CH et al (2022) Short-term supplementation of EPA-enriched ethanolamine plasmalogen increases the level of DHA in the brain and liver of n-3 PUFA deficient mice in early life after weaning. Food Funct 13(4):1906–1920

Gao N, Wu M, Liu S, Lian W, Li Z, Zhao J (2012) Preparation and characterization of O-acylated fucosylated chondroitin sulfate from sea cucumber. Mar Drugs 10(8):1647–1661

Gowda NM, Goswami U, Khan MI (2008a) Purification and characterization of a T-antigen specific lectin from the coelomic fluid of a marine invertebrate, sea cucumber (*Holothuria scabra*). Fish Shellfish Immun 24(4):450–458

Gowda NM, Goswami U, Khan MI (2008b) T-antigen binding lectin with antibacterial activity from marine invertebrate, sea cucumber (*Holothuria scabra*): Possible involvement in differential recognition of bacteria. J Invertebr Pathol 99(2):141–145

Guan R, Peng Y, Zhou L, Zheng W, Liu X, Wang P et al (2019) Precise structure and anticoagulant activity of fucosylated glycosaminoglycan from *Apostichopus japonicus*: analysis of its depolymerized fragments. Mar Drugs 17(4):195

Guo GM (2012) Treadmill training and nutritional supplements on mouse skeletal muscle energy metabolism enzymes and resistance to fatigue. Thesis, Liaoning Normal University

Guo L, Gao Z, Zhang L, Guo F, Chen Y, Li Y et al (2016) Saponin-Enriched Sea cucumber extracts exhibit an antiobesity effect through inhibition of pancreatic lipase activity and upregulation of LXR-β signaling. Pharm Biol 54(8):1312–1325

Guo Y, Han XQ, Che HX, Li ZJ, Dong P, Xue CH et al (2018) Synergistic effect of eicosapentaenoic acid-enriched phospholipids and sea cucumber saponin on orotic acid-induced non-alcoholic fatty liver disease in rats. R Soc Open Sci 5(7):172182

Han XQ, Fu XY, Xu J, Dong P, Wang JF, Xue CH et al (2014) Effects of sea cucumber saponins versus starfish saponins on orotic acid-induced fatty liver in rats. J Food Science 35:268–273

He M, Wang J, Hu S, Wang Y, Xue C, Li H (2014) The effects of fucosylated chondroitin sulfate isolated from Isostichopus badionotus on antimetastatic activity via down-regulation of Hif-1α and Hpa. Food Sci Biotechnol 23(5):1643–1651

Hegde VR, Chan TM, Pu H, Gullo VP, Patel MG, Das P et al (2002) Two selective novel triterpene glycosides from sea cucumber, Telenata Ananas: inhibitors of chemokine receptor-5. Bioorganic Med Chem Lett 12(21):3203–3205

Higuchi R, Inagaki K, Natori T et al (1991) Biologically active glycosides from asteroidea, XXV. Glycosphingolipids from the starfish Asterina pectinifera, 2. Structure of three ganglioside molecular species and a homogeneous ganglioside, and biological activity of the ganglioside. Justus Liebigs Ann Chem 1991(1):1–10

Hu Y (2017) Preparation and functional analysis of bioactive peptides derived from Viscera of sea cucumber. Thesis Jimei University

Hu X, Li ZJ, Xue Y, Xu J, Xue CH, Wang JF et al (2012) Dietary saponins of sea cucumber ameliorate obesity, hepatic steatosis, and glucose intolerance in high-fat diet-fed mice. J Med Food 15(10):909–916

Hu S, Chang Y, Wang J, Xue C, Li Z, Wang Y (2013a) Fucosylated chondroitin sulfate from sea cucumber in combination with rosiglitazone improved glucose metabolism in the liver of the insulin-resistant mice. Biosci Biotechnol Biochem 77(11):2263–2268

Hu S, Chang Y, Wang J, Xue C, Shi D, Xu H et al (2013b) Fucosylated chondroitin sulfate from Acaudina molpadioides improves hyperglycemia via activation of PKB/GLUT4 signaling in skeletal muscle of insulin resistant mice. Food Funct 4(11):1639–1646

Hu S, Chang Y, He M, Wang J, Wang Y, Xue C (2014a) Fucosylated chondroitin sulfate from sea cucumber improves insulin sensitivity via activation of PI3K/PKB pathway. J Food Sci 79(7): H1424–H1429

Hu SW, Tian YY, Chang YG, Li ZJ, Xue CH, Wang YM (2014b) Fucosylated chondroitin sulfate from sea cucumber improves glucose metabolism and activates insulin signaling in the liver of insulin-resistant mice. J Med Food 17(7):749–757

Huang N, Wu MY, Zheng CB, Zhu L, Zhao JH, Zheng YT (2013) The depolymerized fucosylated chondroitin sulfate from sea cucumber potently inhibits HIV replication via interfering with virus entry. Carbohydr Res 380:64–69

Imanari T, Washio Y, Huang Y, Toyoda H, Suzuki A, Toida T (1999) Oral absorption and clearance of partially depolymerized fucosyl chondroitin sulfate from sea cucumber. Thromb Res 93(3): 129–135

Ishikawa J, Takada S, Hashizume K, Takagi Y, Hotta M, Masukawa Y et al (2009) Dietary glucosylceramide is absorbed into the lymph and increases levels of epidermal sphingolipids. J Dermatol Sci 56(3):216–218

Iwata M, Corn T, Iwata S, Everett MA, Fuller BB (1990) The relationship between tyrosinase activity and skin color in human foreskins. J Invest Dermatol 95(1):9–15

Jiang CY (2019) Evaluation of memory enhancing activity of sea cucumber protein peptides and its mechanism. Thesis, Dalian Polytechnic University

Jiang S, Yin H, Qi X, Song W, Shi W, Mou J et al (2021) Immunomodulatory effects of fucosylated chondroitin sulfate from Stichopus chloronotus on RAW 264.7 cells. Carbohydr Polym 251: 117088

Joo KM, Lee JH, Jeon HY, Park CW, Hong DK, Jeong HJ et al (2010) Pharmacokinetic study of ginsenoside re with pure ginsenoside re and ginseng berry extracts in mouse using ultra performance liquid chromatography/mass spectrometric method. J Pharm Biomed 51(1): 278–283

Kalia LV, Kalia SK, Lang AE (2015) Disease-modifying strategies for Parkinson's disease. Mov Disord 30(11):1442–1450

Kaneko M, Kisa F, Yamada K, Miyamoto T, Higuchi R (2003) Structure of a new neuritogenic-active ganglioside from the sea cucumber *Stichopus japonicus*. EurJ Org Chem 2003(6): 1004–1008

Kim YJ, No JK, Lee JH, Chung HY (2005) 4,4'-Dihydroxybiphenyl as a new potent tyrosinase inhibitor. Biol Pharm Bull 28(2):323–327

Kitagawa I, Yamanaka H, Kobayashi M, Nishino T, Yosioka I, Sugawara T (1978) Saponin and sapogenol. XXVII. Revised structures of holotoxin a and holotoxin B, two antifungal oligoglycosides from the sea cucumber *Stichopus japonicus* Selenka. Chem Pharm Bull 26(12):3722–3731

Kitagawa I, Inamoto T, Fuchida M, Okada S, Kobayashi M, Nishino T et al (1980) Structures of echinoside a and B, two antifungal oligoglycosides from the sea cucumber Actinopyga echinites (jaeger). Chem Pharm Bull 28(5):1651–1653

Kuwahara H, Yamasaki T, Hatakeyama T, Aoyagi H, Fujisawa T (2002) Oligomerization process of the hemolytic lectin CEL-III purified from a sea cucumber, Cucumaria echinata. J Biochem 131(5):751–756

Ley K, Laudanna C, Cybulsky MI, Nourshargh S (2007) Getting to the site of inflammation: the leukocyte adhesion cascade updated. Nat Rev Immunol 7:678–689

Li JZ, Bao CX, Chen GZ, Zhang GZ, Fan HZ, Chen JD (1985) Antithrombin activity and platelet aggregation by acid mucopolysaccharides isolated from *Stichopus japonicus* Selenka. Zhongguo Yao Li Xue Bao 6(2):107–110

Li M, Miao ZH, Chen Z, Chen Q, Gui M, Lin LP et al (2010) Echinoside a, a new marine-derived anticancer saponin, targets topoisomerase2α by unique interference with its DNA binding and catalytic cycle. Ann Oncol 21(3):597–607

Li B, Wang JF, An M, Liu Y, Zhang Z, Xue CH (2011) Study on the effect of saponin of Pearsonothuria graeffei on hematopoiesis in myelosuppression mice. J Acta Nutrimenta Sinica

Li S, Jiang W, Hu S, Song W, Ji L, Wang Y et al (2015) Fucosylated chondroitin sulphate from *Cucumaria frondosa* mitigates hepatic endoplasmic reticulum stress and inflammation in insulin resistant mice. Food Funct 6(5):1547–1556

Li X, Luo L, Cai Y, Yang W, Lin L, Li Z et al (2017a) Structural elucidation and biological activity of a highly regular fucosylated glycosaminoglycan from the edible sea cucumber *Stichopus herrmanni*. J Agric Food Chem 65(42):9315–9323

Li J, Li S, Yan L, Ding T, Linhardt RJ, Yu Y et al (2017b) Fucosylated chondroitin sulfate oligosaccharides exert anticoagulant activity by targeting at intrinsic tenase complex with low FXII activation: importance of sulfation pattern and molecular size. Eur J Med Chem 139:191–200

Li Q, Cai C, Chang Y, Zhang F, Linhardt RJ, Xue C et al (2018) A novel structural fucosylated chondroitin sulfate from *Holothuria mexicana* and its effects on growth factors binding and anticoagulation. Carbohydr Polym 181:1160–1168

Li S, Zhong W, Pan Y, Lin L, Cai Y, Mao H et al (2021) Structural characterization and anticoagulant analysis of the novel branched fucosylated glycosaminoglycan from sea cucumber *Holothuria nobilis*. Carbohydr Polym 269:118290

Li Y, Wang T, Shi HH, Wang YM, Xue CH, Huang QR et al (2022) Absorption, pharmacokinetics, tissue distribution, and excretion profiles of sea cucumber-derived sulfated sterols in mice. J Agric Food Chem 70(2):480–487

Lian W, Wu M, Huang N, Gao N, Xiao C, Li Z et al (1830) (2013) anti-HIV-1 activity and structure-activity-relationship study of a fucosylated glycosaminoglycan from an echinoderm by targeting the conserved CD4 induced epitope. Biochim Biophys Acta 10:4681–4691

Liu HH, Ko WC, Hu ML (2002) Hypolipidemic effect of glycosaminoglycans from the sea cucumber Metriatyla scabra in rats fed a cholesterol-supplemented diet. J Agric Food Chem 50(12):3602–3606

Liu CH, Zhu BW, Dong XP, Chen LG (2007a) Isolation of sea cucumber enzymatic hydrolysate and its antioxidative effect in vitro. F & FI 09:50–53

Liu BS, Yi YH, Li L, Zhang SL, Han H, Weng YY et al (2007b) Arguside a: a new cytotoxic triterpene glycoside from the sea cucumber *Bohadschia argus jaeger*. Chem Biodivers 4(12): 2845–2851

Liu X, Sun Z, Zhang M, Meng X, Xia X, Yuan W et al (2012) Antioxidant and antihyperlipidemic activities of polysaccharides from sea cucumber *Apostichopus japonicus*. Carbohydr Polym 90(4):1664–1670

Liu X, Cui J, Li Z, Xu J, Wang J, Xue C et al (2014) Comparative study of DHA-enriched phospholipids and EPA-enriched phospholipids on metabolic disorders in diet-induced-obese C57BL/6J mice. Eur J Lipid Sci Technol 116(3):255–265

Liu X, Xu J, Xue Y, Gao Z, Li Z, Leng K et al (2015a) Sea cucumber cerebrosides and long-chain bases from Acaudina molpadioides protect against high fat diet-induced metabolic disorders in mice. Food Funct 6(11):3428–3436

Liu YJ, Xu H, Xu J, Guo YL, Xue Y, Wang JF, Xue CH (2015b) Vanadium-binding protein from vanadium-enriched sea cucumber *Apostichopus japonicus* inhibits adipocyte differentiation through activating WNT/β-catenin pathway. J Funct Foods 17:504–513

Liu Y, Zhou Q, Zhao Y, Wang Y, Wang Y, Wang J et al (2016a) Enrichment, distribution of vanadium-containing protein in vanadium-enriched sea cucumber *Apostichopus japonicus* and the ameliorative effect on insulin resistance. Biol Trace Elem Res 171(1):167–175

Liu X, Hao J, Shan X, Zhang X, Zhao X, Li Q et al (2016b) Antithrombotic activities of fucosylated chondroitin sulfates and their depolymerized fragments from two sea cucumbers. Carbohydr Polym 152:343–350

Liu X, Liu Y, Hao J, Zhao X, Lang Y, Fan F et al (2016c) In vivo anti-cancer mechanism of low-molecular-weight Fucosylated chondroitin sulfate (LFCS) from sea cucumber *Cucumaria frondosa*. Molecules 21(5):625

Liu S, Lu Y, Kurono M, Matahira Y, Manabe Y, Sugawara T (2022) Oral supplementation of sea cucumber and its hydrolysate mitigates ultraviolet A-induced photoaging in hairless mice. J Sci Food Agric 102(5):1987–1994

Lou QM (2011) Lipid analysis of several marine animals and study on bioactivity of EPA-phospholipid from Cucumaria frondose. Thesis, Ocean University of China

Lu LH, Zhou JY, Yan Y, Hu ZH, Guo YW (2009) Effects of sea cucumber peptide on immune regulation and anti fatigue ability in mice. Shandong Medicine 49(25):35–37

Luo B, Zhong-Bing MA, Wang XF, Wang Y (2008) Study on the antiviral effects of *Stichopus japonicus* Selenka glycosaminoglycans (SJ-GAG) on HSV-I. Chin J Oceanol Limnol

Margalit M, Shalev Z, Pappo O, Sklair-Levy M, Alper R, Gomori M et al (2006) Glucocerebroside ameliorates the metabolic syndrome in OB/OB mice. J Pharmacol Exp Ther 319(1):105–110

Mats MN, Korkhov VV, Stepanov VR, Kupera EV, Oleĭnikova GK, Anisimov MM (1990) The contraceptive activity of triterpene glycosides–the total sum of holotoxins A1 and B1 and holothurin a in an experiment. Farmakol Toksikol 53(2):45–47

Matsuoka Y, Saito M, LaFrancois J, Saito M, Gaynor K, Olm V et al (2003) Novel therapeutic approach for the treatment of Alzheimer's disease by peripheral administration of agents with an affinity to beta-amyloid. J Neurosci 23(1):29–33

Meisen I, Peter-Katalinić J, Müthing J (2003) Discrimination of neolacto-series gangliosides with α2–3- and α2–6-linked n-acetylneuraminic acid by nanoelectrospray ionization low-energy collision-induced dissociation tandem quadrupole TOF MS. Anal Chem 75(21):5719–5725

Meng J, Hu XQ, Zhang TT, Dong P, Li ZJ, Xue CH et al (2018) Saponin from sea cucumber exhibited more significant effects than ginsenoside on ameliorating high fat diet-induced obesity in C57BL/6 mice. Med Chem Comm 9(4):725–734

Mou J, Wang C, Li W, Yang J (2017) Purification, structural characterization and anticoagulant properties of fucosylated chondroitin sulfate isolated from *Holothuria mexicana*. Int J Biol Macromol 98:208–215

Nagase H, Kitazato KT, Sasaki E, Hattori M, Kitazato K, Saito H (1997) Antithrombin III-independent effect of depolymerized holothurian glycosaminoglycan (DHG) on acute thromboembolism in mice. Thromb Haemost 77(2):399–402

Nelson RH (2013) Hyperlipidemia as a risk factor for cardiovascular disease. Prim Care 40(1): 195–211

Nguyen TH, Um BH, Kim SM (2011) Two unsaturated fatty acids with potent α-glucosidase inhibitory activity purified from the body wall of sea cucumber (*Stichopus japonicus*). J Food Sci 76(9):H208–H214

Nguyen BCQ, Yoshimura K, Kumazawa S, Tawata S, Maruta H (2017) Frondoside a from sea cucumber and nymphaeols from Okinawa propolis: natural anti-cancer agents that selectively inhibit PAK1 in vitro. Drug Discov Ther 11(2):110–114

Niu Q, Li G, Li C, Li Q, Li J, Liu C et al (2020) Two different fucosylated chondroitin sulfates: structural elucidation, stimulating hematopoiesis and immune-enhancing effects. Carbohydr Polym 230:115698

Olivera-Castillo L, Davalos A, Grant G, Valadez-Gonzalez N, Montero J, Barrera-Perez HAM et al (2013) Diets containing sea cucumber (*Isostichopus badionotus*) meals are hypocholesterolemic in young rats. PLoS One 8(11):e79446

Pescador R, Diamantini G, Mantovani M, Malandrino S, Riva A, Casu B et al (1980) Absorption by the rat intestinal tract of fluorescein-labelled pig duodenal glycosaminoglycans. Arzneimittelforschung 30(11):1893–1896

Pislyagin EA, Manzhulo IV, Gorpenchenko TY, Dmitrenok PS, Avilov SA, Silchenko AS et al (2017) Cucumarioside A2-2 causes macrophage activation in mouse spleen. Mar Drugs 15(11): 341

Qian W, Tao L, Wang Y, Zhang F, Li M, Huang S et al (2015) Downregulation of integrins in cancer cells and anti-platelet properties are involved in holothurian glycosaminoglycan-mediated disruption of the interaction of cancer cells and platelets in hematogenous metastasis. J Vasc Res 52(3):197–209

Ronca G, Conte A (1993) Metabolic fate of partially depolymerized shark chondroitin sulfate in man. Int J Clin Pharmacol Res 13 Suppl:27–34

Sandhoff R, Schulze H, Sandhoff K (2018) Ganglioside metabolism in health and disease. Prog Mol Biol Transl Sci 156:1–62

Shang F, Gao N, Yin R, Lin L, Xiao C, Zhou L et al (2018) Precise structures of fucosylated glycosaminoglycan and its oligosaccharides as novel intrinsic factor Xase inhibitors. Eur J Med Chem 148:423–435

Shechter Y, Karlish SJ (1980) Insulin-like stimulation of glucose oxidation in rat adipocytes by vanadyl (IV) ions. Nature 284(5756):556–558

Shen Z (1997) Studies on the effects of *Holothuria Leucospilota* on blood coagulation, lipids level and viscosity. J Med Lab Sci

Shi D, Wang JF, Shi-Wei HU, Zhang X, Chang YG, Wang YM et al (2013) Study on the alleviating effects of Thelenata anax sulfated fucan on the insulin resistance of diabetic mice. Chin J Mar Drugs

Shi D, Qi J, Zhang H, Yang H, Yang Y, Zhao X (2019) Comparison of hydrothermal depolymerization and oligosaccharide profile of fucoidan and fucosylated chondroitin sulfate from *Holothuria floridana*. Int J Biol Macromol 132:738–747

Shi HH, Zhang LY, Chen LP, Yang JY, Wang CC, Xue CH et al (2022) EPA-enriched phospholipids alleviate renal interstitial fibrosis in spontaneously hypertensive rats by regulating TGF-beta signaling pathways. Mar Drugs 20(2):152

Song S, Cong P, Xu J, Li G, Liu X, Li Z et al (2016) Absorption and pharmacokinetic study of two sulphated triterpenoid saponins in rat after oral and intravenous administration of saponin extracts of Pearsonothuria graeffei by HPLC-MS. J Funct Foods 25:62–69

Song JJ, Cheng X, Ji XM, Jiang NY, Mao XY (2017a) Protective effect of sea cucumber protein peptide on H2O2-induced oxidative injury in RAW264.7 macrophages. Chinese J Food 17 (09):20–26

Song S, Zhang L, Cao J, Xiang G, Cong P, Dong P et al (2017b) Characterization of metabolic pathways and absorption of sea cucumber saponins, holothurin a and echinoside a, in vitro and in vivo. J Food Sci 82(8):1961–1967

Srihera N, Li Y, Zhang TT, Wang YM, Yanagita T, Waiprib Y et al (2022) Preparation and characterization of astaxanthin-loaded liposomes stabilized by sea cucumber sulfated sterols instead of cholesterol. J Oleo Sci 71(3):401–410

Sugawara T, Kinoshita M, Ohnishi M, Nagata J, Saito M (2003) Digestion of maize sphingolipids in rats and uptake of sphingadienine by Caco-2 cells. J Nutr 133(9):2777–2782

Sugawara T, Zaima N, Yamamoto A, Sakai S, Noguchi R, Hirata T (2006) Isolation of sphingoid bases of sea cucumber cerebrosides and their cytotoxicity against human colon cancer cells. Biosci Biotechnol Biochem 70(12):2906–2912

Sugawara T, Duan J, Aida K, Tsuduki T, Hirata T (2010) Identification of glucosylceramides containing sphingatrienine in maize and rice using ion trap mass spectrometry. Lipids 45(5): 451–455

Sun N, Cui P, Lin S, Yu C, Tang Y, Wei Y et al (2017) Characterization of sea cucumber (*Stichopus japonicus*) ovum hydrolysates: calcium chelation, solubility and absorption into intestinal epithelial cells. J Sci Food Agric 97(13):4604–4611

Svennerholm L, Bråne G, Karlsson I, Lekman A, Ramström I, Wikkelsö C (2002) Alzheimer disease–effect of continuous intracerebroventricular treatment with GM1 ganglioside and a systematic activation programme. Dement Geriatr Cogn Disord 14(3):128–136

Teng LB, Wang JF, Yang YC, Li B, Xue CH (2011) Study on the preventive and therapeutic effects of fucoidan isolated from sea cucumber on acute alcoholism in mice. Chin Pharmacol Bull 27(6):835–839

Thinh PD, Ly BM, Usoltseva RV, Shevchenko NM, Rasin AB, Anastyuk SD et al (2018) A novel sulfated fucan from Vietnamese sea cucumber Stichopus variegatus: isolation, structure and anticancer activity in vitro. Int J Biol Macromol 117:1101–1109

Tong Y, Zhang X, Tian F, Yi Y, Xu Q, Li L et al (2005) Philinopside a, a novel marine-derived compound possessing dual anti-angiogenic and anti-tumor effects. Int J Cancer 114(6):843–853

Tovar AM, Mourão PA (1996) High affinity of a fucosylated chondroitin sulfate for plasma low density lipoprotein. Atherosclerosis 126(2):185–195

Ustyuzhanina NE, Bilan MI, Dmitrenok AS, Borodina EY, Stonik VA, Nifantiev NE et al (2017) A highly regular fucosylated chondroitin sulfate from the sea cucumber Massinium magnum: structure and effects on coagulation. Carbohydr Polym 167:20–26

Ustyuzhanina NE, Bilan MI, Dmitrenok AS, Shashkov AS, Nifantiev NE, Usov AI (2018) Two structurally similar fucosylated chondroitin sulfates from the holothurian species Stichopus chloronotus and Stichopus horrens. Carbohydr Polym 189:10–14

Volpi N (2002) Oral bioavailability of chondroitin sulfate (Condrosulf) and its constituents in healthy male volunteers. Osteoarthr Cartil 10(10):768–777

Wan H, Han J, Tang S, Bao W, Lu C, Zhou J et al (2020) Comparisons of protective effects between two sea cucumber hydrolysates against diet induced hyperuricemia and renal inflammation in mice. Food Funct 11(1):1074–1086

Wang TT (2021) Effects of sea cucumber peptide on blood glucose activity and its mechanism in type II diabetes rats. Dissertation, Guangxi University

Wang Y, Wang JF, Zhang J, Cui FX, Xue (2007) Effect of collagen peptide from *Stichopus japonicus* on melanin synthesis in B16 melanoma cells. J Nutr 04:401–404

Wang Y, Wang JF, Gao S, Zhao Q, Liu ZD, Liang YF et al (2008) Protective effect of collagen polypeptide of *Stichopus japonicus* on skin of ultraviolet induced photoaging model mice. J China Pharm Univ 01:64–67

Wang Y, Kang WL, Xin SG, Xing WQ (2009) Analysis study of trace elements in abalone and sea cucumber. Spectrosc Spectr Anal 29(2):511–514

Wang J, Wang Y, Tang Q, Wang Y, Chang Y, Zhao Q et al (2010a) Antioxidation activities of low-molecular-weight gelatin hydrolysate isolated from the sea cucumber Stichopus japonicus. J Ocean Univ China 1:5

Wang J, Fu J, Wang YM, Liu ZD, Li XL, Xue CH (2010b) Effects of saponins of Pearsonothuria graeffei on immune regulation in mice. J Ocean Univ China 40:28–32

Wang JF, Zhang X, Hui LI, Yang YH, Xue CH (2012a) Inhibitory effects of fucoidan, a sulfated polysaccharides from Cucumaria frondosa on experimental metastasis in mice. Chin J Mar Drugs

Wang Y, Su W, Zhang C, Xue C, Chang Y, Wu X et al (2012b) Protective effect of sea cucumber (*Acaudina molpadioides*) fucoidan against ethanol-induced gastric damage. Food Chem 133(4): 1414–1419

Wang D, Zhang L, Wen M, Du L, Gao X, Xue C et al (2016a) Enhanced neuroprotective effect of DHA and EPA-enriched phospholipids against 1-methyl-4-phenyl-1,2,3,6-tetrahydropyridine (MPTP) induced oxidative stress in mice brain. J Funct Foods 25:385–396

Wang D, Ding L, Dong P, Xue Y, Xue C, Wang Y (2016b) Regulation of cholesterol metabolism by sea cucumber saponins in rats with fatty liver. J Acta Nutr Sin 38:67–70

Wang H, Yang S, Wang Y, Jiang T, Li S, Lv Z (2017a) Immunoenhancement effects of glycosaminoglycan from *Apostichopus japonicus*: in vitro and in cyclophosphamide-induced immunosuppressed mice studies. Mar Drugs 15(11):347

Wang X, Tao S, Cong P, Wang Y, Xu J, Xue C (2017b) Neuroprotection of Strongylocentrotus nudus gangliosides against Alzheimer's disease via regulation of neurite loss and mitochondrial apoptosis. J Funct Foods 33:122–133

Wang T, Cong PX, Cui J, Jiang S, Xu J, Xue CH et al (2019a) Digestion, absorption, and metabolism characteristics of EPA-enriched phosphoethanolamine plasmalogens based on gastrointestinal functions in healthy mice. J Agric Food Chem 67(46):12786–12795

Wang H, Xu L, Yu M, Wang Y, Jiang T, Yang S et al (2019b) Glycosaminoglycan from *Apostichopus japonicus* induces immunomodulatory activity in cyclophosphamide-treated mice and in macrophages. Int J Biol Macromol 130:229–237

Wang X, Wang Y, Xu J, Xue C (2021) Sphingolipids in food and their critical roles in human health. Crit Rev Food Sci Nutr 61(3):462–491

Wang Z, Sun J, Ma X, Liu X, Yin F, Li D et al (2022) Characterization of a synthetic zinc-chelating peptide from sea cucumber (*Stichopus japonicus*) and its gastrointestinal digestion and absorption in vitro. J Sci Food Agric 102(11):4542–4550

Wen M, Wang JH, Liu FH, Wang YM, Xue CH, Xue Y (2013) Hypoglycemic effect of sea cucumber saponins on spontaneous diabetic mice. Sci Technol Food Ind 34(22):149–152

Wen M, Fu XY, Han XQ, Hu XQ, Dong P, Xu J et al (2016a) Sea cucumber saponin Echinoside a (EA) stimulates hepatic fatty acid β-oxidation and suppresses fatty acid biosynthesis coupling in a diurnal pattern. J Nutr Sci Vitaminol 62(3):170–177

Wen M, Xu J, Ding L, Zhang L, Du L, Wang J et al (2016b) Eicosapentaenoic acid-enriched phospholipids improve Aβ1–40-induced cognitive deficiency in a rat model of Alzheimer's disease. J Funct Foods 24:537–548

Wu J (2005) Studies on the bioactive constituents from sea cucumber *Holothuria nobilis* and *Holothuria hilla*. Dissertation, Naval Medical University

Wu M, Huang R, Wen D, Gao N, He J, Li Z et al (2012) Structure and effect of sulfated fucose branches on anticoagulant activity of the fucosylated chondroitin sulfate from sea cucumber *Thelenata ananas*. Carbohydr Polym 87(1):862–868

Wu FJ, Xue Y, Liu XF, Xue CH, Wang JF, Du L et al (2014) The protective effect of eicosapentaenoic acid-enriched phospholipids from sea cucumber *Cucumaria frondosa* on oxidative stress in PC12 cells and SAMP8 mice. Neurochem Int 64:9–17

Wu M, Wen D, Gao N, Xiao C, Yang L, Xu L et al (2015) Anticoagulant and antithrombotic evaluation of native fucosylated chondroitin sulfates and their derivatives as selective inhibitors of intrinsic factor Xase. Eur J Med Chem 92:257–269

Wu N, Zhang Y, Ye X, Hu Y, Ding T, Chen S (2016) Sulfation pattern of fucose branches affects the anti-hyperlipidemic activities of fucosylated chondroitin sulfate. Carbohydr Polym 147:1–7

Xiao QZ, Wang CH, Ai LJ (2012) In vitro antitumor activities of low molecular sea cucumber *Stichopus japonicus* peptides sequentially hydrolyzed by proteases. Adv Mater Res 393-395: 1259–1262

Xiao C, Lian W, Zhou L, Gao N, Xu L, Chen J et al (2016) Interactions between depolymerized fucosylated glycosaminoglycan and coagulation proteases or inhibitors. Thromb Res 146:59–68

Xu HJ, Zhang H, Liu CH, Xue Y, Xue CH, Wang YM (2011) The effect of dietary sea cucumber saponin on hyperuricemia in mice. J Chin Pharmacol Bull 27:1064–1067

Xu H, Wang J, Zhang X, Li Z, Wang Y, Xue C (2015) Inhibitory effect of fucosylated chondroitin sulfate from the sea cucumber *Acaudina molpadioides* on adipogenesis is dependent on Wnt/β-catenin pathway. J Biosci Bioeng 119(1):85–91

Xu H, Zhou Q, Liu B, Chen F, Wang M (2022) Holothurian fucosylated chondroitin sulfates and their potential benefits for human health: structures and biological activities. Carbohydr Polym 275:118691

Yamada K, Hara E, Miyamoto T, Higuchi R, Isobe R, Honda S (1998) Isolation and structure of biologically active glycosphingolipids from the sea cucumber *Cucumaria echinata*. Eur J Org Chem 1998(2):371–378

Yamada K, Harada Y, Miyamoto T, Isobe R, Higuchi R (2000) Constituents of holothuroidea. 9. Isolation and structure of a new ganglioside molecular species from the sea cucumber *Holothuria pervicax*. Chem Pharm Bull 48(1):157–159

Yamada K, Hamada A, Kisa F, Miyamoto T, Higuchi R (2003) Constituents of Holothuroidea, 13. Structure of neuritogenic active ganglioside molecular species from the sea cucumber *Stichopus chloronotus*. Chem Pharm Bull 51(1):46

Yang P, Collin P, Madden T, Chan D, Sweeney-Gotsch B, McConkey D et al (2003) Inhibition of proliferation of PC3 cells by the branched-chain fatty acid, 12-methyltetradecanoic acid, is associated with inhibition of 5-lipoxygenase. Prostate 55(4):281–291

Yang YH, Wang JF, Zhang X, Hui LI, Xue CH (2013) The anti-metastatic effect of sea cucumber fucoidan on human gastric carcinoma cell line Hgc-27 under hypoxia induced by CoCl2. Acta Nutrimenta Sinica

Yang L, Wang Y, Yang S, Lv Z (2018a) Separation, purification, structures and anticoagulant activities of fucosylated chondroitin sulfates from *Holothuria scabra*. Int J Biol Macromol 108: 710–718

Yang W, Cai Y, Yin R, Lin L, Li Z, Wu M et al (2018b) Structural analysis and anticoagulant activities of two sulfated polysaccharides from the sea cucumber *Holothuria coluber*. Int J Biol Macromol 115:1055–1062

Yang W, Chen D, He Z, Zhou L, Cai Y, Mao H et al (2020) NMR characterization and anticoagulant activity of the oligosaccharides from the fucosylated glycosaminoglycan isolated from *Holothuria coluber*. Carbohydr Polym 233:115844

Yang JY, Zhang TT, Dong Z, Shi HH, Xu J, Mao XZ et al (2021) Dietary supplementation with exogenous sea cucumber-derived ceramides and glucosylceramides alleviates insulin resistance in high-fructose-diet-fed rats by upregulating the IRS/PI3K/Akt signaling pathway. J Agric Food Chem 69(32):9178–9187

Yegdaneh A, Mirian M, Dana N, Taheri M (2021) Holothurin B isolated from *Holothuria atra* inhibits angiogenesis more potently than curcumin in vitro. Res J Pharmacogn 8(3):33–40

Yin R, Zhou L, Gao N, Li Z, Zhao L, Shang F et al (2018) Oligosaccharides from depolymerized fucosylated glycosaminoglycan: structures and minimum size for intrinsic factor Xase complex inhibition. J Biol Chem 293(36):14089–14099

Yoshida S, Shimada Y, Kondoh D, Kouzuma Y, Ghosh AK, Jacobs-Lorena M, Sinden RE (2007) Hemolytic C-type lectin CEL-III from sea cucumber expressed in transgenic mosquitoes impairs malaria parasite development. PLoS Pathog 3(12):1962–1970

Yu YH (2021) Study on anti fatigue effect and mechanism of sea cucumber peptide. Thesis, Jiangnan University

Yu SR, Ye XW, Chen L, Xie X, Zhou Q, Lian XY et al (2015) Cytotoxic and anti-colorectal tumor effects of sulfated saponins from sea cucumber *Holothuria moebii*. Phytomedicine: International Journal of Phytotherapy and Phytopharmacology 22(12):1112–1119

Yu YF, Wang T, Wang GF, Wang YZ, Li Y, Gao YL (2019) Effect of sea cucumber peptide on mitochondrial function of skeletal muscle in exercise-induced fatigue rats. Shandong Medicine 59(20):31–34

Zeng BB, Zhang LY, Chen C, Zhang TT, Xue CH, Yanagita T et al (2020) Sea cucumber sterol alleviates the lipid accumulation in high-fat-fructose diet fed mice. J Agric Food Chem 68(36): 9707–9717

Zhang Y, Yi Y (2011) Studies on antitumor activities of triterpene glycoside colochiroside a from sea cucumber *Colochirus anceps*. Zhongguo Zhong Yao Za Zhi 36(4):504–507

Zhang SL, Li L, Yi YH, Liu BS, Sun P, Wu ST et al (2010) Studies on antifungal and antitumor activities of lecanoroside a and C from *Actinopyga lecanora*. J Chin. Pharmacol. Bull 26:839–840

Zhang X, Wang JF, Yang YH, Chang YG, Xue CH (2011) Anti-metastasis effect of SC-FUC on spontaneous metastasis of mouse Lewis lung carcinoma and its mechanism. Chin Pharmaco Bull 27(8):1098–1103

Zhang LY, Song SS, Xu J, Dong P, Xue CH, Wang YM (2015) Effects of sea cucumber saponin on blood pressure in obese mice. Chin Pharmacol Bull 12:1169–1173, 1174

Zhang HJ, Chen C, Ding L, Shi HH, Wang CC, Xue CH et al (2020a) Sea cucumbers-derived sterol sulfate alleviates insulin resistance and inflammation in high-fat-high-fructose diet-induced obese mice. Pharmacol Res 160:105191

Zhang TT, Li R, Han X, Chu YY, Wang CC, Chi NQ et al (2020b) Relationship between structure and efficacy of sea cucumber saponins echinoside a and its derivatives on hemolytic activity and prevention of nonalcoholic fatty liver disease. J Food Sci 85(7):2198–2206

Zhao Q (2008) Study on antioxidant activity of sea cucumber collagen peptides. Thesis, Ocean University of China

Zhao Q, Wand JF, Xue Y, Wand Y, Gao S, Lei M, Xue CH (2008) Comparative study on the main active components and immunomodulatory effects of three kinds of sea cucumber. J Fish Sci China 01:154–159

Zhao Q, Xue Y, Liu ZD, Li H, Wang JF, Li ZJ et al (2010) Differential effects of sulfated triterpene glycosides, holothurin A1, and 24-dehydroechinoside a, on antimetastasic activity via regulation of the MMP-9 signal pathway. J Food Sci 75(9):H280–H288

Zhao Q, Liu ZD, Xue Y, Wang JF, Li H, Tang QJ et al (2011a) Ds-echinoside a, a new triterpene glycoside derived from sea cucumber, exhibits antimetastatic activity via the inhibition of NF-κB-dependent MMP-9 and VEGF expressions. J Zhejiang Univ Sci B 12(7):534–544

Zhao Q, Xue CH, Zhang X, Dong P, Wang YM, Wang JF (2011b) Antiangiogenesis effect of holothurin A1 and 24-dehydroechinosideA from Pearsonothria graeffei. Chin Pharmacol Bull 27:939–943

Zhao Q, Xue Y, Wang JF, Li H, Long TT, Li ZJ et al (2012) In vitro and in vivo anti-tumour activities of echinoside a and dsechinoside a from Pearsonothuria graeffei. J Sci Food Agric 92 (4):965–974

Zhao Y, Zhang D, Wang S, Tao L, Wang A, Chen W et al (2013) Holothurian glycosaminoglycan inhibits metastasis and thrombosis via targeting of nuclear factor-κB/tissue factor/factor Xa pathway in melanoma B16F10 cells. PLoS One 8(2):e56557

Zhao L, Wu M, Xiao C, Yang L, Zhou L, Gao N et al (2015) Discovery of an intrinsic tenase complex inhibitor: pure nonasaccharide from fucosylated glycosaminoglycan. Proc Natl Acad Sci U S A 112(27):8284–8289

Zhao YC, Xue C, Zhang T, Wang Y (2018) Saponins from sea cucumber and their biological activities. J Agric Food Chem 66(28):7222–7237

Zheng W, Zhou L, Lin L, Cai Y, Sun H, Zhao L et al (2019) Physicochemical characteristics and anticoagulant activities of the polysaccharides from sea cucumber pattalus mollis. Mar Drugs 17 (4):198

Zhou XQ, Wang CH, Jiang AL (2012) Antioxidant peptides isolated from sea cucumber Stichopus Japonicus. Eur Food Res Technol 234:441–447

Zhou JJ, Xie LH, Du B (2013) Study on enzymolysis of collagen from sea cucumber intestine and its antioxidant activity. J Food Saf Qual 4(05):1401–1407

Zhou MM, Xue Y, Sun SH, Wen M, Li ZJ, Xu J et al (2016) Effects of different fatty acids composition of phosphatidylcholine on brain function of dementia mice induced by scopolamine. Lipids Health Dis 15(1):1–10

Zhou L, Gao N, Sun H, Xiao C, Yang L, Lin L et al (2020) Effects of native fucosylated glycosaminoglycan, its depolymerized derivatives on intrinsic factor xase, coagulation, thrombosis, and hemorrhagic risk. Thromb Haemost 120(4):607–619

Zhu YZ, Wang JF, Shi D, Xu LL, Xue CH (2012) The protective effect of sea cucumber fucoidan on the liver of mice in chronic alcoholism. Acta Nutrimenta Sinica 34(5):474–477,482

Zhu Q, Lin L, Zhao M (2020) Sulfated fucan/fucosylated chondroitin sulfate-dominated polysaccharide fraction from low-edible-value sea cucumber ameliorates type 2 diabetes in rats: new prospects for sea cucumber polysaccharide based-hypoglycemic functional food. Int J Biol Macromol 159:34–45

Zhu Y, Zhang T, Ding L, Shi H, Xue C, Xie W et al (2021) A comparative study about the neuroprotective effects of EPA-enriched phosphoethanolamine plasmalogen and phosphatidylethanolamine against oxidative damage in primary hippocampal neurons. J Ocean Univ China 20(5):1207–1214

Zoepfl M, Dwivedi R, Taylor MC, Pomin VH, McVoy MA (2021) Antiviral activities of four marine sulfated glycans against adenovirus and human cytomegalovirus. Antivir Res 190: 105077

Zou Z, Yi Y, Wu H, Wu J, Liaw C, Lee K (2003) Intercedensides a–C, three new cytotoxic triterpene glycosides from the sea cucumber *Mensamaria intercedens Lampert*. J Nat Prod 66(8):1055–1060

Zou Z, Yi Y, Wu H, Yao X, Du L, Wu J et al (2005) Intercedensides D–I, cytotoxic triterpene glycosides from the sea cucumber *Mensamaria intercedens Lampert*. J Nat Prod 68(4):540–546

Traditional Processing Techniques and Products of Sea Cucumber: Historical Review

Jing Li and Tiantian Zhang

Abstract The sea cucumber is a high-protein, low-fat, low-cholesterol seafood, containing a variety of substances beneficial for the human body. In China, the sea cucumber has been regarded as a tonic since ancient times, because it is highly nutritious and can be used for medical purposes. The sea cucumber has been part of the Chinese people's diet for over a thousand years. Because of the wide differences among sea cucumber varieties and their morphological diversity, the processing methods vary across different countries. Many traditional processing techniques have continued to this day. The trade in processed cucumbers is growing around the world. With the technological advancement of the twenty-first century, we are seeing a higher level of industrialization in sea cucumber processing and a greater diversity of related products. As we seek to maximize the value of this marine species, we continue to draw from the traditional method of sea cucumber processing. This chapter reviews the recorded ancient processing methods, summarizes the processing techniques of important sea cucumber varieties traded globally, and describes in detail the progress and development of China's sea cucumber products, as well as their processing techniques. The progress in processing techniques, along with more efficient and sustainable use of sea cucumber resources, presents us with both challenges and future advancements. It is also of great significance to the balanced use of marine resources.

Keywords Traditional processing technique · History · Dried sea cucumber · Manual processing

J. Li (✉) · T. Zhang
College of Food Science and Engineering, Ocean University of China, Qingdao, P.R. China
e-mail: lijouc@ouc.edu.cn; zhangtiantian@ouc.edu.cn

C. Xue (ed.), *Advances in Sea Cucumber Processing Technology and Product Development*, Advances in Marine Bioprocesses and Bioproducts,
https://doi.org/10.1007/978-3-031-16512-2_4

1 Introduction

The sea cucumber, an ancient echinoderm class, is distributed in sea areas around the world. Sea cucumbers can be found at various depths, from the intertidal zone to the bottom of the ocean, mostly inhabiting the seabed. Tropical sea areas boast wide varieties of sea cucumbers. The Indo-West Pacific region, with the greatest sea cucumber diversity and the most abundant resources, is home to *Aspidochirotida holothuroidea*, *Bohadschia*, *Astichopus multifidus*, and *Actinopyga mauritiana*. Sea cucumber resources in temperate seas are not as diverse varieties, and the preponderant varieties are mainly distributed along the Pacific coast of the Northern Hemisphere, the coasts of Latin America, and the Arctic Ocean. For example, *Astichopus multifidus* is distributed on the east and west coasts of the Pacific Ocean, and the Mexican *Astichopus multifidus* and *Isostichopus badionotus* are distributed in Latin America, the Caribbean Sea, and Mexico.

In China, the uneven distribution of edible sea cucumber resources is even more obvious. There is only one economically viable variety in the northern waters of Liaodong Bay (*Apostichopus japonicus*), while there are over ten species in the southern seas, such as *Thelenota ananas*, *Stichopus chloronotus*, *Bohadschia*, and *Actinopyga mauritiana*. For more details, please refer to Chapter "Diversity, Distribution, and Biology of Sea Cucumber". However, continuous advancements in aquacultural technologies bring hope to the depleting sea cucumber population caused by overfishing. As Purcell reported in 2012, in China, for example, the annual production of dried, cultured *Apostichopus japonicus* has reached 4000 tons (Purcell et al. 2012). Besides, Mexico and other countries in Central and South America have started to cultivate *Isostichopus fuscus* (Purcell et al. 2012).

Sea cucumbers are highly nutritious, soft in texture, and rich in taste. As a typical high-protein, low-fat seafood, they contain substances such as collagen, sulfated polysaccharides, n-3 polyunsaturated fatty acid, and others that serve good nourishing, known as tonic effects (see Chapter "The Functional Components of Sea Cucumber and their Nutritional and Biological Activities"). Ever since ancient times, the sea cucumber has been considered an important nutritional health supplement tonic in China. Techniques of drying, marinating, soaking, and cooking sea cucumbers have been passed down to this day, from one generation to the next. Sea cucumbers can be eaten fresh. But when taken out of the water for an extended period of time or contaminated by oil, they generally become autolytic, i.e., they decompose. Autolysis also occurs when sea cucumbers die a natural death. Thus, there is a need to preserve them. For prolonged storage time and long-distance transport, sea cucumbers were usually dried or partly dried in ancient times. As demonstrated by historical records, people soak dried sea cucumbers in water before cooking them. Also, other processing techniques have been employed to process sea cucumbers, which will be reviewed in this chapter.

This chapter examines historical records of sea cucumber processing in China and the primary processed products of each period. With a focus on the sea cucumber trade between Japan of the Nara period (CE 710–794) and China under the rule of the Tang Dynasty, as well as the trade between China and Southeast Asia in the Qing Dynasty, this chapter reviews the development of the processing techniques of numerous sea cucumber varieties in different regions, before summarizing the key factors that influence the parameters for sea cucumber processing techniques and their impact on product quality. As the global demand for sea cucumbers grows, so will people's expectations for their quality and that will call for greater diversity in processed sea cucumber products, as well as more innovative processing techniques.

2 Traditional Processing Methods of Sea Cucumber all over the World

Often referred to as a "living fossil" in the ocean, the sea cucumber is a mollusk that has inhabited the seabed for 600 million years (Mao et al. 2015). Its recorded history in China dates back to the Warring States Period (475 to 221 BC), and the first record of sea cucumber consumption can be traced back to the era of the Three Kingdoms (220 to 280 AD). As of the Qing Dynasty (1644–1912), the sea cucumber had become an indispensable part of China's banquet culture, and sea cucumbers were cooked in many ways, such as stir-fried, deep fried, and stewed. Their important role in banquets has remained to this day. Over the course of its long history, this unique echinoderm species has been used by humans in many ways to harness its abundant dietary and medical values (Mao et al. 2015).

Today, more than 70 countries and regions are engaged in sea cucumber capturing, processing, and trade, covering over 60 sea cucumber species (Purcell et al. 2013). Because of their quick autolysis, most sea cucumbers supplied to the Chinese market have been dried or preserved in salt ever since ancient days. A small number of fresh sea cucumbers and preserved sea cucumber organs are supplied to West Pacific countries, constituting a palatable part of these countries' traditional cuisines and an important source of protein. The following section reviews the traditional processing methods of sea cucumber around the world, going back to the very first processed products and tracing their paths in global trade.

2.1 Historical Records of sea Cucumber Processing Methods

The sea cucumber has been part of the Chinese people's diet for over a thousand years (Mao et al. 2015). In the years between 219 and 210 BC, officials were sent off by the First Emperor of Qin to look for an elixir that could endow him with everlasting life (Table 1). Sea cucumbers were then presented as a possible longevity remedy by fishermen. That was how sea cucumbers first entered the historical records of the imperial court. They later became valuable and rare food that could only be enjoyed by the royalty.

Table 1 The Chinese history and related records of sea cucumber processing

Date	Ruling entity	Ruler of the country (reign)	Important event
475–221 BC	Warring states period		The first record of sea cucumber
219–210 BC	Qin dynasty	First emperor of Qin (247–210 BC)	Sea cucumbers were first presented as a possible longevity remedy
220–280 AD	Three kingdoms		The first record of sea cucumber consumption: Shen Ying (*Linhai Shuitu Yiwuzhi*)
1127–1279	Southern song dynasty		*Holothuria scabra* won wide acclaim and became an ingredient of imperial banquets
1271–1368	Yuan dynasty		Jia Ming (*Yinshi Xuzhi*)
1368–1644	Ming dynasty	Zhu Yuanzhang (1368–1398)	Xie Zhaozhe pointed out the origin of the name of sea cucumber (*Wu Za Zu*) Sea cucumber became precious seafood that has an equal reputation as shark fin, bird's nest, and abalone
		Zhu Youjiao (1621–1627)	Yao Kecheng (*Shiwu Bencao*); Liu Ruoyu (*Yinshi Haoshang of Ming gong Shi*)
		Zhu Youjiao (1628–1644)	
1644–1912	Qing dynasty	Shunzhi (1644–1661)	Zhou Lianggong (*Min Xiao Ji*)
		Kangxi (1662–1722)	Nie Huang (*Hai Cuo Tu*)
		Qianlong (1736–1796)	Ding Yizeng recorded the method of soaking dried sea cucumbers; Yuan Mei recorded the cooking method of sea cucumber commonly used at that time; Wu Yiluo (*Bencao Congxin*)
		Jiaqing (1796–1820)	Hao Yixing (*Ji Hai Cuo*)
		Daoguang (1821–1850)	*Annals of Jiaozhou*; Sa Ying'e (*Jilin Waiji*)
		Guangxu (1871–1908)	*Annals of Wendeng County*; Xu Ke (*On Animals of Qingbai Leichao*)

However, the sea cucumber did not gain immediate popularity when it was first used as an ingredient. In the *Treatise on the Anomalous Aquatic and Terrestrial Creatures of Linhai* (*Linhai Shuitu Yiwuzhi*), Shen Ying of the Kingdom of Wu (220 to 280 AD), during the era of the Three Kingdoms described how people of coastal regions captured and roasted sea cucumbers. Sea cucumbers were described as pure black in color, 16 to 17 cm long, with the length of a baby's arm, with a stomach and thirty parapodium, and with no mouth or eye. Admittedly, roasting can harden the body wall of sea cucumbers, which is conducive to storage and transportation, but the tough texture of roasted sea cucumbers makes them not delicious.

Sea cucumbers began to appear again in historical records during the Yuan Dynasty (1271–1368), but there were few mentions of "good taste" or "nutritious quality." Jia Ming (1271–1368), a health expert of the Yuan Dynasty, stated in Volume 6 of his *Dietary Notes* (*Yinshi Xuzhi*) that "Sea cucumbers taste sweet and salty. Loaded with cold energy, they should not enter the diet of people with diarrhea." Xie Zhaozhe (1567–1624) of Ming Dynasty (1368–1644) was the first to explain why the sea cucumber was termed the "ginseng of the sea" in Chinese. According to his book *Wu Za Zu*, the sea cucumbers of the Yellow Sea and the Bohai Sea, as well as those in the coastal waters of Fujian, are a warm tonic with the same medical effect as ginseng, a Panax schinseng that is well-known in China (Mao et al. 2015).

In Ming and Qing Dynasties, people came to recognize the values of sea cucumber as a tonic and delicacy. Zhu Yuanzhang (1368–1398), the first emperor of the Ming Dynasty, had a preference for sea cucumbers. Liu Ruoyu recorded in the *Ancient Book Volumes of Delicate Diet, Court History of the Ming* (*Yinshi Haoshang of Ming Gong Shi*) that the emperor loved sea cucumber, abalone, and shark fin. Sea cucumbers had since come into people's attention as precious seafood that enjoyed an equal reputation as shark fin, bird's nest, and abalone. In the Qing Dynasty (1644–1912), because of the improvement in culinary techniques, sea cucumbers were cooked in various ways. They could be stewed, grilled, or served as a cold dish with dressing. There are as many as 12 dishes, using sea cucumber as the major ingredient, in the 217 dishes of the Manchu-Han Imperial Feast that have been passed on to this day (see *The Manchu Han Imperial Feast*, edited by Zhou Jin). Even the cuisines of provinces far away from the sea such as Hunan and Sichuan include various sea cucumber dishes.

Most of the sea cucumber dishes passed down from ancient times use dried sea cucumbers, which need to be rehydrated before cooking. Abundant historical records can be found about dehydration and rehydration techniques (Wang 2002). Nie Huang, a painter and amateur biologist of the Qing Dynasty, lamented the lack of a complete illustrated book about seafood and started to work on one himself. It took him more than 30 years (1667–1698) to complete the four volumes of the *Hand-Painted Pictures of Seafood* (*Hai Cuo Tu*). In addition to the colored paintings of over 300 marine animals, the works also introduced ways of processing seafood. He wrote about the processing of *Actinopyga mauritiana* and *Apostichopus japonicas*, two types of sea cucumbers: "*Actinopyga mauritiana* grows in the sea mud of Guangdong. Large ones can be as long as 16 to 20 cm. They have greenish blue backs, white abdomens, and no pricks. Collectors cut open their backs, coat them with ground oyster shells, and dry them on bamboo chips. Dried sea cucumbers are about the size of a man's palm. Before cooking, people soak them in water to clear away the mud and sand. Stewed in meat broth, their surfaces become as smooth as leather. Sea cucumbers harvested in the waters of Liaodong and Japan, also 16 to 20 cm long, are as black as ox horns. Their backs are hunched and their abdomens flat. They have pricks all over their bodies, and tiny pricks that resemble the feet of a silkworm along the sides of their abdomens. Collectors hollow out the sea cucumbers without cutting them open. The sea cucumbers are then dried in their original shapes and cooked in the same way as *Actinopyga mauritiana*. Soft and tasty, they are even better than *Actinopyga mauritiana*. The prices are thus higher. These sea cucumbers constitute an integral part of banquets. As the demand for sea cucumbers grows, so does their production. Only through stewing with meat broth will the aroma of sea cucumbers be brought out. Because of the warm climate in Guangdong, sea cucumbers must be processed with ground oyster shells there. Otherwise, they will easily rot and cannot acquire a tender texture when stewed. However, the climate in Liaodong is cold. As a result, there is no need for ground oyster shells. Sea cucumbers dry on their own and maintain their original features. When stewed, they turn soft and tender, and taste great. We can thus tell that the latter variety is of better quality than the first."

During the rule of Emperor Jiaqing of the Qing Dynasty (1796–1820), Hao Yixing compiled information in his book *On Seafood* (*Ji Hai Cuo*), which describes in detail the process of dehydrating sea cucumbers: "Sea cucumbers are dried under the sun and marinated in salt before being washed to reduce the salty taste. Covered in coal ash, they harden and turn black. Dried sea cucumbers are still 15 to 20 cm long. They are transported to distant destinations and enjoyed by gourmets. The sea cucumber is referred to as ginseng in the sea because of its rarity and tonic function." In addition, *Annals of Jiaozhou* under the rule of Emperor Daoguang (1821–1850) and *Annals of Wendeng County* under the rule of Emperor Guangxu (1871–1908) also record the dehydrating techniques. They emphasize that sea cucumbers should be disemboweled before being boiled with seawater and marinated in salt. They are then boiled again in saline water, covered in plant ash, and dried. Today in China's coastal areas, people still process sea cucumbers this way. Our ancestors realized the importance of salt in processing sea cucumbers. To minimize the saltiness, soaking is a necessary step before cooking sea cucumbers, and whether they are soaked properly determines the acceptance of the dish. Thus, there basic food processing techniques to preserve and prepare sea cucumber have already been used by Chinese people.

According to Ding Yizeng of the Qing Dynasty (1736–1796), "dried sea cucumbers were soaked in water, scrubbed to remove the rugged parts on the surface, and then rinsed and cut open to take out the intestines. They were then cut into slices, boiled in saline water, and stewed in a meat broth before fully steamed in a container." Soaking in water and reboiling in meat broth was fairly popular at the time. Yuan Mei, a writer in Qing Dynasty, stressed that if one intended to serve his guests sea cucumber dishes, he must simmer the ingredients the day before so that the sea cucumber would soften and become fit for use. Fully soaked sea cucumbers swell and acquire a soft texture. Only then can they be cooked. Yuan also wrote about the way gourmets of the time cooked sea cucumbers: "Without a delicious taste of their own, sea cucumbers carry mud and sand, and a salty smell of the sea. They are not fit for simmering in light soup. We should pick prickly sea cucumbers of small sizes, soak them in water to clear away the sand, and boil and soak them in meat broth several times before fully braising them in chicken and meat broth."

From these written traditions, we can tell that ancient Chinese people already understood the following properties of sea cucumbers: (1) When taken out of the water, sea cucumbers become autolytic. They must be dehydrated before being stored and transported. (2) The acceptance of sea cucumber dishes highly depends on whether the ingredient is properly soaked. (3) Sea cucumbers do not have much taste of their own. They must be boiled in meat broth or seasoned with spices so that they become palatable dishes.

Ancient Chinese people accumulated a great body of experiences in the processing and cooking of sea cucumbers. We are left with abundant records on sea cucumber varieties and processed quality. For example, in the Southern Song Dynasty (1127–1279), *Holothuria scabra*, a prickless sea cucumber variety produced in Wenzhou, won wide acclaim and became an ingredient of imperial banquets. But it was soon replaced by *Astichopus multifidus*, a prickly variety that was considered a delicacy of the sea. As described in Volume 1 of *Shiwu Bencao* compiled by Yao Kecheng toward the end of the Ming Dynasty (1621–1644), "Sea cucumbers, produced in the coastal waters of eastern and southern China, are black, with the shape of a silkworm. One variety, 16 to 20 cm long, is clean inside out and delicious in taste. With an excellent tonifying effect, it is one of the most precious varieties. Another variety, 6 to 10 cm long, carries mud and sand that cannot be thoroughly cleaned even after scraping. It does not have a rich taste and is of poor quality." Another record can be found in *Bencao Congxin* of the Qing Dynasty: "Sea cucumbers produced in the waters of the Liaodong Peninsula are of excellent quality. Those with pricks on the back are named *Apostichopus japonicus*, while those without pricks are called *Holothuria scabra*." Sa Ying'e wrote in Volume 7 of *Jilin Waiji*: "Sea cucumbers have the shape of a worm and pricks over their bodies. Those produced in Hunchun are of the best quality." Hunchun refers to the vast area from the northwestern coast of the Sea of Japan to Peter the Great Gulf. As pointed out by Zhou Lianggong in his *Min Xiao Ji* at the beginning of the Qing Dynasty, "Sea cucumbers harvested in Fujian are unique in color and as large as a man's palm when spread out on bamboo chips. They look different from those harvested in Jiaozhou and Liaohai, and their taste cannot match the latter in richness." According to *On Animals* of *Qingbai Leichao*, "Sea cucumbers are echinoderms, found in offshore areas. Dried sea cucumbers can be an ingredient. Those presented to emperors are *Apostichopus japonicus* and of the best quality. They are black in color with many pricks. Sea cucumbers produced in Guangdong are yellow and of lesser quality. Those produced in Ningbo are white, prickless, and of the worst quality. Fujian produces *Holothuria scabra*, another type of sea cucumber that is white and prickless." From the works cited above, it can be concluded that as early as in Ming and Qing Dynasties, people had come to realize that sea cucumbers in the Bohai Sea and Liaodong Gulf were of the best quality. Sea cucumbers were classified on the basis of their quality into *Apostichopus japonicus* in Liaodong Peninsula of China, *Actinopyga* in Guangdong, *Acaudina leucoprocta* in Zhejiang, and *Holothuria scabra* in Fujian. Sea cucumber production in China was not only limited to the Yellow Sea and the Bohai Sea but also reached the tropical waters of the East China Sea and the South China Sea.

Over the course of China's ancient history, people's understanding and demand for sea cucumbers kept growing and reached a peak in the Qing Dynasty (1644–1912). Sea cucumbers imported from half of the Pacific countries appeared in the Chinese market. China's commercial ties with Australia began with the trade of sea cucumbers (trepang, bèche-de-mer), which is one of the earliest exported goods from Australia. In the eighteenth century, the Makassan of Indonesia who depended on maritime trade and the Yolngu of Australia traded sea cucumbers with the Qing government (Dai 1998). The Australian aboriginals learned through the trade the Chinese techniques of sea cucumber processing (boiling disemboweled sea cucumbers before drying and smoking). The Museum and Art Gallery of the Northern Territory in Australia display artifacts, images, and written records related to sea cucumber processing. The extensive exchanges between China and Australia today find their origin in the sea cucumber trade of over 200 years ago.

A lot has been recorded about sea cucumber processing in Japanese history as well (Jun 2015). Fresh sea cucumbers are often cut into small slices and marinated with vinegar. The intestines and gonads taken out during processing are also salted and cured and served as konowata and konoko (heavily salted, fermented viscera dishes). Dried sea cucumbers were exported to China. The Nara period saw the earliest record of drying sea cucumbers. Inscribed wooden slips, excavated from the relics of the Heijo-kyo Palace (the political center of the period), recorded how sea cucumbers were processed in the area of today's Noto Peninsula to the north of Ishikawa prefecture. At that time, people would boil disemboweled sea cucumbers with saline water before leaving them to be sun-dried. Records from the Heian period show that dried sea cucumbers were paid to China as tributes. The tributes from the Noto Peninsula also included konowata. The *Drawings of Famous Japanese Mountain and Sea Products*, printed toward the end of the eighteenth century, depicts the scenes of sea cucumber processing. People would boil and stir disemboweled sea cucumbers in big pots, fix them on nets, and leave them to dry and take shape in the sea wind and sun. The *Book of Japanese Cuisine*, published in 1643, includes recipes for dried sea cucumbers. High-grade food materials including dried sea cucumber, dried abalone, and shark fin appeared in the trade records between Japan (the Edo period) and China (Qing Dynasty). These goods were packed in straw bales and exported to China. Because of this, Japan saw an increase in sea cucumber capturing and dried processing. Such sea cucumbers were mostly exported to China and seldom entered the diet of most Japanese people.

2.2 Processing Techniques Used for Global Trade of Sea Cucumbers

The Chinese people consider the sea cucumber one of the four delicacies of the sea. Statistics from the official website of the International Trade Center (ITC) show that 90% of the world's sea cucumber harvest is made into dried products (smoked, salt-dried, and pure-dried sea cucumbers, etc.). A small percentage supplied to Asian markets is in salted, frozen, or fresh form. The major Asian markets include the mainland, Hong Kong, and Taiwan of China, as well as Singapore and Malaysia. According to the ITC, the global imported volume of dried and salted sea cucumbers (product category 030819) reached USD 306 million in 2020, with 13,824 tons imported, nearly doubling that of 2017. Despite the fact that only 7099 tons were exported globally in 2020 due to the impact of COVID-19, the export volume amounted to USD 209 million. In the 3 years before the outbreak of the pandemic, the export volume of sea cucumbers rose from USD 218 million in 2017 to USD 258 million in 2019. As shown by the statistics of the Food and Agriculture Organization (FAO), the global sea cucumber harvest in 2019 registered 59,300 tons, up from 35,300 tons in 2010, a 60% increase within a decade.

The commercial catch of sea cucumbers has had a history of over a thousand years, with more than 70 wild species utilized for food. As many as 50 species were used in Asia and over 30 in the Pacific region (Purcell et al. 2018; Toral-Granda et al. 2008). Asia-Pacific has thus become a major supplier of sea cucumbers, with Indonesia being the largest tropical sea cucumber producer (Aprianto et al. 2019). The Sembilan Islands in South Sulawesi is a major producing area, with a wide variety of commercial sea cucumbers. The locals have developed targeted processing techniques for each variety based on the differences in the thickness of sea cucumber body wall and body wall texture (Aprianto et al. 2019). For example, the minimum harvest length for *Bohadschia vitiensis*, *B. similis*, *Holothuria edulis*, and *Actinopyga miliaris* is 20–25 cm. Since the body walls of these species are thin, directly boiling them would distort their surfaces. As a result, they must be salted beforehand. Their processing includes five steps: disemboweling, marinating with saline water, salt drying, boiling, and sun drying. *Thelenota ananas* and *Stichopus vastus* are species that are larger in size. Their wet weights are usually between 1 and 3 kilograms, and their wet lengths exceed 30 cm. Since their body walls are thick enough for direct boiling, their processing includes eight steps: disemboweling, boiling, cooling, dry-salting, second boiling, sun-drying, third boiling, and final sun-drying. *Holothuria scabra* is the most common variety in the Indo-Pacific region and has the most economic value. Because of the pigmentation and chalk-like sediments on the surface, several steps are needed to remove their thick and tough leather-like surface. The processing covers six steps: boiling, disemboweling, scraping of a rough surface with shell, dry-salting, second boiling, and sun-drying. Trepong with/without smoking is the major sea cucumber type produced today.

The Labbakkang region in Indonesia mainly deals with the processing of *Holothuria scabra*. With a smaller scale of production than that of the Sembilan Islands, this region relies on manual labor and boasts a more sophisticated and science-based surface removal technique. Disemboweled sea cucumbers are boiled at a temperature of 95–100 °C, cooled down in the refrigerator, and boiled for a second time. Their rough surfaces are softened with shredded papaya leaves and scraped off, and the viscera that is left are removed. The sea cucumbers are boiled a third time, salted, boiled one last time, and sent to dry under the sun. The sea cucumbers go through a total of 11 steps to be fully processed. Most tropical production areas in the Indo-Pacific still adopt the traditional method of surface removal, dissolving and removing the sediments on the sea cucumber surface with the inherent bacteria of the local sand beaches (Sachithananthan et al. 1975). Of course, after removal of the surfaces, the sea cucumbers have to be heat-treated through boiling, so that potential pathogens are inactivated.

Tropical sea cucumbers are of important economic value, and their main species include *Holothuria, H. fuscogilva, H. scabra, H. scabra versicolor, Actinopyga* echinites, A. *miliaris, Thelenota ananas, Stichopus chloronotus*, and *S. variegatus* (Conand 2018). Since tropical sea cucumbers vary greatly, most of them are harvested in different fisheries (Purcell et al. 2013). To reduce the differences in sea cucumber processing at an artisanal scale, the South Pacific Commission and the National Fisheries Administration of Papua New Guinea summarized four major methods for tropical sea cucumber processing: (1) Boil fresh sea cucumbers for 2 to 5 min until they swell. Their bodies are squeezed, and the viscera taken out. Then, the cucumbers are boiled for a second time until their body walls become tough and elastic. They are buried in sand pits for 12 to 18 h, and the redundant and dissolved parts removed (2) The sea cucumbers are washed with seawater and boiled with clear water three times. They are dried in heated air or smoked before leaving them to be sun-dried, which takes between 4 days and 2 weeks, based on their water contents. (3) The fresh sea cucumbers are boiled for ten minutes until they swell. Two incisions are created, 3 cm each, on their backs to let out the viscera, but the five longitudinal muscle layers are kept. They are then washed with seawater before boiling them for a second time in fresh water until the texture turns tough and elastic. The remaining viscera are removed. Wooden sticks are inserted through the cuts to spread sea cucumbers' bodies. They are dried in heated air or smoke for 12 to 48 h. Then, the sea cucumbers are also dried under the sun, with their cuts downward for 1 to 2 days. The wooden sticks are taken out, and the sea cucumbers are tied up with strings or rattans. They are dried again, under the sun, for 4 days to 2 weeks, based on the water contents. The strings or rattans are removed before packaging. (4) Two incisions are created, 3 cm each, on the abdomen of the sea cucumbers. The remaining steps are exactly the same as those in the previous method.

The traditional processing techniques above can also be applied to industrialized sea cucumber processing in local companies, where the processing equipment would usually be more sophisticated. For example, the equipment includes pulp steamers with automatic temperature control, as well as drying and smoking ovens (Purcell et al. 2013). Much larger than personal workshops, these companies enjoy great advantages in aspects such as product quality and graded packaging.

High-quality sea cucumber species with relatively high market prices mostly grow in the coastal waters of the Pacific in the Northern Hemisphere, Latin America, and the Arctic Ocean. *Astichopus multifidus, S. herrmanni, S. chloronotus, T. ananas*, and *Holothuria nobilis* are the main species of temperate production areas. East Asian countries, such as China, Japan, and South Korea, mostly process *Apostichopus japonicus*, small-sized sea cucumbers with thin and tender body walls. Iceland, the United States, and Canada mainly process *Astichopus multifidus* and *T. ananas*, large-sized sea cucumbers with thick body walls that are more suitable for

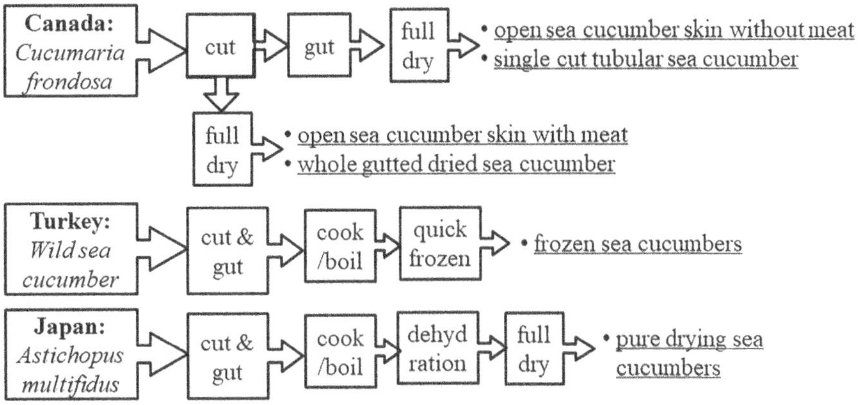

Fig. 1 Processing methods used for different species of cold-water sea cucumbers in different countries

mechanized processing. With specialized equipment that ensures efficiency, the cost of labor and production is significantly reduced. Large quantities of affordable sea cucumbers thus enter the Chinese market, generating profitable returns. For example, *Cucumaria frondosa* sea cucumbers are mainly processed by a simple drying technique that includes cutting, disemboweling, and drying in heated air. Based on the differences in cutting techniques, the dried products can be divided into open sea cucumber skin with or without meat, single-cut tubular sea cucumber, and whole gutted dried sea cucumber, as shown in Fig. 1. Canada boasts a high level of mechanization and automation with advanced technologies in sea cucumber processing (Purcell et al. 2013). As better freezing technologies are increasingly applied in the field, frozen sea cucumbers, as a primary product, account for a growing share in the global trade of sea cucumbers. For example, Turkey's total export volume of sea cucumbers in the 5 years between 2017 and 2021 reached USD 683 million, 18.37% higher than that of Canada, which stood at USD 577 million (International Trade Center, ITC). Turkey has gradually taken the place of Canada as the largest exporter of frozen sea cucumbers. Its processing plants of wild sea cucumbers are mostly small- and medium-sized, with monthly production rates of less than 10 tons. As the processing plants are mainly held by Chinese people in Turkey, fresh sea cucumbers are often transported and sold to Chinese communities after they are disemboweled, boiled, and quick-frozen. Japan, a major exporter of high-quality cold-water *Astichopus multifidus* since ancient times, has an annual import volume of only 3 to 4 tons. Primary products such as frozen and pure-dried sea cucumbers dominate its production. The steps for pure drying include disemboweling fresh sea cucumbers, boiling them in fresh water, dehydrating them at low temperatures, and drying them in heated air.

From studying the above processing techniques of tropical and cold-water sea cucumbers, it can be seen that today's dried sea cucumber processing still follows the four major traditional steps of disemboweling, salting, boiling, and drying. The size and position of the incision are the keys to maintaining the shape of processed sea cucumber. The cutting positions of *H. fuscogilva*, *H. whitmaei*, and *T. ananas*, for example, have a greater impact on the prices of final products, as compared to those of *B. argus* and *S. herrmanni* (Purcell 2014a). Salting removes much of the water within sea cucumbers, prevents them from being decomposed by bacteria, and improves the quality of final products. The duration and order of salting are thus essential. Parameters such as the order and frequency of boiling and drying are also major factors behind product quality (Purcell et al. 2016; Mangubhai et al. 2016; Aprianto et al. 2019). For instance, the temperature, duration, and the number of times boiling are often specifically designed to achieve optimum product quality. As reported by Aprianto reported, the grade A dried sea cucumbers, produced in Fiji, are boiled three times, while those of grade C are only boiled once (Aprianto et al. 2019). Sea cucumbers produced in the Sembilan Islands are also boiled once. They are thus of relatively low quality and can only be graded C. Drying is also a key factor that determines product quality. In addition to duration, the technique of drying is also important. Smoking enables better product quality than drying in heated air chambers, as it adds to the aroma and prolongs the shelf life of sea cucumbers. It is thus the most recommended technique for drying high-quality tropical sea cucumbers in the Pacific region. However, the smoked flavor has not gained popularity among Chinese consumers. For traditional Chinese techniques of sea cucumber processing, please refer to Sect. 3.2.

There are abundant commercial sea cucumber varieties, about which the processing techniques vary across the globe. The steps to take, the order of the steps, and the necessity of repeating certain steps are determined in light of the following factors: (1) features of the body walls, especially the thickness and roughness of surfaces; (2) quantity of sea cucumbers to be processed; and (3) requirement for water content after boiling. For sea cucumbers of the same variety, the grade of the final products varies, due to the processing techniques, which depend on geographical, cultural, and technological factors. Inappropriate handling leads to lower quality, grade, and price of the final products, resulting in the waste of sea cucumber resources and reductions in the country's export revenue. This would be detrimental to the development of sea cucumber-related industries. To avoid such mishandlings, the Australian Centre for International Agricultural Research has issued a set of standard handling and processing methods for the commercialized processing of sea cucumbers (Purcell 2014b), with the aim to transform sea cucumber processing from the traditional manual model to a scientifically based and efficient commercialized model.

3 Traditional Sea Cucumber Products and their Processing Techniques Developed in China

China enjoys an abundance of sea cucumber resources, with 134 identified species, accounting for 14.9% of the world's total (Liao 1997). There are about 20 edible species, among which *Astichopus multifidus*, *Thelenota ananas*, *Stichopus herrmanni*, and *Holothuria nobilis* are the most popular and widely consumed species by Chinese people. In 2017, China's sea cucumber production reached 200,000 tons, and the total output of sea cucumber products was valued at USD 4.44 billion (FAO). According to the *Almanac of Fisheries in China* (2021), China's sea cucumber output in 2020 amounted to 196,500 tons, 24,800 tons more than that of 2019, up by 14.5%. Using a per ton price of RMB 150,000 for fished sea cucumber products to calculate, the output value of China's sea cucumber raw materials in 2020 was RMB 29.17 billion, and the whole industrial chain totaled an output value of RMB 60 billion. Despite China's long history of sea cucumber processing, it only has a few processing techniques and a small number of processed products. The following section will trace the development of China's traditional sea cucumber processing techniques, with a focus on the most common and representative processed products such as semi-dried and dried sea cucumbers.

3.1 Semi-Dried Sea Cucumber

Compared with dried sea cucumbers, fresh ones have a better taste and a higher nutritional quality. But most Chinese people do not have easy access to fresh sea cucumbers because of their short shelf life. Preservation in salt or salt water is one of the oldest processing and storage methods, with a history of over a thousand years. It effectively prolongs the shelf life while retaining the quality and nutritional attributes of sea cucumbers. Fresh sea cucumbers are kept in salt water tanks until the moisture content of sea cucumber is below 50%. In temperatures of 0 °C–4 °C or under wet and cold conditions, salted sea cucumbers can be preserved for 3 to 6 months. They are not suitable for storage in the freezer. The steps for making salted sea cucumbers are simple: Create an incision of 2 cm on the abdomen of the sea cucumber. Let out the viscera, but keep the intestines and gonads inside. Put the sea cucumber in hot water, pre-boil for 3 to 5 min, and take it out. The body of the sea cucumber will contract and turn black, and the pricks will be tough, straight, and pointed. Put the pre-boiled sea cucumber in a container and preserve it in salt or saturated salt water. The mass ratio of fresh sea cucumber to salt should be 5:1. Preservation in salt or salt water will leave an amount of high-salinity liquid inside the sea cucumber, which maintains the flavor of the sea cucumber. Currently, the salt-preserving techniques adopted by businesses often have flaws, such as uneven distribution of salt and excessive salinity in products (also known as "Lagang Yan" sea cucumber). As for sea cucumbers preserved in saline water (also known as pickled sea cucumber), their market prices are relatively low because of the high water content and short processing period (Hou et al. 2015). Dried sea cucumbers account for a larger share of the market.

3.2 Dried Sea Cucumber

Dried sea cucumber (trepang) is the most popular form of processed fresh sea cucumber in the international market. Though not as nutritious as fresh or salted sea cucumber, its advantages in storage and transport make it very popular. Salt, plant ash, or mild-salt pretreated dried sea cucumbers, as well as freeze-dried sea cucumbers, are the most common types in China. This section will focus on the traditional Chinese sea cucumber drying techniques with the examples of plant ash-pretreated drying and salt-pretreated drying, techniques invented in ancient times and passed on to this day. Mild-salted drying and freeze-drying are modern techniques that will be discussed in another chapter.

3.2.1 Plant Ash-Pretreated Dried Sea Cucumber

Plant ash-pretreated dried sea cucumber is a kind of dried sea cucumber with a layer of plant ash on the surface. It can be preserved for 5 to 10 years without any additives. It also boasts greater taste and texture, as well as better nutrition and easier alimentation for the human body. Plant ash-pretreated drying is the oldest technique for sea cucumber processing, which carries the wisdom and experience from generations of craftsmen. According to historical records, the Empress Dowager of the Qing Dynasty (1861–1908) had the habit of eating plant ash-pretreated dried sea cucumbers. Plant ash-pretreated drying has also been the standard technique for the publicly purchased sea cucumbers since the founding of the People's Republic of China.

In Jiaodong Peninsula, an ancient processing technique for preparing plant ash-pretreated dried sea cucumber is still in use today. This centuries-old technique has been included in the third group of the intangible cultural heritage of the city of Longkou in 2020. The processing steps are shown in Table 2, and its key control points include the following:

1. Pick female sea cucumbers, which have a sound, fleshy quality.
2. Cut open, clean, and disembowel the sea cucumbers.
3. Boil with clear water. Use charcoal to make a fire, as the fire temperature it creates is lower than coal fire. The control of the water temperature is very important. Constantly stir and control the duration and degree of heating to prevent it from reaching the boiling point, and stir the sea cucumbers to make sure that they are boiled evenly. The boiling normally lasts 40 min. Scoop up the sea cucumbers immediately after their skin tightens and their pricks stiffen.

Table 2 Different processing methods used for plant ash-pretreated dried sea cucumber

Processing steps	Specific requirements/standards Ancient processing technique used in Jiaodong Peninsula	Processing technique commonly used in the 1970s
Harvest	Female Sea cucumbers before spawning	Fresh Sea cucumber
Gut	Cut open Disembowel guts Clean	Clean with seawater Cut an incision one third the size of the body Disembowel guts Stir until their mouths fully retract
First boil	Use charcoal to set fire Boil with clear water for about 40 min Constantly stir Ensure the temperature below the boiling point Scoop up right after the body walls tightens and pricks stiffen	Boil with either seawater or freshwater for about 30 min Constantly stir Skim the scum Scoop up right after the body walls tighten, the pricks stiffen, and the incisions turn golden
Marination	Put in clay vats Marinate in saline water for about 10 days	Cool down the sea cucumbers Add in 30% (w/w) of salt and fully stir Pouring in salt brine Sealing the top with salt Keep for 2–3 months or longer
Second boil	Boil in water Add cold water to control the temperature below the boiling point Observe the changes in the color and softness of the sea cucumbers to stop the boiling accordingly	Boil in saturated saline water for 30–50 min Observe the surface of sea cucumber to stop the boiling accordingly
Plant ash treatment	Straw ash	Chinese oak ash
Sun-drying	Put in clean gauze cages Sun-dry Remove at least 98% of the sea cucumbers' water content	Sun-dry Flip once in a while to be evenly dried Move to shady spots occasionally Seal for a few days
Soaking before eating	Soak in boiled water for 2–3 h Change the water and repeat previous steps six times Simmer for 45 min Soak again for 20 h	Soak in water until they soften Remove the meat on the inner side of the body walls Boil for 2 h Soak in cold water Repeat previous steps once a day for three consecutive days

4. Put the sea cucumbers in clay vats and marinate them in saline water for about 10 days. Observe the changes in the shape of the sea cucumbers, and adjust the time of marination and the concentration of the saline water accordingly. This step determines the taste and texture of the sea cucumbers after soaking.
5. The second boiling is of utmost importance. Sea cucumbers are sterilized in boiling water. Add cold water to control the water temperature. Observe the changes in the color and softness of the sea cucumbers, and adjust the intensity

of the fire and the time to stop the boiling accordingly. This determines the taste and texture of the finished products.

6. Dehumidify and dehydrate the sea cucumbers by wrapping them in plant ash. Straw ash is the best option.
7. Put the sea cucumbers in clean gauze wrappings and dry them under well-ventilated conditions and adequate sunlight. If it is humid outside, take the sea cucumbers indoors right away. Repeat the step until at least 98% of the sea cucumbers' water content is removed.
8. Soak plant ash-pretreated dried sea cucumbers in boiled water for 2 to 3 h, and change the water. Repeat the step six times. Put the sea cucumbers in a pot and pour in purified water. Boil until the water bubbles, turn the heat down, and simmer for 45 min. Turn off the heat and soak the sea cucumbers in the same pot for 20 h. Then, they will be ready to serve or used as ingredients. Fully soaked sea cucumbers prepared by this ancient method can be ten times larger than dried ones, which are typically larger than that prepared by other methods. As a result, this ancient processing technique is unique and famous in Jiaodong Peninsula, especially in the Qimu Island of Longkou.

Obviously, ancient processing method holds strict requirements for each step of the process, including capture, salting, boiling, time of fermentation, ash grinding, and drying. The quality of its product greatly depends on the experience of craftsmen. The processing methods of plant ash-pretreated sea cucumber, especially the processing requirements and standards vary in different regions and periods (Table 2). Zong He detailed the sea cucumber processing technique commonly used in China in the 1970s (Zong 1976). Sea cucumbers are processed in two stages. The first stage includes the following steps.

1. Clean fresh sea cucumbers with seawater; create an incision one third the size of their bodies, on their heads, or tails; and remove the viscera. Stir the sea cucumbers in a wooden barrel to make them contract and discharge the water inside their bodies until their mouths fully retract.
2. Put the sea cucumbers in boiling water (either seawater or freshwater). Boil them for about 30 min, stirring constantly, and skim the scum until the body walls tighten, the pricks stiffen, and the incisions turn golden.
3. Take out the sea cucumbers and put them in a ceramic vat after they cool down. Add in 30% of the salt and fully stir before pouring in salt brine and sealing the top with salt. Keep the sea cucumbers in clean brine for 2 to 3 months or longer. Make sure that their insides are yellow. This stage produces salted sea cucumbers. The second stage involves the following steps.

   a. Boil the salted sea cucumbers in saturated saline water for another 30–50 min. Take out a sea cucumber from time to time and observe its surface. If its surface dries immediately after it leaves the water and salt particles crystalize on the surface, the sea cucumbers are ready and boiling could be stopped.
   b. Stir the sea cucumbers in ash. Chinese oak ash is the best option.

c. Sun-dry the sea cucumbers, flipping them once in a while to make sure that they are evenly dried. From time to time, move them to a shady spot and leave them in a sealed condition for a few days. The final result is fully dried sea cucumbers are obtained with hard pricks.
d. Grade the sea cucumbers. Usually, there should be less than 35 dried sea cucumbers per catty of top-quality sea cucumbers. Those that are in full size, flawless, and ink black, with hard and straight bodies and pricks, closed incisions, and retracted mouths are the best.
e. Soak the sea cucumbers in water until they soften. Remove the meat on the inner side of their body walls. Boil the sea cucumbers for 2 h before soaking them in cold water. Repeat the step once a day for three consecutive days.

To summarize the two stages above, fresh sea cucumbers are cleaned, disemboweled, boiled, salted, boiled for a second time, stirred in ash, and sun-dried. Boiling is a step that requires experience accumulated over many years. Sea cucumbers are normally boiled in batches with lidless pots. Nuances in water type, water-sea cucumber ratio, and starting temperature would lead to different outcomes. Research data regarding the abovementioned key processing parameters will be discussed in Chapter "The Pretreatment Technology of Raw Sea Cucumber and new Processing Technology of Salted Sea Cucumber". The benefits of using plant ash are as follows. First, plant ash contains botanic and mineral elements. Light and alkaline, it is easily removed by the wind when the sea cucumber is dry, or by water when it is wet, without fouling the sea cucumber or adding to its weight. Second, it is a natural desiccant that protects the sea cucumber from becoming dampened or rotten. Third, plant ash stiffens when dried, creating a protective layer on the tender surface of the sea cucumber, so that the sea cucumbers are not be easily damaged in transport. Though as hard as rocks, the sea cucumbers would absorb water and soften when soaked in water, as water removes the plant ash.

3.2.2 Salt-Pretreated Dried Sea Cucumber

Salt-pretreated drying, which involves repeated boiling, processing, and sun-drying, has a history of over a thousand years. Salt-pretreated dried sea cucumbers are white, as a result of being covered in a thin layer of salt particles or powder. Their surfaces and tiny parapodium cannot be clearly distinguished. Compared with the plant ash-pretreated drying technique, salt-pretreated drying features easier steps and lower costs. Sea cucumbers processed with either technique are suitable for long-term storage and long-distance transport. But salt-pretreated drying leads to a heavy loss of active substances, and the steps for soaking and rehydrating salt-pretreated dried sea cucumbers are more complicated. Moreover, salting adds to the weight of the sea cucumber and creates a white layer of salt over its surface, making it more difficult to judge the quality of the product. Some unscrupulous traders increase the weight of dried products by salting and drying them multiple times, garnering more profit at the expense of nutrition loss. The huge profit generated by the hidden use of additives has seen the introduction of sugar-dried sea cucumbers into the market in the guise of pure-dried sea cucumbers. This production involves repeated soaking with highly concentrated sugar solutions, boiling, and drying. Caramelization over the course of processing takes a heavy toll on the tonic effect and quality of the sea cucumber. The products have a shorter shelf life and have a poor quality after soaking. To ensure the quality of dried sea cucumber products, safeguard the legitimate right and interests of consumers, and prohibit such practices as adding sugar or salt, China has issued the *National Food Safety Standard for Dried Sea Cucumbers* (GB 31602–2015). It stipulates that the salt content of dried sea cucumbers be lower than 40%, the water content not exceed 15%, and the content of water-soluble sugar be under 3%.

4 Conclusion and Prospects

Sea cucumbers have been part of the Chinese people's diet for over a thousand years. The Chinese people have long recognized the value of sea cucumbers. From drying to soaking, the variety in changes in the shape of sea cucumbers demonstrates the exhaustive techniques developed by Chinese craftsmen. The traditional primary products, especially dried sea cucumbers, which effectively address the seasonal and regional differences of sea cucumbers, will remain dominant in the Chinese and global markets of sea cucumber products for a long period of time.

China is the world's largest producer in terms of the aquaculture and processing of sea cucumbers. In the past, China's sea cucumber processing industry is labor intensive, as it still relies on mature traditional techniques. Traditional techniques such as air and hot blast drying were carried out solely based on experience, without science-based optimization and improvements. It was difficult to guarantee product quality, as processing would seriously damage the nutrition of sea cucumbers. Limited to dried, salted, and rehydrated sea cucumbers, the products were far from diversified.

In the future, sea cucumber manufacturers should further invest in improving the quality of sea cucumber products with modern technologies. The optimization and automation of processing steps should be further improved, as the traditional mechanized production merely includes the manual cleaning of sea cucumbers in water sinks, manual disembowelment assisted by mechanic rollers, removal of intestines with grid screening conveyors, and boiling with jacket steamers. Major sea cucumber processing companies in China have started to introduce continuous and automated technologies in the screening of raw materials, washing with soft air bubbles, mechanized disembowelment, continuous and automatic boiling, and reshaping (Mao et al. 2015). These technologies are critical to transforming the traditional landscape of sea cucumber processing and to the standardized, mass production of sea cucumbers. With the development and introduction of processing techniques, new products like instant, canned, high- or ultrahigh pressure processed, and deep-processed sea cucumbers (capsules, oral liquid, and enzyme hydrolysis products) will meet the different consumption needs of Chinese people.

References

Aprianto R, Amir N, Kasmiati M, Fahrul S, Tresnati J, Tuwo A, Nakajima M (2019) Economically important sea cucumber processing techniques in South Sulawesi, Indonesia. Earth Environ Sci 370:1–9. https://doi.org/10.1088/1755-1315/370/1/012082

Conand C (2018) Tropical Sea cucumber fisheries: changes during the last decade. Mar Pollut Bull 133:590–594. https://doi.org/10.1016/j.marpolbul.2018.05.014

Dai Y (1998) Sea cucumber trade between China and Southeast Asia in the 18th and 19th centuries. J Chinese Soc Econ His 4:71–81. Link: https://kns.cnki.net/kcms/detail/detail.aspx?FileName=ZSJY199804010&DbName=CJFQ1998

Hou Z, Wang M, Zhang J, Zhao Y, Wang Y, Wang G (2015) The research progress of sea cucumber processing technology. The Food Industry 36:248–252. Link: https://kns.cnki.net/kcms/detail/detail.aspx?FileName=SPGY201504076&DbName=CJFQ2015

Jun A (2015) *Apostichopus japonicus*: fisheries, trade, and foodways in Japan. Dev Aquac Fish Sci 39:399–421. https://doi.org/10.1016/B978-0-12-799953-1.00022-2

Liao Y (1997) *Fauna sinica*: *phylum Echinodermata*: class *Holothuroidea*. Science Press, Beijing

Mangubhai S, Nand Y, Ram R, Fox M, Vodivodi T (2016) Value chain analysis of the wild caught sea cucumber fishery in Fiji. Wildlife conservation society and Fiji department of fisheries. Report No. 02/16. Suva, Fiji

Mao Y, Yao H, Quan L (2015) Processing and cuisine. Dev Aquac Fish Sci 39:367–382. https://doi.org/10.1016/B978-0-12-799953-1.00020-9

Purcell S (2014a) Value, market preferences and trade of beche-de-mer from Pacific Island sea cucumbers. PLoS One 9:1–8. https://doi.org/10.1371/journal.pone.0095075

Purcell S, Ngaluafe P, Aram KT, Lalavanua W (2016) Variation in postharvest processing of sea cucumbers by fishers and commercial processors among three Pacific Island countries. SPC Bêche-de-mer Information Bulletin 36:58–66. Link: https://www.researchgate.net/publication/299535039

Purcell S, Samyn Y, Conand C (2012) Commercially important sea cucumbers of the world. Food and Agriculture Organization of the United Nations, Rome

Purcell S, Williamson D, Ngaluafe P (2018) Chinese market prices of beche-de-mer: implications for fisheries and aquaculture. Mar Policy 91:58–65. https://doi.org/10.1016/j.marpol.2018.02.005

Purcell SW (2014b) Processing Sea cucumbers into bêche-de-mer: a manual for Pacific Island fishers. Southern Cross University, Lismore, and the secretariat of the Pacific community, Noumea

Purcell SW, Mercier A, Conand C, Hamel J-F, Veronica Toral-Granda M, Lovatelli A, Uthicke S (2013) Sea cucumber fisheries: global analysis of stocks, management measures and drivers of overfishing. Fish Fish (Oxf) 14(1):34–59. https://doi.org/10.1111/j.1467-2979.2011.00443.x

Sachithananthan K, Natesan P, Alagaratnam C, Thevathasan A, Philip LB (1975) De-scummer for beche-de-mer processing. Bulletin of the Fisheries Research Station, Sri Lanka 26:11–15. URL: http://hdl.handle.net/1834/32942

Toral-Granda V, Lovatelli A, Conand C, Mercier A, Purcell S, Uthicke S (2008) Sea cucumbers. A global review on fisheries and trade. FAO Fisheries and Aquaculture Technical Paper. No. 516. Rome

Wang S (2002) A research on the origins of sea cucumber. J Sichuan Tourism Univ 1:8–9. Link: https://kns.cnki.net/kcms/detail/detail.aspx?FileName=SCPR200201005&DbName=CJFQ2002

Zong H (1976) References on sea cucumbers and their processing methods. Fish Sci Technol Inform 3:14–16. Link: https://kns.cnki.net/kcms/detail/detail.aspx?FileName=SCKJ197603009&DbName=CJFQ1979

The Pretreatment Technology of Raw Sea Cucumber and New Processing Technology of Salted Sea Cucumber

Xiaoming Jiang and Changhu Xue

Abstract The preprocessing and salting of fresh sea cucumber is the most traditional and fundamental process for sea cucumber processing. The preprocessing includes cleaning, slaughtering, and eviscerating, which is not only the first step of processing but also a key to ensuring the quality of the sea cucumber in subsequent processing. The cooking process denatures proteins, inactivates enzymes, and dehydrates the sea cucumber. The salting process further dehydrates for better preservation. The preprocessing and salting process results in salted sea cucumber, the most important intermediate raw material of the sea cucumber industry. The salted sea cucumber is then further processed into various sea cucumber products. This chapter introduces the methods and techniques of sea cucumber pretreatment, thermal processing, and salting, as well as the related equipment, and finally, the trend of research and development for relevant technology and equipment.

Keywords Preprocessing · Salting · Cooking · Machine · Cleaning · Slaughtering

1 Introduction

There are approximately 1200 known species of sea cucumber worldwide (McElroy 1990), and more than 60 species are currently in established commercial trade in over 70 countries and regions worldwide (Purcell et al. 2012). The most commercially valuable species in the world is the *Apostichopus japonicus*, which is mainly distributed in the West Pacific Ocean, the Yellow Sea, the Sea of Japan, and the Okhotsk Sea. The northern boundary of its geographical distribution is the coast of Sakhalin Island, the Russian Federation, and Alaska (USA). The southern boundary is the Japanese island of *Tanegashima*. In China, it is commonly found in the coastal areas of Liaoning, Hebei, and Shandong Provinces, covering Yantai, Qingdao, and

X. Jiang (✉) · C. Xue
College of Food Science and Engineering, Ocean University of China, Qingdao, P. R. China
e-mail: jxm@ouc.edu.cn; xuech@ouc.edu.cn

C. Xue (ed.), *Advances in Sea Cucumber Processing Technology and Product Development*, Advances in Marine Bioprocesses and Bioproducts,
https://doi.org/10.1007/978-3-031-16512-2_5

other places in Shandong Province, and its southern boundary is Xiapu County in Fujian Province in China (Purcell et al. 2012). In 2020, the aquacultural output of *Apostichopus japonicus* in China was 196,000 tons (Liu 2021). Therefore, the largest scale of processing and utilization for sea cucumbers worldwide occurs in China.

With the expanding aquaculture for sea cucumbers, there is a growing concern about how to make full use of sea cucumber resources. The sea cucumber is a very unique biological material with many special properties compared to general aquatic and meat products. On the one hand, the sea cucumber contains autoenzymes (Zhu and Han 2004), leading to a strong autolysis ability. Under the stimulation of certain external conditions, the sea cucumber can easily turn into liquid by itself through the process of epidermal destruction, guts spitting, and dissolution. This brings many technical problems to the preservation, storage, transportation, and processing of sea cucumbers. On the other hand, the sea cucumber also has a strong ability to rehydrate, as the body wall of fresh sea cucumber shrinks significantly after being heated, while salted sea cucumber, after processing, can swell up to close to its original size through steaming. The process can even be repeated several times.

2 Processing Techniques of Sea Cucumber Preprocessing

The pretreatment of sea cucumber is generally defined as the process from capture to heat treatment, and the main operations involved in this process are cleaning and slaughtering. The unique characteristic of autolysis for the sea cucumber dictates that captured sea cucumbers need to be thermally processed as quickly as possible to inactivate their autoenzymes for the maintenance of quality, so the pretreatment should be efficient and rapid enough to minimize stimulation to sea cucumbers.

2.1 Cleaning Method of Sea Cucumber

In the natural environment, sea cucumbers usually live on reefs and gravels under the sea, and freshly caught sea cucumbers often carry a large amount of mud and attached marine impurities in their skinfolds. If such foreign matters are not cleaned out in time, the viscous liquid secreted on their body surface will often cause a deterioration in quality during the subsequent evisceration and steaming process or even secondary contamination of the finished products.

After being caught by professional divers at sea, sea cucumbers undergo a preliminary manual cleaning in seawater and are then loaded into a waterless container for temporary storage. After a certain amount of the catch is completed, the container needs to be transported to shore and then to a nearby centralized processing site via vehicles. During this process, sea cucumbers are exposed to external stimuli and are generally packed in containers loaded with 20–50 kg each, encountering the problem of mutual extrusion. After the sea cucumbers are transported to the processing site, they are dumped into a cleaning tank for quick cleaning. The general procedure is to rinse off the obvious mud and sand on the surface of sea cucumbers with water and then put them into cool freshwater (below 10 °C) to remove the residual mud, sand, sea salt, and other impurities. The process in most cases adopts manual cleaning, which requires attention to prevent damage to the body wall of sea cucumber and avoid overstimulation of the sea cucumber for prevention of autolysis. At present, some companies have also adopted a more automated cleaning method by dumping sea cucumbers from the temporary storage container onto a conveyor belt, which transports sea cucumbers to the slaughterhouse and sprays to sea cucumbers using spraying equipment. However, this method also faces the problem of dead-end spraying and the intensity of spraying that stimulates the sea cucumbers, which is often not effective enough and easily leads to epidermal damage. A more sensible solution is to use the flexible bubble cleaning technology, which utilizes compressed air to form small bubbles in the water and create air cells with a negative pressure zone that adsorb body surface attachments. Flow layers are created through air injection into the water, applying friction to the surface of the sea cucumber body to remove the viscous liquid. Sea cucumbers could be cleaned in the water and air bath in a flexible, all-round way, so that there are no dead corners on the surface of the sea cucumber, satisfying the process requirements for clean sea cucumbers with minimal or even no damage. The water and air bath process are also suitable for the rapid cleaning of large quantities of sea cucumbers.

According to the observation of the cleaning process, sea cucumbers were generally located at the bottom of the cleaning water tank due to their weight and gravity. The water was filled with pressurized gas to produce a large number of rising bubbles, and sea cucumbers were thus driven by these bubbles to tumble and sink in the water. After the cleaning process, the water tank had some slightly reddish viscous liquids and tiny impurities floating on the upper layer of the water. A certain number of sediments could be collected through filtration of the discharged wastewater. Cleaned sea cucumbers were visually inspected to ensure that there was no mud or sand on the surface. This was contributed to the fact that the objects being cleaned formed a very complex solid-liquid-gas three-phase flow during the bubble cleaning process. The cleaning mechanism could be explained by the following hydrodynamic: When the gas was injected into the liquid medium at a certain speed, it drove the liquid to flow; faster flow layers drove slower flow layers, and slower flow layers inhibited faster flow layers. Flow layers at different speeds contained each other, resulting in the internal friction between layers; under this action, each contact surface of the objects rubbed with the water flow, leading to the gradual separation of the impurities attached to the objects. At the same time, the gas injected into the liquid also produced tiny bubbles on the surface of sea cucumbers, and the bubble cavitation phenomenon compressed the liquid surrounding bubbles to produce high pressure, while the liquid in the center of each bubble stretched to form a negative pressure zone, absorbing impurities. Air bubble cleaning means the use of internal friction and the bubble cavitation phenomenon to effectively remove impurities on the surface of the target object. This flexible, intermediate, and seamless cleaning method is a perfect fit for the cleaning process of fresh sea cucumber (Xu et al. 2009a, b).

A traditional cleaning method of sea cucumber is manual rinsing, is labor-intensive, and requires a higher labor cost. High-pressure roller spray cleaner and brush roller scrubber are more suitable for the cleaning of fish and shellfish other than sea cucumber. At present, some companies use spraying equipment to clean sea cucumber surfaces, but there are also problems associated such as monotonous functionality, poor cleaning effect, and easy damage to sea cucumber surfaces. The actual effect is far from ideal. Xu Wenqi et al. used the flexible bubble cleaning technology that applies compressed air to form small bubbles in the water, generating air cells with a negative pressure zone to adsorb body surface attachments and producing changes in the flow layer through gas injection, which process eventually applies friction on the body surface to remove viscous liquid.

Using an abluent ratio to judge the effectiveness of sea cucumber cleaning encompasses both a quality measurement for the extent of sea cucumber cleaning and a quantitative measurement for sea cucumbers meeting the cleanliness requirements in a single batch of cleaning. Damage rate is used to quantify the quality of the cleaned product because the absence of damage to the skin and the absence of skin autolysis are important indicators for the sensory quality of sea cucumbers. Gutting is a normal physiological phenomenon for the sea cucumber under stimulation, while skin autolysis is a symptom that marks the beginning of the decay of the sea cucumber surface. Therefore, sea cucumbers with skin damage and skin autolysis should be strictly counted during processing. The ideal state for fresh sea cucumber cleaning should be characterized by a high abluent ratio with minimal loss of sensory quality due to skin damage, or even with no damage.

Cleaning intensity and cleaning time are important process parameters for sea cucumber cleaning. This is due to the internal friction produced by bubbles in the liquid and the intensity of the bubble cavitation phenomenon, which are both determined by the airflow into the water body. The greater the cleaning intensity, the better the cleaning effect on the sea cucumber. The change of damage rate is consistent with the abluent rate, which increases with the growth of cleaning intensity. The higher the cleaning intensity, the higher the cleanliness of the sea cucumber and the higher the probability of damage to the sea cucumber, so the differences between the abluent rate and damage rate for each group of sea cucumbers are remarkable.

Cleaning time also has a dual effect of simultaneously increasing the abluent rate and the damage rate for the sea cucumber. The abluent rate and the damage rate are both directly proportional to the cleaning time, so the two influencing factors should be reasonably deployed in the sea cucumber cleaning process. Cleaning should be completed at the most appropriate cleaning intensity and within the shortest time. It should also be noted that in the cleaning process, the abluent rate should not be blindly pursued at the expense of the damage rate nor should the requirements for sea cucumber cleaning be lowered for the sake of less damage, with the consequence of the cleaning result being substandard.

2.2 Slaughter Method of Sea Cucumber

There are many ways of slaughtering and eviscerating sea cucumbers; most of them are manual labor. The most common method is to make a long incision with a sharp knife in the middle of one of the three equal parts between the anus and the tentacles on the ventral side of the sea cucumber, squeeze out all the guts and silt, etc., leaving only the body wall of the sea cucumber, which is rinsed clean for subsequent processing. The process requires a high level of operator skill and the viscera is easily cut during the dissection process, which affects the reuse of sea cucumber intestines. The fine sand in the viscera also adds to the difficulty of cleaning, and the viscera are generally discarded. An evisceration method popular in the Kansai region of Japan uses a special evisceration syringe, which resembles an injector hollow inside, to pierce the body cavity through the anus of a sea cucumber. The guts remain intact within the cavity of the evisceration syringe and are then removed from the cavity after penetration through the mouthpiece of the head. This achieves the purpose of eviscerating the sea cucumber and obtaining complete intestines (Xu et al. 2009a, b). Sea cucumber intestines are further processed into salted products in Japan. In Canada, Iceland, and Russia, there is a slaughtering method for *Cucumaria frondose*, which adopts a circular cut to remove the sea cucumber's tentacles and anus, squeeze out the guts from both ends, and then clean the cylindrical body wall of the sea cucumber for further processing. The tentacles obtained by this method have a high protein content, which can serve as a food ingredient after drying and soaking, mostly in Chinese restaurants and called "flower of sea cucumber" (not sea cucumber gonad). In the Americas, *Parastichopus californicus*, a variety of slender sea cucumber, is processed in a similar way, with the intestines cleaned and frozen and the radial muscular band stripped separately and processed in Chinese restaurants as a catering ingredient called "Osmanthus Band."

In response to the trend of modernized and mechanized processing for sea cucumbers, Chinese researchers have made more attempts by utilizing the unique physiological characteristics of the sea cucumber, which are easily stimulated by external factors for "gutting." Biomimetic principles have been adopted to simulate external stimulation, which applies rolling and vibration to continuously stimulate the sea cucumber, so that the sea cucumber can gut autonomously, with a gutted rate of up to 90%. The intensity and duration of stimulation during the eviscerating process functioned as core parameters and were directly proportional to the final gutted ratio and gutted efficiency and inversely proportional to the sensory quality. This technique has not been widely used in sea cucumber processing yet (Xu et al. 2009a, b).

3 Processing Techniques of Sea Cucumber Cooking

Fresh sea cucumbers have a strong autolysis ability and are not easy to store and transport. After being harvested, sea cucumbers need to be cooked to deactivate enzymes and are then processed into various related products. Cooking is a key step of sea cucumber processing, which directly affects the yield, appearance, taste, nutrient content, and other indicators and has a significant impact on product quality.

For fresh sea cucumbers, producers usually use boiling water for pretreatment. For example, natural cooking does not require special processing equipment and has an easily controlled process. Natural cooking is popular among ordinary consumers for processing sea cucumbers. Autoclaving requires less time and helps to improve productivity but is dependent on production equipment, and the high intensity of heat treatment can also adversely affect product quality. Low-pressure vacuum cooking is a relatively mild processing approach that reduces the boiling point of cooking liquor by using sealed equipment and vacuum devices to reduce the boiling point of the cooking liquid boil, thus reducing the impact on thermally sensitive substances in the cooking material (Liu et al. 2015).

3.1 Traditional Cooking Technique

The autolytic enzyme within sea cucumbers is very active, and collagen, the main component of the sea cucumber's body wall, is difficult for humans to masticate and digest. The traditional cooking process is the easiest and most convenient way to inactivate the sea cucumber's autolytic enzymes and achieve the desired dehydration and formation. Meanwhile, the cooking process also facilitates the transformation of collagen to gelatin, extending a palatable taste.

The traditional processing method for the sea cucumber is to clean, slaughter, eviscerate, and then boil the sea cucumber, usually using an open pot for intermittent cooking, with the sea cucumbers put into the pot when the cooking water reaches a certain temperature. The ratio of water to sea cucumber is usually more than 4:1. Sea cucumbers are put into hot water at different starting temperatures, leading to different cooking effects during the cooking process.

During the traditional cooking process, the nutritional ingredients of sea cucumber suffer a loss, and the body wall of sea cucumber is mainly composed of dermis connective tissues. The intercellular space of the dermis connective tissues of sea cucumber abounds in collagens and other fiber compositions and proteoglycans. The proteoglycans and collagen fibers in the intercellular matrix are bound in an orderly manner via non-covalent bonds (Scott 1988). The cooking process of fresh sea cucumber is a process of converting the originally rigid, ordered, and hydrophobic collagens in the body wall into flexible, dispersed, and hydrophilic gelatinous proteins, accompanied by disaggregation of proteoglycan molecules and collagen fibers in the course of conversion. Meanwhile, glycosides, which are lipophilic (especially the cell membrane lipids), are released as the cell membrane is disrupted by heat. The nearly simultaneous flow of collagens, polysaccharides, and glycosides from sea cucumber tissues to cooking liquor indicates that the loss of these three functional components is inevitable. However, if the cooking time is strictly controlled with complete conversion, it might be possible to reduce the loss and is worthy of further investigation (Jiao and Kang 2010).

The histological changes and their rheological properties under different heating conditions (40, 60, 70, 80, 90, and 100 °C) for fresh sea cucumber have been investigated with the contents and types of collagens verified. It was found that the muscle structure of *Stichopus japonicus* experienced obvious changes under different heating temperatures, and there was a certain regularity between rheological properties and heating temperature. These changes were mainly caused by the different states of collagen at different temperatures. Collagen accounted for the highest percentage of crude protein in *Stichopus japonicus*, which was 71.45%, and collagen was the type I collagen with three $\alpha 1$ chains (Xue et al. 2006). Similar studies have also been conducted from the perspectives of texture characteristics, enzyme activity, and histological changes, and it was found that the texture characteristics of sea cucumber varied significantly under different heating temperatures, especially in the temperature range of 40–70 °C. These changes mainly came from the changes in collagen, which could be clearly observed via scanning electron microscopy, and most enzymes in sea cucumbers maintained high activity after being heated for one hour at 40–60 °C (Dong et al. 2014). Researchers also investigated the effect of temperature on the change of collagen structure and molecular weight in *Stichopus japonicus*, and the extracted collagens had a high purity and an intact triple-helical structure. The collagen was not only destroyed in the secondary structure but also partially destroyed in the primary structure under heat treatment at 40 °C. The α- and β-chain bands disappeared, and the molecular weight decreased (<100 kDa). At 70 °C, the thermal contraction was exhibited on the superimposition of peptide chains, which aggregated and contracted with an increased molecular weight; at 80 °C, the peptide chains further contracted with a molecular weight up to about 120 kDa, and the collagen still maintained the secondary structure to a certain extent. When the temperature exceeded 100 °C, the primary structure of the collagen was severely damaged (Wang et al. 2011).

Researchers studied the weight loss, morphological change, weight loss pattern, and water absorption phenomenon of sea cucumbers under different heating temperatures and found that the masses of sea cucumbers decreased uniformly at 50 °C with a gradual change in appearance. The tissue of the body wall remained full within 1 h of heating, with little change compared to fresh sea cucumbers. The sea cucumber tissues started to become soft and sticky for an extended period of time, which might be contributed to the autolysis of sea cucumbers, because 50 °C was close to the optimum reaction temperature of cathepsin. Therefore, sea cucumbers should not be processed at 50 °C for a long period of time.

When the heat processing temperature was 60 °C, 70 °C, 80 °C, 90 °C, or 100 °C, most of the water flowed out at the beginning of heat treatment. With the extension of the processing time, the quality loss of the sea cucumbers gradually leveled off after 5 h, 3 h, 1.5 h, 1 h, and 0.5 h, respectively, when the collagen of the sea cucumber's body wall was basically denatured and the moisture content essentially reached equilibrium. When the heating temperature was greater than 70 °C, as the heating time is extended, the water absorption phenomenon would appear in the body wall of the sea cucumber. The higher the temperature, the earlier the water absorption appeared, with 70 °C corresponding to 24 h, 80 °C corresponding to 6 h, 90 °C corresponding to 3 h, and 100 °C corresponding to 2 h. This was due to the fact that the collagen in the body wall of sea cucumber was denatured to a certain extent when heated and gradually turned into gelatin and showed water absorption.

Heat treatment also has a big influence on the morphology of sea cucumbers, as manifested by reduced sea cucumber volume. With the increase in heating temperature, the time required for sea cucumber to shrink to a certain degree will also decrease. During the whole heating processing, the trend of sea cucumber mass and volume changes remained basically the same.

Changes in the heat treatment for sea cucumbers can also be clearly observed through sensory evaluation. During the processable time at 60 °C, sea cucumbers maintained good shape. With the increase of heating time, the elasticity increased, and the hardness showed a trend of changing from brittle hard to tough hard, to relatively hard, and eventually to solid hard. However, there was no obvious fragrance at all times. After the heating time exceeded 24 h, an obvious fishy smell was noticed and the processing could not continue. When the temperature exceeded 70 °C, with the extension of heating time, the body wall of sea cucumber gradually decreased in hardness, and the tissues gradually became softer due to the denaturation of collagen and the formation of gelatin. After heating exceeded a set amount of time and a certain temperature, the sea cucumber body wall became soft and sticky, and with the increase of temperature, the duration of the above phenomenon shortened gradually. This was due to the denaturation and degradation of collagen in sea cucumber body wall under heating, generating gelatin that started to disintegrate and lose, resulting in the tissues becoming fragile and difficult to further process.

3.1.1 Continuous Cooking

Currently, sea cucumber processing is adopting the method of continuous cooking in order to improve processing stability by standardizing operating parameters, thus reducing the constraints of experience-related matters.

Continuous cooking resorts to a conveyor belt to continuously transport sea cucumber body wall materials for continuous material placement, completing a series of actions such as sending sea cucumber's body walls to the cooking liquor, operating in the liquor, and removing body walls out of the liquor. The processing method has thus been changed from batch processing to uninterrupted delivery through optimization.

In addition, sea cucumber's body walls traveling in the cooking liquor produce a relative displacement movement with the liquor, eliminating the need for stirring operation and avoiding the problem of uneven heating. After sea cucumbers are eviscerated and preliminarily cleaned, they are sent to the cooking equipment by a stainless steel mesh belt conveyor and then removed from the equipment after 10 to 20 min of blanching. The conveying and cooking speed can be adjusted based on the demand to control the weight loss ratio of sea cucumber in order to meet the development requirements of subsequent products.

In the process of continuous cooking, issues on temperature control and the elimination of a large amount of foam produced by cooking need to be resolved, and there are already technical solutions to achieve temperature control of the cooking water via a programmable controller (PLC) for temperature sensing test and feedback regulation. Heating could be supplied in a variety of ways, such as electric heating rod heating, steam heating, and conduction oil heating. Steam heating is currently the most widespread heating method, but the accuracy of temperature control is also the most unreliable, so a solution of regulation by proportional valve or steam bypass has been proposed to achieve small overshoot, small hysteresis and reliable ability to adjust and control the temperature. The large amount of foam produced by the sea cucumber saponin can be removed through mechanical scraping, compressed air blowing, or collecting and pumping.

3.1.2 Batch Cooking

Currently, batch cooking is still a popular cooking method because of its minimal requirements for processing equipment and ability to use ordinary cooking devices, of which the most commonly used is the jacketed cooking pot. During processing, after raw sea cucumbers are placed into the pot in batches, workers need to continuously stir the raw materials in the pot with a stirring stick to prevent the sea cucumber body walls from sticking to the bottom and to keep the materials evenly heated. The temperature of the cooking liquor also needs to be controlled for a stable temperature. After cooking for a certain amount of time, normally 10 min, it is necessary to remove the froth agglutinated by the precipitates of the body wall using tools during cooking, so as to ensure the cleanliness of the cooking liquor and the quality of sea cucumber body wall.

According to the technique observed by the author at a sea cucumber processing site, workers responsible for cooking sea cucumbers added tap water to an iron pot, first of all, and used coal for heating (since the preliminary processing site was on a sea island, the facilities were relatively rudimentary with no liquefied gas or electricity). When the water was heated to about 70 °C, gutted sea cucumbers were put into the pot for cooking and were stirred with a spatula to ensure that the sea cucumbers were heated evenly. After the sea cucumber body walls were added, the heating rate was increased immediately, and the airflow of the coal stove was adjusted through the blower until the cooking liquor started to boil. Workers kept skimming the floating foam with tools during cooking while adding enough cold water to the pot to stop boiling; when the hot liquid boils again, the sea cucumber's body walls were fetched out. The whole process took about 15 min, during which sea cucumber body walls shrank significantly, the weight decreased to about 20% of the original weight, and the moisture content dropped to about 75%. After all the sea cucumbers were removed, the amount of hot liquid in the cooking pot was reduced, and enough cold water was then added to reduce the temperature of the new cooking liquor to about 70 °C. The next batch of sea cucumber body walls was added. After processing three batches of sea cucumbers, each pot of cooking liquor was completely replaced. In the whole process, there were no specific quantified and measured control parameters, and the operators were the only ones that determined when to put in the sea cucumber body walls, how to control the stove blower to adjust airflow, when to add cold water for cooling, how much cold water to add, and how frequently to turn and stir the sea cucumbers. All of the operations were purely based on operators' experience. Therefore, the intermittent cooking of sea cucumber required long-term experience, and no one had been able to translate the process into a standard operation specification for a fairly long period of time.

3.2 New Cooking Technique

3.2.1 Steam Curing

Steaming is a new method of heat treatment for the sea cucumber, which changes the original method of completely immersing sea cucumbers in the cooking liquor, using hot steam as the heat transfer medium instead to achieve high-temperature curing of the sea cucumbers. By controlling the temperature of the hot steam, various indicators, such as yield, appearance, taste, and nutrition, are regulated. This method is advantageous in that the loss of water-soluble components is somewhat reduced. However, a major drawback is that the temperature of the steam needs to be controlled when fresh sea cucumbers are being processed, and superheated steam directly contacting fresh sea cucumbers' body wall will lead to partial shrinkage due to uneven heating. This results in the bending of sea cucumber and partial bulging, which affect the quality of subsequent products.

3.2.2 Low-Temperature Cooking

Currently, low-temperature heating is widely used in the processing of meat products, such as minced fish products, in which the low-temperature gelation process results in the better elasticity and quality of minced fish products. In sea cucumber processing, low-temperature cooking has less impact on the bioactive components within sea cucumbers, which retains the sea cucumber's nutrition to the maximum extent. Moderate denaturation of collagen in the body wall of sea cucumber through low-temperature curing produces a water loss rate that is different from that in high-temperature steaming. A study was conducted to prepare ready-to-eat sea cucumbers by two methods: boiling traditionally salted sea cucumbers and vacuum low-temperature steaming of fresh sea cucumbers. Through the analysis of the quality loss and changes in nutrients, functional components, and mineral elements, it was found that there were significant differences in the nutritional and functional components of the edible sea cucumbers obtained from the traditional boiling of salted sea cucumber and from low-temperature steaming of sea cucumber without rehydration for 3 h. The mass-loss rates of low-temperature steamed ready-to-eat sea cucumber and boiled salted ready-to-eat sea cucumber were 43.21% and 80.71%, respectively. The contents of nutrients, functional components, and mineral elements (Mg, Cr, Zn) of low-temperature steamed ready-to-eat sea cucumber were higher than those of boiled salted ready-to-eat sea cucumber. Compared with fresh sea cucumber, the loss rate (except for fat) of low-temperature steamed ready-to-eat sea cucumber was significantly lower than that of boiled salted ready-to-eat sea cucumber, and the losses of mineral elements (Cu, Fe, Mg, Al, Mn, Cr, Zn) of the two ready-to-eat sea cucumbers both exceeded 50%.

3.2.3 High-Pressure Cooking

High-pressure cooking requires less time, is beneficial to production efficiency, but is dependent on production equipment. Moreover, the high intensity of heat treatment can also adversely affect product quality. The traditional consumption method for sea cucumber focuses on water boiling, which requires repeated boiling, and is lengthy and difficult to control. Processing sea cucumbers at a temperature of 100 °C or higher not only shortens the time of heat treatment but also significantly improves the water absorption capacity of sea cucumber products. The same processing time (36 h) results in twice the weight of rehydrated sea cucumbers. It is concluded that the higher the temperature, the shorter the time is required. Currently, the process of preparing the body wall of fresh sea cucumber with an autoclave in the market is simple and easy to operate.

3.2.4 Low-Temperature Vacuum Cooking

The low-temperature vacuum cooking technique includes two concepts. One is sous-vide processing, a new type of processing, in which sea cucumbers are vacuum packed and then heated at a relatively low, constant temperature for a set time for full curing (Schellekens 1996). The other is vacuum cooking processing, which involves lowering the boiling temperature of the cooking liquor through negative pressure to ensure the cooking liquor in the system at a low boiling temperature, while the sea cucumber body wall is in direct contact with the cooking liquor (Jiang 2013; Xue et al. 2013). The difference lies in the distinction between indirect heating and direct heating.

Regardless of the type of low-temperature vacuum cooking, two factors are necessary; the first of which is a constant temperature and the second is a definite period of time. Precise temperature control allows for more options than traditional heat processing methods provide. With stable heat treatment, low-temperature vacuum processing is more beneficial to the retention of nutrients and flavor of sea cucumbers. It keeps all parts of the sea cucumber heated at a constant temperature, reducing the loss of heat-sensitive components, such as polysaccharides, amino acids, and vitamins, and retaining nutrients of the sea cucumber to the greatest extent possible.

3.2.5 Low-Temperature Steam Curing

The low-temperature steam curing technique is based on the sea cucumber steaming technique, added with a negative pressure environment to generate relatively low-temperature steam for the heat treatment of sea cucumber in a high vacuum environment. The processing technique is highly dependent on special equipment, which sets up a temperature control program for low-temperature curing and generates low-temperature steam through the control of a vacuum system and steam generation system. Meanwhile, the equipment uses an ultrasonic transducer to generate multifrequency oscillation, which transmits instantaneous pulses of low-temperature steam to the sea cucumber's body wall to degrade the collagen structure rapidly and moderately. At the same time, the high-efficiency nano-infrared radiation plate supplements the steam energy to form a circulating vortex in the work chamber and ensures the formation of an atomized water layer on the body wall of the sea cucumber under micro-pressure, so that the sea cucumber is heated more evenly and the nutrients are not easily lost, thus maintaining a better texture.

3.2.6 Microwave Curing

The microwave curing technique, with its fast-heating speed and low heat loss, can shorten the processing time while better retaining both the nutrients and the original flavor of sea cucumbers. The traditional method of curing fresh sea cucumber requires at least 30 min of atmospheric pressure heating to achieve a suitable taste for consumption. High-temperature and high-pressure heating also require at least 15 min, while the vacuum cooking method requires 3 to 4 h to achieve an appropriate taste. Microwave-assisted heating can achieve rapid heating of sea cucumber. Existing solutions include a high-temperature and high-pressure water bath combined with microwave-assisted heating, as well as electromagnetic heating in an enclosed space combined with microwave-assisted heating. Both options enable the cooking of sea cucumber in about 1 min to achieve a palatable taste for consumption.

3.3 Effects of Different Cooking Methods on the Quality of Sea Cucumber

How do different cooking methods affect the quality of *Stichopus japonicus*? What are the pros and cons of thermal cooking methods? How do the appropriate thermal cooking methods vary according to the needs of practical product development?

Researchers compared the changes in the mass loss rate, textural characteristics, tissue morphology, rehydration effect, and nutrient composition of sea cucumbers after low-pressure, atmospheric, and high-pressure cooking treatments. It was found that the mass loss rate of sea cucumbers treated with low-pressure cooking was significantly lower than that of sea cucumbers treated with atmospheric pressure and high-pressure cooking and that sea cucumbers treated with different cooking pressures all maintained good morphology, but the hardness varied sharply. The rehydration speeds and maximum rehydration ratios of sea cucumbers after cooking also varied significantly and showed a pattern of negative correlation with hardness values. The low-pressure treatment caused relatively lesser damage to the tissue structure of the sea cucumber's body wall, resulting in a larger hardness value, slower rehydration speed, and smaller rehydration ratio. In addition, compared with atmospheric and high-pressure cooking processes, low-pressure cooking could significantly reduce losses of proteins and sea cucumber polysaccharides (Liu et al. 2015).

Researchers also investigated the changes in the weight loss rate, protein loss rate, cross-sectional morphological structure, fresh body wall water content, and maximum transformation temperature for the body wall of *Stichopus monotuberculatus* at different cooking temperatures (50 °C, 60 °C, 70 °C, 80 °C, and 90 °C) and durations (10 min, 20 min, and 30 min). The moisture content of fresh sea cucumber body walls was about 93.67%. With increasing temperature and duration, the body wall started to shrink and lose weight at 50–60 °C, and the rate of weight loss gradually increased. The protein loss rate gradually increased from 70 to 80 °C, and the maximum transformation temperature of the fresh body wall was 73.5 °C, which was consistent with the temperature at which the shrinkage and protein loss of the sea cucumber's body wall occurred. Scanning electron microscopy (SEM) was used to observe the cross-sectional morphological structure of the body wall, and it was found that the density of the collagen fiber network in the body wall increased significantly from 70 °C, and cavities appeared. Therefore, water cooking for *Stichopus monotuberculatus* should start at 70 °C for 20–30 min, or 80 °C for no more than 10 min (Zhong et al. 2016).

In addition, researchers studied the changes in the nutritional quality of sea cucumbers from industrialized aquaculture subject to different heat treatment methods (blanching, blanching-atmospheric water boiling, and blanching-high pressure steaming). It was discovered that the mass loss rates of sea cucumber were basically the same after constant pressure boiling and atmospheric steaming, but both were significantly higher than the rate in the blanching process. The loss rates of crude protein, crude fat, and total sugar were greater in high-pressure cooking than those in atmospheric cooking, while the opposite was true for ash and collagen contents. The saponin content of fresh sea cucumber was only 0.97%, and the loss occurred mainly in the blanching process during heat treatment. The eicosapentaenoic acid (EPA) and docosahexaenoic acid (DHA) contents of atmospheric boiled sea cucumbers were slightly higher than those of high-pressure boiled sea cucumbers. After heat treatment, the essential amino acids, flavor amino acids, and total amino acids of sea cucumber were remarkably increased, but the rate of amino acid loss was much higher in the atmospheric cooking process than that in the high-pressure boiling and blanching. The hardness, springiness, cohesiveness, and chewiness of the sea cucumbers were significantly higher in atmospheric and high-pressure cooking than those in blanching. The different nutritional components were affected by heat treatment to various degrees.

4 Processing Technique of Sea Cucumber Salting

Salted sea cucumber, also known as Lagang Yan sea cucumber, represents a common processing method used by coastal residents in North China. The processing method can be simply summarized: fresh sea cucumbers are gutted, washed, and then boiled for 10–15 min. They are then mixed thoroughly with salt and stored in cold preservation. The products can be sold as an end product or further processed into dried sea cucumbers, ready-to-eat sea cucumbers, and other forms of sea cucumbers. Currently, this product is the primary intermediate raw material for the sea cucumber industry.

The consumption method of salted sea cucumber consists of first soaking and desalinating the sea cucumber in water, steaming it, soaking it in clean water, and refrigerating it. The product is characterized by its relatively low price and wide consumption. Compared to the current products that are processed directly from fresh sea cucumber to edible ones, repeated steaming and soaking lead to the loss of some water-soluble nutrients. However, compared to dried sea cucumber, the processing method of salted sea cucumber has the advantages of maintaining the original flavor and nutrition and being convenient for transportation, storage, and consumption, but the salted one does not maintain a long shelf life as it needs to be frozen for storage and the rehydration ratio is limited, usually only 2 to 4 times.

High-quality salted sea cucumber is in black or brownish gray color, with tight and elastic tissue, intact body shape, plump flesh, straight spine, and a neat incision. The salinity and moisture content should not exceed 22.0% and 65.0%, respectively. Due to the high salinity of salted sea cucumbers, desalination is a necessary step in the preparation of ready-to-eat sea cucumbers and mild-salted dried sea cucumbers for consumption or further processing. Temperature, time, material-water ratio, and the frequency with which water is changed will all make a difference in the desalination of the sea cucumbers.

Processing of salted sea cucumber: fresh sea cucumber → evisceration and washing → cooking → salting → cold preservation.

Sea cucumber intestine is a by-product in the processing of sea cucumber, which can also be pickled to make full use of sea cucumber by-products and can also be processed into salted sea cucumber intestine. Salted sea cucumber intestine is a salted product made from sea cucumber intestine, a by-product of sea cucumber processing, which improves the comprehensive development and utilization of sea cucumber. The process of salted sea cucumber intestine includes sea cucumber → temporary cultivation → intestine collection → cleaning → drainage → weighing → salted dehydration → packaging → preservation → final product.

4.1 Salting Method

Salted sea cucumbers are divided into dry salted sea cucumbers and pickled sea cucumbers preserved in salt brine. The dry salted sea cucumber has salt crystals on the body surface; in contrast, the pickled sea cucumber is preserved in saturated saltwater after being heated and cooked. The processing cycle of salted sea cucumber is shorter than that of dry salted sea cucumber, and the market price of the former is relatively lower due to the lesser dry matter content of its products.

4.1.1 Liquid Immersion and Salting

The liquid immersion and salting method immerses the sea cucumber directly in the salt brine for preservation after eviscerating and cooking. In this process, the cooked sea cucumber body wall is immersed in the salt (NaCl) solution, and the salt is transferred into the flesh. The salt concentration of the solution is higher than the liquid concentration within the cell, as well as between the cells in tissues that constitute the flesh. Therefore, salt ions pass through the flesh via diffusion and osmosis, and water and some water-soluble substances in the sea cucumber's body wall leach into the solution. This method controls the moisture content of the salted sea cucumber by regulating the salt concentration of the salt brine. The moisture content of the salted sea cucumber obtained by this method is generally higher than that of the product obtained by the dry salting method. The flesh is relatively loose and elastic, and the sea cucumber is complete in shape, with plump flesh, straight thorns, and neat incision and with no peculiar smell or impurities and no salt crystallization inside or outside. The protein content of this product is relatively low, about 8%, and the moisture content is high, up to 68%. The price of this kind is slightly higher than that of fresh sea cucumbers. Requiring cryopreservation, this kind of sea cucumber keeps an even lower yield of dried sea cucumber or rehydrated sea cucumber processed from it.

4.1.2 Salt Particles Covering and Salting

The traditional dry salting method for sea cucumber is to eviscerate and cook the sea cucumber, put it into a basin or a wooden trough, add the same weight of salt as the sea cucumber when the sea cucumber is hot, mix well, transfer the sea cucumber into another tank after 3 days cooling, wash out the dirt and salt on the sea cucumber with the water in the original basin, and add as much salt as 60% of the weight of the sea cucumbers. After the water in the original basin is clarified, pour it into another tank and seal it tightly. After 7 to 8 days of salting, remove the sea cucumber and pour it into a large pot with saturated salt brine at a temperature of 100 °C for sea cucumber reheating. Then, add as much salt as 60% of the weight of the sea cucumber, continually stirring it gently from the bottom with a spatula while picking the insoluble salt. After 20 to 30 min, or when the surface of the sea cucumber is dry and covered with white frost when contacted with air, the production of salted sea cucumber is finished. This kind of product keeps tight flesh and tissue elastic and intact, with plump texture, straight thorns, and neat incision, with no or mixed impurities, and with no salt crystallization inside or outside. The protein content stays high, about 12% or more, the salinity is about 25% or less, and the total content of moisture and salt is relatively small, about 85% or less.

At present, the best way to work with the tunnel-type continuous cooking equipment is to put sea cucumbers after 15 min of cooking into a foam box with good thermal insulation and pile up one layer of sea cucumbers and then one layer of raw salt, and repeat this process until the box is full. Then, cover the box and stew for 12 h. On the following day, dump out the water that has precipitated from the wall of the sea cucumber, obtaining the finished salted sea cucumber product. Depending on the duration of cooking, the weight loss of sea cucumbers during cooking, as well as the difference in water precipitation during salting, the ultimate moisture content of salted sea cucumbers is 50 ~ 65%. The products with high moisture content are called "high moisture" salted sea cucumber, while the products with low moisture content are called "low moisture" salted sea cucumber.

4.2 Quality Control of Salted Sea Cucumber

This section provides a specific process case to illustrate the key points of quality control for salted sea cucumber processing.

The process flow of salted sea cucumber processing in a salted sea cucumber producer is as follows:

1. Raw material acceptance: The production manager or quality inspector conducts sensory and specification acceptance for fresh sea cucumbers delivered to the facility and performs cost accounting based on the processing body wall yield and reject ratio.

2. Raw material handling: Raw material entering the factory is handled in a timely manner, and ice is added to cool down the temperature during intermediate processing. The staff is required to wear pink latex gloves when processing the raw material. Firstly, pour sea cucumbers into a white basin, hold the scissors in one hand (scissors need to be managed by a designated person), hold the sea cucumber with the other hand with the abdomen facing upward, and then cut a slit of 1/4 of the body length toward the head at 3/4 of the body length of each sea cucumber. The slit should be kept neat and not too big, and only one slit can be made for each sea cucumber. Then, slowly squeeze the sea cucumber from both ends with both hands to clean out the mud and guts from the cavity.
3. Rinsing: After removing the dirt and sand, put the body wall of the sea cucumbers into a colander, and rinse them well with clean water, and then weigh and calculate the body wall yield.
4. Water boiling: Use a jacketed pot or iron pot for boiling, and put the sea cucumber into a pot in proportion according to pot capacity. Pour in the right amount of water, boil until the edge of the pot is bubbling, and then pour in the corresponding proportion of sea cucumber body wall and start heating up; stir continuously with a long wooden spatula during heating, and maintain the temperature until it reaches a certain degree. From the time the body walls are poured into the completion of the cooking process, the softness of sea cucumber is mainly tested by hand, and the cooking duration is determined by the conditions of sea cucumber.
5. Salt stewing: Quickly take the boiled sea cucumbers out with stainless steel strainer, pour them into a foam box, and then weigh them, filling each box with a set amount of sea cucumber. Add salt according to the weight of the sea cucumbers, stir well with high-efficiency acid and alkali-resistant industrial gloves, and then seal the box and stew the sea cucumbers. The exact duration is based on the yield (ratio to raw material) after cooking. If the yield is low, shorten the stewing time; if the yield is high, increase the stewing time.
6. Cleaning and cooling: Take out the stewed sea cucumbers with stainless steel strainer, pour them into cold saturated salt brine for rapid cooling and cleaning, and occasionally replace the saturated brine with a colder one when the temperature rises.

   (i) Simmering of saturated salt brine: Boil the water and slowly pour in salt until salt crystals surface on top of the liquid or thawless salt deposits appear at the bottom of the pot. Keep it for about 6 min and then collect the salt brine.
   (ii) Cleaning of plastic bucket: Wash with detergent to remove the oil before use, disinfect with 100 ppm sodium hypochlorite solution, and rinse with clean water for reuse.

7. Brining: Put the drained sea cucumbers and the boiled saturated brine into a plastic bucket in proportion; label the bucket with the product batch, calibration weight, and actual weight; and then store it with the lead seal at 0 to 4 °C for 5 to 10 days. During storage, the products should be kept away from oily substances

and items with fishy smells such as fish and shrimp. Do a good job in product protection to ensure its safety.

8. Primary sorting: Pour the sea cucumbers that have been salted with saturated brine, at 0 to 4 °C for 5 to 10 days out of the bucket, drain and weigh them within 12–20 h, and then conduct primary sorting according to the specification. After sorting, fill the sea cucumbers into the bucket and add fresh saturated salt brine for further storage and salting. Label the bucket with the batch number, per catty amount, calibration weight, and other processing information of the sea cucumbers per batch.
9. Salting with changed saturated brine: After the sorted sea cucumbers are loaded into the bucket, add new saturated brine (calibrated ratio of sea cucumber to saturated brine is 1:1), and then put them into a − 18 °C cold storage for about 5 to 10 days. During this period, check whether the brine in the barrel has frozen. If there is ice over, change the saturated brine promptly, and replace it with new saturated brine after 6 months of refrigeration in the −18 °C cold storage.
10. Take the sea cucumbers that have been salted for more than 20 days out of storage, pour the sea cucumbers from the bucket into a colander, refrigerate and drain them for 20 h, and then reheat the sea cucumbers. Add 180 kg of saturated brine into the jacketed pot, heat it to boiling, and pour in 90 kg of drained sea cucumbers after calibration. Turn on the steam at 0.25 mpa to quickly raise the temperature to about 85 °C, keeping that temperature for about 3 min. Take the sea cucumbers out and put them into the colander, and then after putting the colander into the prepared cold saturated brine at 15 °C at the same time, sort and pack them. The same pot of brine can be used continuously, but the brine must be saturated. The saturated brine for cooling needs to be changed promptly when the temperature rises, and a designated person is responsible for the reheating operation.
11. Secondary sorting: Sort the drained sea cucumbers according to the product specification, and pick out inferior sea cucumbers with bulging necks, melted body walls, or seriously broken body walls.
12. Packaging: Sea cucumbers should be kept in a sealed package according to the packaging specifications in a refrigerated environment. Sea cucumbers inside the packing bag must be kept dry with no obvious trace of water on the body surface.
13. Warehousing: Put the packaged salted sea cucumbers into cold storage for inventory and shipment.

Processing steps with significant hazards include raw material handling, rinsing, draining and weighing, secondary sorting, and packaging. There are biological hazards (bacterial reproduction and introduction of pathogenic bacteria); thus, effective preventive measures need to be taken to reduce the hazards to an acceptable level.

4.3 Relevant Standards of Salted Sea Cucumber

During the evolvement of salted sea cucumbers, some businesses/individuals, in the pursuit of longer shelf life and higher economic benefits, adopt excessive amounts of salt and even illegally use chemical agents, such as formaldehyde in the processing. This seriously affects the consumption safety of salted sea cucumber products and has caused widespread concern in society.

In order to secure the quality of salted sea cucumber and protect the interests of consumers, the Ministry of Agriculture of the People's Republic of China enacted the industry standard SC/T 3215–2007 *Salted Sea Cucumber* in 2007. However, it was no longer in compliance with reality due to the continuous development of the industry. Therefore, the standard was revised in 2012, and a new standard SC/T 3215–2014 *Salted Sea Cucumber* was issued in 2014.

The standard SC/T3215–2014 *Salted Sea Cucumber* applies to products made from fresh or live *Stichopus japonicas* through the process of eviscerating, cleaning, precooking, and salting. It covers all the requirements of salted sea cucumber products, including raw and auxiliary materials, product specifications, sensory requirements, physical and chemical indicators, net content, and contaminants and veterinary drug residues. The test methods encompass product specifications, sensory, protein, salt, moisture, attached salt, net content, contaminants and drug residues, etc. The inspection rules and specific requirements for labeling, packaging, transportation, and storage are stipulated according to market needs and business types.

The new standard for *Salted Sea Cucumber* has scientific and reasonable requirements, according to the principle of guaranteeing a fair and reasonable market, as well as protecting consumers' rights and interests. The promulgation and implementation of the new standard will regulate the production of salted sea cucumbers and help to improve the quality of related products. The standard also facilitates the management and supervision of products by technical supervision and management authorities, effectively protects the legitimate interests of producers and consumers, and promotes the standardized and scientific development of salted sea cucumber processing.

5 Processing Machinery for Sea Cucumber Preprocessing and Salting

5.1 Processing Machinery for Sea Cucumber Preprocessing

At present, a large amount of dedicated sea cucumber processing equipment is launched in China, such as the highly automated sea cucumber preprocessing equipment that can conduct pretesting and impurity-removing, flexible cleaning, eviscerating and sorting, grading and measuring, high-temperature cooking, extrusion and shaping, as well as low-temperature cooling. The pretesting and the impurity-removing process includes wetting and rinsing the sea cucumber by the spraying of clean water; sea cucumbers are filtered by holes and slits with certain geometries. In the cleaning process, the sea cucumbers are put into a water reservoir with pressurized gas input from the bottom upward, and the pressurized water is used to transport the sea cucumber. The high-temperature cooking process drives sea cucumbers to move forward at a set speed in the boiling water of the cooking tank, collect the floating foam on the water surface via guided aggregation and/or diversion, and then drive the sea cucumbers out of the cooking tank. The squeezing and shaping process consists of cooling the sea cucumber by spraying it with low-temperature water and then squeezing by elastic and flexible forces while the water is draining. The machinery is designed to provide pretesting, cleaning, cooking, and shaping devices suited to the sea cucumber pretreatment process, in a highly efficient manner and guarantees the high quality of the final product.

Only in the field of sea cucumber cleaning are there various equipment solutions that apply a variety of principles to achieve the process. These include equipment that uses mechanical friction for deep cleaning to remove the viscous lining on the surface of the sea cucumber for a better cleaning effect; automatic conveying and unloading during the cleaning process have been applied for a high level of automation, which can greatly improve the cleaning efficiency (Zhang et al. 2018a, b). A multi-nozzle water jet device (CN201910611931.8) is used to remove the cuticle from the body surface of sea cucumber, which can achieve deep cleaning of the mucous membrane on the surface of sea cucumber as well as accomplish the cleaning of sea cucumber with a better cleaning effect (Zhang et al. 2019). The processing equipment for cleaning sea cucumbers using ultrasonic waves (CN201810403106.4) is highly automated, green, and energy-efficient, with high cleaning efficiency, good cleaning effect, and less damage to sea cucumbers, and can improve sea cucumber quality (Zhang 2018). There is also cleaning equipment for sea cucumber processing by-products, such as the low-pressure shower for cleaning the intestinal sediment of sea cucumbers (CN201821760314.1), which adopts low-pressure spraying to clean sea cucumbers. It uses the pressure difference to force sea cucumbers to discharge the intestinal sediment, with high cleaning efficiency and good cleaning effect, effectively improving the processing efficiency of sea cucumbers (Yang and Zhang 2018).

In regard to slaughter and evisceration, the semiautomatic sea cucumber gutting and cleaning machine (CN202010613056.X) can be used to manually place the sea cucumber in the curved slot, so that the L-shaped shift lever restricts the sea cucumber in the curved slot. When the rectangular rotor can be rotated to drive the movable lifting frame to move reciprocally, under the elasticity of the spring and the pilot fit between the straight slot frame and the guiding plate, the blade breaks the abdomen of the sea cucumber. In conjunction with the motor, the first cylinder and the water pipe stir and clean the sea cucumber innards. The L-shaped shift lever swings by gravity, and the sea cucumber falls into the collection box eventually. This process can effectively break the abdomen and eviscerate, clean, sort, and collect the sea cucumber, thereby reducing the labor and improving the evisceration efficiency of sea cucumbers (Li et al. 2020).

5.2 *Processing Machinery for Sea Cucumber Salting*

Processing machinery related to salted sea cucumbers has also been extensively developed, such as the square tabletop (CN 214555475 U) for grading and selecting salted sea cucumbers and the rectangular sealing door in its middle, between which the interoperability facilitates the selection of salted sea cucumbers and improves the efficiency and saves manpower for manual selection. The rest of the sea cucumbers are left in the corresponding rectangular collection box; the thermal insulation coating inside the storage box and the low temperature of the transport box can adequately provide a suitable storage environment for sea cucumbers, avoiding death and nutritional loss (Liu et al. 2020).

The efficient and convenient cleaning device for salted sea cucumber (CN 215270325 U) is driven by a driving motor to rotate the impeller, stirring the water flow and sea cucumbers under the action of the water impact force and the cleaning brush head inside the cleaning barrel. This way, the impurities and sea cucumber skin film are peeled off from the sea cucumber and cleaned, and the brush head can be quickly disassembled for cleaning by turning the threaded rod a great convenience (Wang et al. 2021).

5.3 *Other Machinery and Instruments*

Other related equipment and instruments and technology for the quality identification of salted sea cucumbers include machine vision for detecting the adulteration of qualified salted sea cucumbers (CN 111597973 A). This triggers the deformation of salted sea cucumbers under the action of contact pressure, and computer vision that tracks the change in the contour of sea cucumbers during the recovery process after the contact pressure is withdrawn. Through the use of image processing and machine learning, automatic, rapid, and nondestructive identification of out-of-limits salted sea cucumbers is achieved. The samples do not require pretreatment and can be analyzed quickly and reliably. With no consumption of organic reagents in the process nor damage will be caused to the sea cucumber, representing a non-invasion measurement method that is both accurate and stable (Wang et al. 2020).

The texture quality of sea cucumber during salting is detected via a low field NMR (CN 105445307A). The transverse relaxation spectrum T2 on salted sea cucumber quality and an MRI information database were created, as well as a TPA texture information database to conduct correlation analysis of the two databases. This method maintains the integrity of samples, and the operation was simple and fast with accurate results. The results were not affected by the size or the appearance of the samples. And there were also advantages, including easy sample preparation, rapid determination, high precision, and sound reproducibility (Song et al. 2015).

The fingerprint of volatile organic compounds has been utilized to rapidly identify the quality and origin of salted sea cucumbers. A database of VOCs from salted sea cucumbers of different origins has been established using a FlavourSpec gas chromatography-ion migration combined instrument. A database of VOCs from salted sea cucumbers at different storage times has been created. Sea cucumbers have been tested for volatile organic compounds and compared with the established databases to determine the origin and quality of the salted sea cucumbers. The detection was fast and accurate, and no pretreatment was required for the samples (Zhang et al. 2018a, b).

6 Future R&D Trends in Sea Cucumber Processing Techniques for Preprocessing and Salting

In terms of the cleaning and slaughtering of sea cucumber, highly automated processing equipment based on visual recognition and intelligent control represents the trend of future development. Pretreatment equipment that can provide flexible cleaning, accurate positioning, and precise eviscerating will especially become a hotspot for research and development. Such equipment can replace the current large amount of manual labor and achieve automation and standardization of sea cucumber processing. It can also reduce the limitation of manpower on the production of sea cucumbers in the harvest season and further expand the scale of sea cucumber aquaculture.

The major role cooking plays in the heat treatment of sea cucumber is the inactivation of enzymes, curing, and dehydration. However, the current cooking methods have a significant impact on sea cucumber quality, leading to the loss of nutrients and changes in texture, especially in protein properties and water retention. Heating will lead to protein denaturation, which accelerates as the temperature rises. In the process of sea cucumber cooking, heating can cause the denaturation of proteins, especially in collagen. Excessive heating time or over-temperature can cause denaturation of collagen and gelatin production. This results in thermal contraction of the sea cucumber from severe dehydration, which causes significant changes in sea cucumber structure and loss of water-soluble components between the tissues, including polysaccharides, saponins, and other minerals. Future research and development in heat treatment technology will focus on rapid curing and low-temperature curing, achieving the palatable taste of sea cucumbers with moderate protein denaturation and minimizing the loss of solids. In-depth research will also be conducted into the recycling of solid content from the heat treatment cooking liquor, with the aim to improve the utilization of the biological resources.

Salting (pickling) of sea cucumbers. The background of the salting (pickling) technique is to preserve sea cucumbers for a longer retention period and reduce the water activity and moisture content of sea cucumbers for the convenience of drying. With the popularization of cold chain transportation and the development of new dehydration technologies and equipment, salted sea cucumber products will be gradually replaced, and its status as an intermediate raw material in the sea cucumber industry will gradually fade away. In the future, salted sea cucumbers will be replaced by ready-to-eat or fresh sea cucumbers in the end consumer market. While the slightly processed cooked and frozen sea cucumber as raw material will replace the salted sea cucumber as the intermediate raw material of the sea cucumber industry, the complex and lengthy desalination process will be no longer required before subsequent processing into dried sea cucumber or other products.

7 Conclusion

In summary, research on sea cucumber processing has been reported frequently in recent years, and the processing techniques and research methods adopted have become more and more scientific. However, key processes such as boiling and salting still lack research. Moreover, some processing methods are still far from mature, and cost issues have subjected them to certain limitations, not to mention the lack of research on low-cost and high-quality product processing methods. Therefore, companies are lagging behind in terms of processing technologies and techniques of sea cucumber and the adoption of scientific processing methods.

As living standards improve, so do people's requirements for the quality of sea cucumbers, and healthy, green, and original foods are receiving increasing popularity. With the recent improvement of technology and scientific research, scholars have discovered a variety of improved and innovative techniques in the processing of sea cucumber, which has effectively raised the product quality. However, the cost of those new techniques makes it difficult to employ them in sea cucumber processing in the field. In the future, a trend to apply new technologies will continue, but methods and techniques with low cost that improve the sea cucumber quality are the most effective way to improve sea cucumber processing in businesses in the near future. During processing, when a large amount of salt is added to improve the storage period and the appearance of sea cucumbers, direct consumption is very harmful to humans, so sea cucumbers must be desalted before eating. Furthermore, the predominant sea cucumber products in the market are primarily dried sea cucumbers, which have lost a lot of nutrients during processing. Therefore, higher-quality sea cucumber products with low salt, less processed, and easy for storage will become more and more popular. They will also increase the variety of sea cucumber products, thereby enriching the sea cucumber market.

References

Dong J, Xue Y, Xue CH (2014) Texture properties and histology changes of sea cucumber *Stichopus japonicas* during heat treatment. Abstracts of 2014 academic annual meeting of China Fisheries Society 392 https://kns.cnki.net/kcms/detail/detail.aspx?FileName=OGSB201410001390&DbName=CPFD2015

Jiang XM (2013) The sea cucumber vacuum cooking technology and product development. Dissertation. Ocean University of China

Jiao J, Kang HY (2010) Studies on losses of some functional components during traditional processing trepang from *Apostichopus japonicus*. Chin J Mar Drugs 29:46–49. https://doi.org/10.1016/j.carbpol.2017.06.128

Li PC, Zhao XF, Kong DP, Li Y, Zhang JQ (2020) Semi-Automatic Sea cucumber intestine removing and cleaning machine. CN Patent Appl 202010613056.X

Liu XZ (2021) China fishery statistical yearbook. China Agricultural Press, Beijing

Liu Q, Cao R, Guo YY, Li ZC, Zhao L (2015) Effects of sea cucumber processing on the quality of its products. Mod Food Sci Technol 31:313–317. https://doi.org/10.13982/j.mfst.1673-9078.2015.12.047

Liu XT, Che J, Zuo RT, Li SS, Che YQ (2020) Classified screening table for salted sea cucumbers. CN Patent Appl 202022997097.1

Mcelroy S (1990) Beche-de-mer species of commercial value–an update. Beche-de-mer Information Bulletin 2:2–7

Purcell SW, Samyn Y, Conand C (2012) Commercially important sea cucumber of the world. FAO, Rome

Schellekens M (1996) New research issues in sous-vide cooking. Trends Food Sci Technol 7:256–262. https://doi.org/10.1016/0924-2244(96)10027-3

Scott JE (1988) Proteoglycan fibrillar collagen interactions. Biochem J 252:313–323. https://doi.org/10.1042/bj2520313

Song L, Tan MQ, Zhao J, Li Y, Dong XP, Zhou DY, Jiang PF, Li XL (2015) LF-NMR detection method for texture quality of sea cucumber during salting. CN Patent Appl 201510896913.0

Wang ZP, Liu Q, Cao R, Yin BZ (2011) Effect of temperature on the changes of structure and molecular weight of *Apostichopus japonicus* collagen. Progress Fish Sci 32:80–84

Wang XH, Liu W, Li SS, Wang RW (2021) Salted Sea cucumber cleaning device. CN Patent Appl 202120670500.1

Wang HH, Yan PT, Qiu XJ, Zhu XY, Li PP, Tao XH (2020) Method for detecting adulteration conditions of qualified salted sea cucumbers by utilizing machine vision. CN Patent. Appl. 202010407815.7

Xu WQ, Shen J, Cai SJ, Wang CQ (2009b) Study on mechanical evisceration technique of sea cucumber. Fishery Modernization 36(52–54):69

Xu WQ, Shen J, Xue CG, Cai SJ, Lin W (2009a) Study on washing process technique of fresh sea cucumber. Fishery Modernization 36:42–45

Xue D, Gao X, Cui FX, Dong P, Li BF, Xue CH (2006) Rheological and structure properties changes of *Stichopus japonicus* during heat treatment. J Chin Inst Food Sci Technol 6:161–166. https://doi.org/10.16429/j.1009-7848.2006.01.035

Xue Y, Xue CH, Jiang XM, Li ZJ, Wang YM, Wang JF, Xu J, Tang QJ, Chang YG (2013) Low-temperature vacuum heating cooking method of sea cucumber. CN Patent Appl 201310100683.3

Yang T, Zhang T (2018) Novel low-pressure sea cucumber intestinal sediment cleaning device. CN Patent Appl 201821760314.1

Zhang J (2018) Ultrasonic Sea cucumber cleaning and processing device. CN Patent Appl 201810403106.4

Zhang J, Jing YX, Wang MJ, Zhao YP, Wang GM, Liu F (2018a) Method for rapidly identifying quality and producing places of salted sea cucumbers. CN Patent Appl 201810721166.0

Zhang X, Wen BG, Yang L, Wang HD, Jiao J, Wang HH, Tao XH, Dong XP (2018b) Deep cleaning equipment for sea cucumber surfaces. CN Patent Appl 201811448409.4

Zhang X, Wen BG, Zhao SB, Xia WW, Tao XH, Wang HH, Lu Y, Dong XP (2019) Multi-nozzle water jet type removing device of cuticle on body surface of sea cucumber. CN Patent Appl 201910611931.8

Zhong M, Zhang LP, Hu CQ (2016) Influence of water cooking condition on morphology and protein loss of body wall from sea cucumber (*Stichopus monotuberculatus*). South China Fish Sci 12:23–29. https://doi.org/CNKI:SUN:NFSC.0.2016-01-004

Zhu BW, Han B (2004) Characterization and purification of holothurians autoenzyme. Food Ferment Ind 30:132–134, 137

The New Processing Technology of Dried Sea Cucumber Products

Yong Xue and Changhu Xue

Abstract The dried sea cucumber does not need to be a part of the cold chain in the process of circulation and storage, with its extended shelf life. This advantage makes dried sea cucumbers occupy a dominant position with consumer goods. In order to avoid the decomposition of sea cucumbers by the action of the autolytic enzyme, the sea cucumber usually requires preprocessing before drying, using processes such as water boiling, brine boiling, salting, ultrasonic treatment, enzyme treatment, soaking, etc. The preprocessing can change the internal structural state of the sea cucumber's body wall, which is conducive to the formation of a special product structure and to the characteristics of rehydration. Additionally, every drying method has both pros and cons when it comes to single drying methods such as hot air drying, freeze-drying, microwave drying, and heat pump drying. In order to make up for the deficiency of single drying methods and obtain better quality dried sea cucumbers at a lower processing cost, combined drying technology can be adopted, such as heat pump-hot air combined drying, heat pump-microwave combined drying, and freeze-microwave combined drying. According to these characteristics, two or more drying methods are used simultaneously in stages based on the principle of complementary advantages, reducing production costs and energy consumption, improving product quality, and preventing nutrient loss. This chapter provides a reference for the selection and optimization of sea cucumber processing technology and provides a theoretical basis and scientific basis for the scientific consumption and promotion of dry sea cucumber.

Keywords Autoenzyme · Cold air drying · Freeze-drying · Hot air drying · Microwave drying · Sea cucumber

Y. Xue (✉) · C. Xue
College of Food Science and Engineering, Ocean University of China, Qingdao, Shandong, P. R. China
e-mail: xueyong@ouc.edu.cn; xuech@ouc.edu.cn

C. Xue (ed.), *Advances in Sea Cucumber Processing Technology and Product Development*, Advances in Marine Bioprocesses and Bioproducts,
https://doi.org/10.1007/978-3-031-16512-2_6

1 Introduction

The harvesting of sea cucumbers has obvious seasonal characteristics, and fresh sea cucumbers are rich in autolytic enzymes. The sea cucumber will decompose under the action of autolytic enzymes soon after leaving its growth environment, so it is difficult to store and transport sea cucumbers. Therefore, sea cucumbers must be processed quickly after fishing and then sold after being processed into different product forms. Dried sea cucumber is the most important sea cucumber product in the market. The dried sea cucumber does not need to be a part of the cold chain in the process of circulation and storage, with its extended shelf life. Consumers can easily identify the strengths and weaknesses of sea cucumber from its appearance and edible quality compared with other products. These advantages give the dried sea cucumber a dominant position among sea cucumber consumer goods.

The traditional way of drying sea cucumbers focuses on "hanging salt at high temperature." Although the dried sea cucumber retains its full shape after processing, the salt content is in excess, and repeated pickling, heating, and salt hanging lead to the loss of a large number of nutrients in the sea cucumber. Modern methods of dried sea cucumber processing have already experienced significant improvements over traditional processes. The preprocessing and drying methods for fresh sea cucumbers are more diversified, and the products enjoy richer characteristics. The distinctions among different techniques of dried sea cucumber processing are mainly reflected in the preprocessing and drying methods, which have a great impact on the loss of nutrients, rehydration process, and sensory quality of the products.

2 Predrying Processing of Sea Cucumbers

The purpose of pretreatment before drying sea cucumber is to destroy the enzymes within the sea cucumber, improve the stability of raw materials, reduce the water content of the sea cucumber, change the body wall structure, and adjust the product characteristics. The preprocessing methods can be divided into two categories. One is the simple preprocessing methods used in a majority of dried sea cucumber processing, including water boiling, salt water boiling, and salting; another is the preprocessing methods used for freeze-drying or rapid rehydration of dried sea cucumber products, including enzyme treatment, salting, soaking, etc. Such preprocessing methods mentioned above can change the internal structural state of the sea cucumber's body wall and then form a special product structure and rehydration characteristics in the drying process.

2.1 Regular Predrying Processing

During the processing of most dried sea cucumber products, only a simple preprocessing is required for the sea cucumber. Regular preprocessing of sea cucumber includes water boiling, salt water boiling, and salting. According to the salt content of dried sea cucumber and market demands, several methods can be used individually or in combination. The difference in preprocessing conditions mainly lies in the salt content of the water used in the boiling process, boiling time, the salt content of the osmotic solution, osmotic dehydration time, etc. After preprocessing under different conditions, the initial moisture content, salt content, shrinkage, drying process, and rehydration properties upon drying of sea cucumbers will differ. The authors' research team preprocessed sea cucumbers under seven different conditions and then studied the effects of the different preprocessing conditions on the drying process and product quality of sea cucumbers.

The different preprocessing approaches were presented as follows: Group (1) boiled in 3.5% sodium chloride solution for 15 min; Group (2): boiled in fresh water for 15 min; Group (3): boiled in 3.5% sodium chloride solution for 15 min and then boiled in supersaturated sodium chloride solution for 15 min; Group (4): boiled in 3.5% sodium chloride solution for 15 min and then soaked in supersaturated sodium chloride solution for 6 h; Group (5): boiled in 3.5% sodium chloride solution for 15 min and then soaked in supersaturated sodium chloride solution for 12 h; Group (6): boiled in 3.5% sodium chloride solution for 15 min and then soaked in supersaturated sodium chloride solution for 24 h; and Group (7): boiled in 3.5% sodium chloride solution for 15 min and then soaked in supersaturated sodium chloride solution for 48 h.

The salt contents of the sea cucumbers dried in the seven groups were significantly different. The salt content of the dried sea cucumber was less than 1% through the preprocessing of boiling with clean water (Group (2)). The dried sea cucumber treated with this preprocessing method is exactly the pure mild-salted dried sea cucumber on the market. In practical production, the sea cucumber is often boiled with 3.5% salt solution (Group (1)). After drying, its salt content is 13.2%. This kind of sea cucumber is still classified as mild-salted dried sea cucumber because of its low salt content. When the sea cucumber is boiled or soaked in saturated sodium chloride solution, the salt content is significantly increased to about 40%. After the sea cucumber is processed into dried sea cucumber, it is commonly known as the salted dried sea cucumber. The salt content of sea cucumber (Groups (4)–(7)) subjected to soaking showed a growth trend in the first 12 h of soaking. When the soak time exceeded 12 h, the salt content of sea cucumber stayed basically unchanged. When the soak time reached 24 h, the difference in salt contents between different sizes of sea cucumber became negligible. Therefore, as for salted dried sea cucumber, the soak time should reach at least 24 h.

The mass, volume, and moisture content of sea cucumber also changed greatly after different pretreatment. The sea cucumber boiled in clean water for 15 min had the highest initial moisture content and shrinkage coefficient, followed by that boiled in 3.5% sodium chloride solution for 15 min. When the sea cucumber was treated with saturated brine, it was in a state of continuous water loss, and the initial water content of the sea cucumber decreased significantly. In the first 12 h of immersion, the salt content of sea cucumber was increasing, while the volume contraction changed little. After soaking for 24–48 h, the changes of sea cucumber mass, volume and moisture content tended to be stable. Based on this point, the soaking time should be more than 24 h when the sea cucumber is treated by salt water immersion.

During the hot air drying process at 60 °C after different types of preprocessing for sea cucumbers, the variation trend for the water content of sea cucumbers on a dry basis over drying time stayed similar, but the drying time differed sharply. The drying times of sea cucumbers in Groups (1), (2), (7), and (3) were 13 h, 23 h, 35 h, and 40 h, respectively. The drying process of sea cucumbers in each group showed certain similarities. The drying process did not experience any stage of increasing rate or constant rate drying, but entered the stage of the falling rate at the very beginning of the drying process. The water diffusion in sea cucumber was the dominant factor, which directly determined the drying rate of sea cucumber. The difference between the migration rate of water from the inside of the material to the surface and the evaporation rate of water from the surface of the material to the outside intensifies with the increase in drying time. When the drying process reached 5 h, the drying rate decreased significantly, and the progress of the drying process tended to slow down; after drying, a large amount of salt precipitated on the surface of samples from Groups (3) and (7) with high salt content and the surface shrinkage were serious, while the surface of samples from Groups (1) and (2) with low salt content experienced no salt precipitation, and the surface shrinkage was negligible.

2.2 *Preprocessing Before Special Drying*

Freeze-drying or rapid rehydration of dried sea cucumber products requires some special types of preprocessing, including ultrasonic processing, enzyme treatment, salting, and soaking. Those types of preprocessing can change the internal structure and hydration state of the sea cucumber body wall, so as to increase the volume of dried sea cucumber products and improve the rehydration speed.

2.2.1 Predrying Processing of Freeze-Dried Sea Cucumbers

Freeze-drying is a drying method in which frozen materials are placed in a vacuum state, and the ice crystals within them directly sublimate into water vapor. Due to the formation of a significant amount of holes in the materials after the sublimation of frozen ice crystals, the materials become porous. Therefore, when the materials are rehydrated, water can directly and quickly enter the food, and rehydration becomes faster. The freeze-dried sea cucumber usually needs to be boiled, salted, or treated with an enzyme and then soaked to make the sea cucumber fully absorb water and expand. After drying, a large volume of dried sea cucumber takes shape, and a significant amount of porous structures are formed within the body wall. Therefore, the preprocessing cannot only give the freeze-dried sea cucumber products a full shape, but can also improve the drying and rehydration efficiency.

Researchers have developed a variety of approaches to preprocess freeze-dried sea cucumbers. The preprocessing methods used in industrial production are mainly the boiling or soaking of sea cucumber after enzyme treatment. Yuan et al. (2010) adopted the preprocessing of soaking and absorbing water in low-temperature water after boiling. After cleaning, fresh *Apostichopus japonicus* was boiled in 95–100 °C water for 10 min and then soaked in 4 °C pure water to absorb water and expand. The soaking and water expansion of sea cucumber before freezing has a great impact on the quality of the final product, and the soaking and water absorption time should be moderate. If the soaking and water absorption time is too short, the rehydration will be lengthy, although the freeze-dried sea cucumber takes a full form; if the soaking and water absorption time is too long, the body wall of the dried sea cucumber features large pore spaces without aesthetic form, although the freeze-dried sea cucumber rehydrates quickly. Approximately 48 h is an appropriate soaking time; Yun et al. (2006) used the preprocessing method of soaking after enzyme treatment for freeze-dried sea cucumbers. Fresh sea cucumbers were washed and treated with protease and then soaked in water to absorb water. Based on the comparison of the processing effects of neutral protease, papain, and composite protease, it was found that the sea cucumber treated with papain had a lighter fishy smell and better taste than the sea cucumber treated with enzymes. The optimum conditions of papain treatment were as follows: the amount of enzyme used was 0.07% of the weight of the sea cucumbers, the temperature was 25 °C, and enzymolysis lasted for 30 min. The best processing effect could be achieved through soaking and water absorption for 8 h after enzyme treatment of sea cucumber.

In addition to the above methods of boiling and enzymatic hydrolysis, researchers found that physical field treatment using an electric field, ultrasound, etc. could also change the structural state of sea cucumber's body wall, so as to improve the quality of freeze-dried sea cucumber products. Yanara et al. (et al. 2020) preprocessed sea cucumbers before freeze-drying with the electrohydrodynamic (EHD) method of electric field treatment and processed the sea cucumbers at the voltages of 20 kV, 30 kV, and 45 kV for 30–60 min. The EHD treatment changed the water content of the sea cucumbers, and the initial drying rate increased with the increase of treatment voltage. 30 kV–30 min preprocessing yielded a higher drying rate and a higher rehydration rate, achieving better results than freeze-drying without preprocessing did; Duan et al. (2008) preprocessed sea cucumbers using ultrasound before microwaving freeze-drying to strengthen the mass transfer in the penetration process. Higher ultrasonic power could significantly improve the water loss rate of sea cucumbers, but it would also lead to a higher salt content of the product. When the ultrasonic power exceeded 240 W, the change in water loss rate was not obvious, but the growth rate of solid content was very prominent. Ultrasonic-assisted penetration treatment could shorten the drying time by a little more than 2 h at most, but the drying energy consumption was reduced by nearly 20%. The sensory evaluation value, preservation rate of active ingredients, and rehydration ratio were not significantly different from those of untreated sea cucumber.

2.2.2 Other Special Predrying Processing of Sea Cucumbers

In order to improve the rehydration rate of dried sea cucumber, some special types of preprocessing can be carried out before drying, so as to change the dense structure of the sea cucumber body wall and ensure the dried sea cucumber can rehydrate quickly. Sui and Yin (2016) preprocessed dried sea cucumber for quick rehydration by double-boiling, salting, and soaking. The optimum processing conditions of rapid rehydration for dried sea cucumber consisted of boiling for 5 min, salting for 24 h, secondary boiling for 10 min, and soaking for 14 h. The dried sea cucumbers were in dark brown color after rehydration, with complete body wall and good elasticity. The comprehensive score of the sensory quality was higher than that of regular preprocessed dried sea cucumber. The rehydration time of dried sea cucumber subjected to this preprocessing was 5 h, and the weight of the sea cucumber after rehydration was 20.8 times that before rehydration. The soak time of dried sea cucumber subjected to regular preprocessing was 12.7 h, and the weight after soaking is 16.8 times that before soaking. The contents of protein, total sugar, and crude fat of sea cucumbers were higher than those of dried sea cucumbers subjected to regular preprocessing.

In addition, Li et al. (2005) proposed that curing and enzyme deactivation for sea cucumbers under high pressure and subsequent drying by microwave or far-infrared and vacuum could further reduce the nutritional loss in the drying process of sea cucumbers.

3 Sea Cucumber Processing by Single Drying Method

3.1 Hot Air Drying of Sea Cucumbers

Hot air drying is a simple and economical drying method, which uses heated air as the medium to heat and dry the materials. Hot air drying is advantageous in simple operation and low cost and has moderate requirements for equipment and environment, but it also has a certain impact on food quality, resulting in the loss of heat-sensitive nutrients or active ingredients. Hot air drying has been widely used in the industry for drying sea cucumbers.

The authors' research team studied the characteristics of hot air drying for sea cucumbers at different temperatures. Under the drying conditions with different hot air temperatures, the water content change of sea cucumber on a dry basis showed that the higher the hot air temperature, the faster the water loss rate and the shorter the drying time; but when the temperature was greater than or equal to 45 °C, there was little difference in drying time. Drying with hot air at 30 °C (23.5 h), 45 °C

(12 h), 60 °C (11.5 h), 75 °C (12.5 h), and 90 °C (11 h), respectively, resulted in a dramatic decrease of the sea cucumber water content, from 308% to 13%. The whole hot air drying process did not experience any stage of an increasing rate or constant rate of drying, but directly entered into the stage of a falling rate of drying. In the initial stage of drying, the drying rate increased with the increase in temperature. When the temperature was greater than 45 °C, the difference in the drying process decreased. A drying rate at 90 °C is more than twice that at 30 °C. The reason might be that in the initial stage of drying, the contraction amplitude of the sea cucumber's body wall fibers was different, and the extent of the reduction for the tissue voids was also different. After 2 h of drying, the drying rate tended to slow down. The sea cucumber dried at a temperature above 45 °C has uniform morphology and no obvious bending after shrinkage.

3.2 Sea Cucumber Freeze-Drying

Like many other aquatic products, the drying quality of sea cucumber depends partly on the drying temperature. Drying at a relatively low temperature results in better sensory quality and nutrition retention. The vacuum freeze-drying technology has been increasingly applied in the processing of sea cucumber products. Freeze-drying can maintain the original shape of the sea cucumbers, and the sea cucumber after freeze-drying is large and full, with a fast rehydration speed and long shelf life. However, vacuum freeze-drying also has its disadvantages, including high energy consumption, high cost, and a long freeze-drying time.

The entire vacuum freeze-drying process can be divided into three stages: pre-freezing, sublimation drying, and desorption drying. The pre-freezing condition of sea cucumbers has a significant influence on the drying process. Since the freeze-drying process of materials is carried out in a vacuum, only after the material is completely frozen can it be sublimated in a vacuum. Otherwise, if some liquid exists, it will evaporate rapidly under a vacuum and cause material deformation. Therefore, the pre-freeze temperature should first of all be 5–10 °C lower than the eutectic point temperature of the material (i.e., the temperature at which the free water in the material is completely frozen). Generally, the pre-freeze temperature of sea cucumbers is set to −30 °C. Secondly, in addition to freezing the material completely before freeze-drying, freeze-drying should also ensure that the original characteristics of the material should be maintained as much as possible when frozen. Therefore, the freezing rate should make it so that the ice advance speed of material tissue is greater than the water movement speed, the ice crystal distribution is similar to the distribution of liquid water in the material before freezing, and the ice crystals are as fine as possible and therefore do not damage the cell tissue. The faster you freeze, the better you keep your original properties. It is worth mentioning that it is very important to pass through the maximum ice crystal generation zone quickly for the freezing quality of the material. However, if the freezing rate is too fast and the fine ice crystals are too small, the gas escape channels during sublimation and drying will be too small, and the resistance to the escaping gas will increase, which will also reduce the sublimation rate and the rehydration rate of dried products. The commonly used rapid freezing method is to reduce the pre-freeze cabinet temperature to below −25 °C, load and weigh the preprocessed sea cucumbers evenly, and then put them in the pre-freeze cabinet for freezing. Considering that the material temperature will rise during the transfer to the drying bin, in order to prevent the material from thawing, it is necessary to stay at the temperature spot of −30 °C for more than 1 h after the sea cucumbers reach the pre-freeze temperature, which ensures that the material is completely frozen.

The sublimation drying means that the frozen materials are heated in a closed vacuum container, and the ice crystals will sublimate into water vapor, so as to dehydrate the material. Drying starts from the outer surface of the material and gradually moves to the interior. The space left by the sublimation of ice crystals becomes an escape channel for further sublimation of water vapor. Sublimation drying is almost complete when all ice crystals have been removed, with about 90% moisture removed. The following basic conditions must be met to realize sublimation drying: (1) The vacuum pressure within the vacuum container must be lower than the saturated vapor pressure of water corresponding to the temperature of the eutectic point of the material; otherwise, the ice crystals in the material will partially thaw. (2) The latent heat of sublimation for ice crystal sublimation must be provided to the material. The more heat that is supplied, the faster the sublimation. A large amount of sublimated water vapor cannot be completely removed from the container via the vacuum pump alone. The vacuum pump is only used to extract non-condensable gas, and the sublimated water vapor must be condensed into ice (or frost) on its surface via a cold trap to maintain the required vacuum degree. The sublimation drying process is actually the sublimation of the ice in the material into water vapor, which then condenses into ice on the surface of the cold trap, with as much sublimation as condensation. Since the latent heat of sublimation is the same as that of condensation, stable vacuum pressure can be maintained only when the amount of cooling actually provided by the cold trap for condensation is equal to the amount of heat added by the amount of heating for sublimation. The vacuum pressure should generally be kept within the range of slightly less than 1/2 of the saturated vapor pressure corresponding to the highest sublimation temperature, generally between 30 and 90 Pa. At the same time, when the material in the sublimation drying process is heated, the temperature of the dried part of the material should not exceed the maximum allowable temperature of the material. For the sea cucumber, dried products will be damaged or shriveled and deformed when the temperature exceeds 55 °C. When the sensor temperature on the sea cucumber surface approaches 50 °C, the sublimation drying process is basically completed. In order to eradicate the residual ice in the sea cucumber, the operation of the heating plate needs to be extended for more than 30 min.

At the end of sublimation drying, part of the moisture is absorbed on the capillary wall and the polar group of the dry substance, which is not frozen. When such moisture reaches a certain content, it provides conditions for the growth and reproduction of microorganisms and for some chemical reactions, so it is necessary to remove such adsorbed moisture, which is the purpose of desorption drying. Due to the high adsorption energy of adsorbed water, this water is difficult to secrete from the adsorption without sufficient energy. Therefore, the temperature of the material at this stage should be kept high, as long as it is controlled below the maximum allowable temperature. At the same time, in order to make the secreted water vapor maintain a driving force high enough to escape from the material, it is necessary to create a considerable vapor pressure difference inside and outside the material, so a high degree of vacuum must be maintained in the container at this stage. At the beginning of desorption drying, the temperature is relatively low. When the material center temperature rises to the surface temperature, desorption drying is basically over. In order to eliminate the unevenness of drying caused by uneven loading, inconsistent particle sizes of materials and other adverse factors, after the center temperature of sea cucumber is basically the same as the surface temperature, the sea cucumber should be continuously heated for about an extra hour to ensure that the water content of the dried sea cucumber products is less than 5%, and then the vacuum can be removed; now it is the time to open the hatch for vacuum packaging, ensuring the integrity of the sea cucumbers.

The change in the water content and the drying rate for sea cucumbers during freeze-drying were studied by the authors' research team. It took 26 h for freeze-dried sea cucumbers to change their initial dry basis moisture content of 426% to the dry basis moisture content of less than 5% at the end of the drying process. The freeze-drying process of sea cucumber only experienced a falling rate during the drying stage, which was consistent with the hot air drying that also experienced the falling rate during the drying stage only. The difference was that the freeze-drying rate curve had a higher rate in the first 2 h, while the rate decreased significantly at 3 h onward. This showed that the internal migration of water was obvious in freeze-drying. After drying, the size of the sea cucumber was similar to that before drying, the color of the body surface was pale gray, and the sea cucumber's body was fragile. The elastic modulus, stress relaxation time, viscosity modulus, and breaking force of the freeze-dried sea cucumbers decreased significantly after rehydration, while the elasticity increased, and the hardness and viscosity decreased.

The process parameters of freeze-drying sea cucumber were discussed in detail in some studies. The influence of the different factors on the quality of the sea cucumbers in the vacuum freeze-drying process was explored, and it was found that the pre-freeze time had the most obvious effect on the quality of the sea cucumbers. Optimal process conditions were obtained through response surface optimization: pre-freeze time of 4.3 h, pre-freeze temperature of −34.6 °C, and maximum heating plate temperature of 80.3 °C (Zhang et al. 2017). A vacuum freeze-drying test was performed when the sea cucumbers expanded after water absorption. The process of the conditions of freeze-drying were as follows: freezing temperature of −25 ± 1 °C, cold trap temperature of −29–31 °C, vacuum degree of 10–20 Pa, and a final freeze-drying temperature of 60 °C (Yun et al. 2006). The research findings for the vacuum freeze-drying process showed that the sea cucumbers were prefrozen through quick freezing. In the vacuum freeze-drying process, the vacuum degree was under 10 Pa for 0.5 h before heating, and then the sea cucumbers were sublimated at 50 °C for 3 h; the temperature was raised to 70 °C until the end. The final freeze-dried sea cucumbers took a full and appealing form (Yuan et al. 2010).

Freeze-drying (FD) can yield the best-dried sea cucumbers, but at the cost of high energy consumption, high economic cost, and a long freeze-drying time. In order to achieve fast drying time and high-quality sea cucumber products, microwave freeze-drying technology was adopted, and microwave heating was used as the heating source for freeze-drying. To avoid corona discharge during the microwave freeze-drying process, the cavity pressure was kept within the range of 50–100 Pa (Duan et al. 2010). Microwave freeze-drying can shorten the freeze-drying time to about half of that of the conventional freeze-drying process and keep the product quality basically unchanged. Although microwave freeze-drying can significantly improve the drying rate, there are still many problems to be solved. One of the problems is that the uneven distribution inherent to the microwave field leads to uneven temperature distribution of the dry materials, which may lead to overheating. Another problem is the corona or plasma discharge under a high vacuum and the possibility of ice thaw in the product afterward.

3.3 Microwave Drying of Sea Cucumbers

Microwave drying of sea cucumber can be carried out under atmospheric pressure or vacuum conditions. For microwave drying under vacuum conditions, the material is heated rapidly, and the steam generated within the material creates the "pumping effect," which drives the moisture to the surface in the state of water vapor; the surface water vapor is then removed by the vacuum pump, thus gradually completing the drying of the material. Microwave vacuum drying not only overcomes the disadvantage of a low heat transfer efficiency in vacuum drying but also makes up for the deficiency of environmental moisture that cannot be eliminated during microwave drying. It can heat up the surface and the interior of sea cucumbers at the same time, speeding up the drying process. However, in the later period of continuous microwave vacuum drying, due to the low moisture content and uneven water distribution of materials, it is easy to lead to local overheating, excessive drying of the tip and surface, as well as a reduced sensory quality.

The authors' research team studied the microwave drying process of sea cucumbers under different pressure conditions and measured the drying curve of sea cucumbers under microwave heating conditions of 1.3 kPa, 51.3 kPa, and 101.3 kPa. Microwave drying was carried out under the condition of a high vacuum (1.3 kPa), and the duration of the entire drying process was 15 min. When the pressure was 51.3 kPa and 101.3 kPa, the drying durations were 20 min and 21 min, respectively. Microwave drying of sea cucumbers under different pressures underwent three drying stages: preheating, constant (this stage was very short, so the figure showed a peak), and falling rate. The microwave vacuum drying process from the preheating stage to the falling rate stage all appeared in the position of moisture content of about 70%. When drying was about to end, the drying rate curve of sea cucumber showed a peak value, and sea cucumber volume expanded rapidly, which was the phenomenon of puffing, giving the dried products a unique plump appearance. The causes of puffing consist of two aspects: first, the dense structure of sea cucumber makes it difficult to provide an effective channel for water migration from within and, second, during microwave heating, the temperature of the sea cucumber body increases rapidly, and the internal heat gathers easily. The constant accumulation of heat makes a lot of moisture in the sea cucumber body vaporize quickly, and the moisture cannot be fully discharged, so the high-pressure difference occurs between the inside and the outside of the sea cucumber, causing the rapid expansion of the volume of the sea cucumber. The phenomenon of puffing also occurs in the microwave drying process of other materials, but it does not appear in the process of hot air drying, freeze-drying and so on. At the end of the drying process, there was no significant amount of salt precipitation and visible depression on the surface of the sea cucumber. With the increase of pressure during microwave vacuum drying, the elasticity of dried sea cucumbers after rehydration increased, and the hardness and viscosity decreased.

During microwave drying, the continuous microwave heating, intermittent microwave heating, and microwave power density (microwave power/dry material quality) affected the drying speed and sensory quality of sea cucumber. In the continuous microwave drying, the microwave power densities were 0.5 W/g, 0.65 W/g, and 0.8 W/g, and the moisture contents of sea cucumber reached 45% at 90 min, 100 min, and 110 min, respectively (Zhu et al. 2015). When the microwave power density was 0.8 W/g, it only took 180 min to reach the wet basis moisture content of 13%. During intermittent microwave drying, the drying curve was relatively flat. When the initial water content of sea cucumber was 60% and the microwave power densities were 0.5 W/g, 0.65 W/g, and 0.8 W/g, the water contents of the sea cucumbers reached 22% at 900 min, 600 min, and 480 min, respectively. The intermittent drying process could promote the dynamic balance between internal water migration and surface water evaporation and effectively alleviate the accumulation of salt and crusts on the body's surface. Finally, the best drying effect could be obtained under the condition of the vacuum degree of 90 kPa, the intermittent ratio of 10s-on/20s-off, and the microwave power density of 0.65 W/g.

As long as similar conclusions were obtained, it was also found that the salt concentration of cooking water during the preprocessing also had an impact on dried sea cucumbers (Zhang et al. 2012). At different microwave power densities (1.0, 1.5, 2.0, 2.5, and 3.0 W/g), vacuum degrees (0.087, 0.090, and 0.093 MPa), and levels of pre-boiled water salinity (80, 160, and 240 g/L), the sea cucumbers were subjected to microwave drying. The microwave power density had a significant effect on the drying rate and product quality of sea cucumber; whenever the power density increased by 0.5 W/g, the drying time decreased by 22.5 min on average. The salinity of pre-boiled water had a significant influence on the appearance of dried sea cucumbers. The optimal process parameters of microwave vacuum drying for sea cucumber were as follows: microwave power density of 2 W/g, vacuum degree of 0.090 MPa, and pre-boiled water salinity of 80 g/L. The color and shape of dried sea cucumber remained intact, the shrinkage was relatively low (32.20%), the rehydration rate was relatively high (266.32%), and the drying time was only 110 min.

3.4 Cold Air Drying of Sea Cucumbers

Cold air drying is an approach that dries the material under the condition of low temperature, low humidity, and high air speed. The equipment for cold air drying of sea cucumbers is the low-temperature heat pump drying equipment, which is essentially an energy transfer device. The operating principle of the equipment is similar to that of the refrigerator, following the Carnot cycle, which is a thermal cycle process consisting of evaporation → compression → condensation→ throttling → re-evaporation, so as to absorb energy from low-temperature heat sources and then apply such heated energy in high-temperature environments. The heat pump drying equipment mainly consists of a compressor, evaporator, expansion valve, and condenser. When the humid air from the drying room passes through the evaporator, the heat is absorbed by the refrigerant inside, reducing the temperature and discharging the condensate, so the moisture content of the air decreases and the relative humidity approaches saturation. Then, the condenser absorbs the condensation heat discharged from the refrigerant, the temperature rises, and the relative humidity decreases and enters the drying room as the drying medium. At the same time, the refrigerant in the evaporator is converted from liquid to gas due to heat absorption in the evaporator, compressed by the compressor, and then sent to the condenser, which condenses the refrigerant to liquid after the release of latent heat in the condenser. The liquid passes through the expansion valve and declines in pressure to move to the next refrigeration cycle. By controlling the temperature and humidity of the drying environment, heat pump drying can effectively avoid the oxidation of material fat, color change, and nutrient loss caused by hot air drying and better maintain the characteristics of the material while saving energy.

In the process of heat pump drying, the relative humidity of air and air speed significantly influence the drying rate. Mu et al. (2007) studied the heat pump drying process of sea cucumber and found that the drying time was 360 min and 420 min when the relative humidity of the air was 28% and 32%, respectively; when the airspeed increased from 0.85 m/s to 1.80 m/s, the drying time for reducing sea cucumber moisture content from 69.4% to 13.0% was shortened by 60 min. Different drying conditions on the shrinkage of sea cucumber were not notable, but the rehydration rate was higher when the airspeed was larger (1.80 m/s). Compared with traditional drying methods, heat pump drying of sea cucumber simplified the operation steps and improved the sensory quality of the product. Chen et al. (2017) also found that as the speed of cold air increased, the relative humidity in the drying room and the humidity difference between the materials increased, which accelerated the dehydration rate of the drying process. When the cold air speed was 2–3 m/s, the drying characteristics were different from those under 1–1.5 m/s. The effect of the speed of the cold air on the drying rate was higher than that of cold air temperature, and airspeed control should be made stricter. The speed of the cold air had the most significant effect on the sea cucumbers' appearance and had a limited effect on the characteristic odor after rehydration. The sensory index scores were higher with a cold air speed of 1–2 m/s than those at 2.5–3 m/s, so the cold airspeed should preferably not be higher than 2.5 m/s. The optimal process conditions for drying were determined by a single-factor test and response surface test: cold air temperature of 19 °C and cold air speed of 1.70 m/s.

The temperature of the heat pump drying also affects drying. Chen et al. (2017) studied the heat pump drying process of sea cucumbers in the temperature range of 8–28 °C and found that the falling rate process of sea cucumbers drying at 18–28 °C was significantly shorter than that at 8–13 °C, and the drying time could be shortened at a temperature above 18 °C; the drying rate was a little higher at a reduced temperature than at a constant temperature. Therefore, centering on the cold air temperature corresponding to the moderate dehydration rate, drying could be carried out at a temperature higher than the center temperature at the early stage of drying; with the increase of cold air temperature, the sensory index for the texture of dried sea cucumber after rehydration showed a trend of increasing before decreasing, and a better texture could be achieved at 13–23 °C. The cold air temperature had less effect on the flavor of *Apostichopus japonicus*, but had more effect on the appearance. At 18–28 °C, the higher the temperature, the more obvious the collapse of sea cucumber folds and the lower the sensory score. Therefore, a cold air temperature of around 18 °C helped to save drying time and achieve a better quality of *Apostichopus japonicus* after rehydration, with less nutrition loss, moderate shrinkage, and a high sensory score.

3.5 *Quality Comparison of Dried Sea Cucumbers Processed by Different Single Drying Methods*

Due to the significant differences in drying principles and drying conditions, the quality of dried sea cucumbers processed by different single drying methods varies significantly. Studies have compared the quality of dried sea cucumbers processed from different single drying methods, respectively. Zhang et al. (2018) used hot air drying (AD), microwave drying (MD), vacuum freeze-drying (VFD), and microwave vacuum freeze-drying (MVFD) for *Stichopus japonicas* and the differences in product quality, drying rate, and energy consumption for different drying methods. The drying conditions of the four drying methods are presented as follows.

1. Hot air drying (AD): drying temperature of 70 °C, relative humidity of 55–60%, air speed of 1.5 m/s.
2. Microwave drying (MD): microwave power of 500 W.
3. Vacuum freeze-drying (VFD): pre-freeze temperature of −35 °C, pre-freeze time of 4 h, drying vacuum degree of 60 Pa, heating plate temperature constantly

increasing from 0 °C to 80 °C, set temperature rise time of 6 h, cold trap temperature of −40 °C.

4. Microwave vacuum freeze-drying (MVFD): pre-freeze temperature of −35 °C, pre-freeze time of 4 h, drying vacuum degree of 60 Pa, microwave power of 480 W, cold trap temperature of −40 °C.

Comparing the edible quality of *Stichopus japonicas* revealed that the VFD *Stichopus japonicas* had the best quality in terms of rehydration ratio, appearance, and flavor, the MVFD *Stichopus japonicas* was slightly worse, and the AD *Stichopus japonicas* was the worst. The VFD *Stichopus japonicas* enjoyed the highest rehydration ratio (10.13), much higher than products from other drying methods. The AD and MD products had low rehydration ratios; especially the AD products had the worst rehydration ratio (5.58). In terms of sensory quality of dried *Stichopus japonicas*, the AD and MD *Stichopus japonicas* had more severe shrinkage and color changes, and their appearance scores were very low; the VFD kept the best appearance, with only slight shrinkage overall and spongy tissue patterns visible on the surface, and its color was the same as before drying. The MVFD *Stichopus japonicas* had a worse appearance and darker color than the VFD *Stichopus japonicas*. The VFD *Stichopus japonicas* maintained the best elasticity after rehydration, with an elasticity value of over 90%; the MVFD was slightly worse than VFD, with an elasticity value of about 5%. The worst was the AD, with an elasticity value below 70%. In terms of smell, the VFD *Stichopus japonicas* enjoyed an intense delicate smell of *Stichopus japonicas*, while the MVFD had a lighter smell; the AD and MD *Stichopus japonicas* did not give the smell of *Stichopus japonicas* because of the formation of the hard shell of the surface.

On the side of nutrient loss, the VFD *Stichopus japonicas* had the highest acidic mucopolysaccharide content (5.02 g/100 g), whereas the MD *Stichopus japonicas* had the lowest acidic mucopolysaccharide content (4.66 g/100 g); the effects of the four drying methods on the acidic mucopolysaccharide content of *Stichopus japonicas* were not significant. It was observed via electron microscopy that the VFD product formed a uniform porous spongy structure with fine and even pores. The MFVD product had poorer pore uniformity than the VFD product did, and the MD and AD products did not have a better pore structure.

Although VFD yielded higher quality products, it had the highest energy consumption and the longest drying time, requiring about 16.5 h to achieve a moisture content of less than 12% in the wet basis of *Stichopus japonicas*. MVFD shortened the drying process by about 8 h compared with VFD because of its higher heating efficiency. The MD rate was the highest and MD was completed within 1.5 h. VFD featured the highest energy consumption, whereas MVFD could save about 60% energy compared to VFD.

Chen et al. (2017) compared the quality of *Apostichopus japonicus* treated with cold air drying (CAD), vacuum freeze-drying (VFD), hot air drying (HAD), and vacuum microwave drying (VMD) processes. The drying conditions used for the different drying methods were as follows.

1. Cold air drying (CAD): vacuum degree of 90 kPa, cold air drying temperature of 19 °C, air speed of 1.7 m/s.
2. Vacuum freeze-drying (VFD): drying vacuum degree of 200 Pa, cold trap temperature of −40 °C, the total drying time of approximately 36 h.
3. Vacuum microwave drying (VMD): microwave power of 2 W/s, drying pressure of 90 kPa, the total drying time of approximately 120 min.
4. Hot air drying (HAD): drying temperature of 45 °C, air speed of 1 m/s, the total drying time of approximately 10 h.

The dry weight rate and rehydration ratio of dried *Stichopus japonicas* after rehydration varied significantly among different drying methods, with VMD having the largest dry weight rate and the lowest rehydration ratio and CAD having the largest rehydration ratio after rehydration. During the rehydration process, all four types of dried *Stichopus japonicas* experienced nutrient losses, and the nutrient losses of VFD and CAD were significantly smaller than those of HAD and VMD. The difference in protein loss between the CAD and HAD *Stichopus japonicas* was significant ($P < 0.05$). HAD had the highest total sugar loss, and VFD had the lowest loss; VFD and CAD were significantly better than HAD and VMD in terms of texture. In conclusion, the CAD and VFD *Stichopus japonicas* were more in line with people's requirements in terms of quality. However, the VFD *Stichopus japonicas* was more costly, whereas CAD was more practical in industrial production.

4 Combined Drying Method of Sea Cucumbers

Every method has its own advantages and disadvantages when a single drying method is used, such as hot air drying, freeze-drying, microwave drying, and heat pump drying, to dry sea cucumbers. Freeze-drying is an excellent method for obtaining high-quality products; its shortcoming is on the economic side, i.e., the initial capital investment and the high operating costs. Microwave vacuum drying is characteristic of high speed and efficiency, but it is still difficult to determine the drying endpoint and uneven heating. With the heat pump drying method, the falling rate phase of drying lasts too long, and the heat transfer efficiency at this phase of drying is limited. The excessive drying temperature, the product's rehydration nature, and the product appearance from hot air drying are all far from satisfactory. In order to make up for the shortage of any single drying method and obtain higher quality dried sea cucumbers at a lower processing cost, the combined drying technology can be used. Based on material properties, two or more drying methods are utilized in the combined drying technology following the principle of complementary advantages, drying materials in stages. This technology can achieve lower production costs, reduce energy consumption, improve product quality, and prevent nutrient losses.

4.1 Combined Heat Pump-Hot Air Drying Technology of Sea Cucumbers

The authors' research team studied the drying process of the combined heat pump-hot air drying for sea cucumber. The dried sea cucumbers obtained from the processing exhibited better appearance quality, and the drying time was shortened compared with that in heat pump drying, so the combined drying was better than any of the single drying methods. The sea cucumbers were successively dried in the heat pump and hot air drying equipment with a gradient temperature rise. The air speed of heat pump drying was 1.5 m/s, and the sea cucumbers were dried at 13 °C, 18 °C, 23 °C, 28 °C, 34 °C, and 39 °C for 1 day, respectively. Then, they were transferred to the hot air drying oven under the drying temperatures of 45 °C, 50 °C, and 55 °C until the moisture content of sea cucumber reached (13 $\pm$ 1) %. Within the 1-day drying cycle, the sea cucumbers were taken out after drying for a period of time, sealed, and packed for "moisture balance" treatment, which made the moisture in the sea cucumbers evenly distributed inside and outside.

The water loss rate throughout the drying process was relatively low. At 13 °C, the sea cucumbers were dried through a preheating stage, a short constant rate stage, and a falling rate stage. Due to the "water balance" treatment on the first day of drying, the water within the sea cucumbers was redistributed evenly, so the peak rate appeared again on the second day of drying; this cycle continued until the end of drying. The programmed heating and drying enabled the sea cucumbers to maintain the same rate from the third day of drying onward. Under this condition, the sea cucumbers had the longest possible time in the preheating and constant rate stages, making it possible to eliminate the residual stress generated during the drying process in time. The sea cucumbers were not easily deformed through stretching, twisting, stressing, etc. and shrank uniformly in all directions without the phenomenon of local undercutting.

However, some studies have also shown that the quality of dried sea cucumber processed with heat pump-hot air was lower than that of sea cucumber processed with a heat pump. Ma et al. (2015) used cold air drying and combined cold air-hot air drying methods to dry sea cucumbers and found that although the drying time required by the combined cold air and hot air drying method was 20% shorter than that by cold air drying, and the drying rate was significantly higher; the sensory quality of the former was significantly worse than that of cold air drying; the lower the cold air drying temperature, the better the sensory quality of the sea cucumber; and the sensory quality of sea cucumber was best when the cold air drying temperature was 8 °C. The study applied heat pump drying temperatures of 8, 15, and 25 °C, respectively, and the cold air speed was 6.5 km/h; during the cold air-hot air combined drying, cold air drying was first performed for several hours, and then hot air drying was conducted until the end of drying, with experimental conditions of the combined drying as follows:

1. Cold air temperature of 8 °C, air speed of 6.5 km/h, the drying time of 264 h; hot air temperature of 45 °C, air speed of 9.0 km/h, the drying time of 120 h.
2. Cold air temperature of 15 °C, air speed 6.5 km/h, the drying time of 240 h; hot air temperature of 45 °C, air speed of 9.0 km/h, the drying time of 96 h.
3. Cold air temperature of 25 °C, air speed of 6.5 km/h, the drying time of 168 h; hot air temperature of 45 °C, air speed of 9.0 km/h, the drying time of 96 h.

When the cold air drying was used alone, the drying rate was higher at the early stage of drying; after 144 h of drying, drying entered the falling rate stage. The drying rate was directly related to the drying medium temperature. When the combined cold air-hot air drying was used, the rate of the hot air drying stage was significantly higher than that of cold air drying. Given the same hot air temperature, increasing the cold air temperature could significantly shorten the drying time. When the cold air drying temperature was 8 °C, the total drying time of combined drying toward the final moisture content was 384 h. In contrast, when the cold air drying temperature was 25 °C, the total drying time of combined drying toward the final moisture content was approximately 280 h. The drying time was shortened significantly. Although drying efficiency increased with rising drying temperature, the product quality declined among the dried sea cucumbers after combined drying; the sample's sensory quality was best when the cold air drying temperature was 8 °C after rehydration.

4.2 Combined Hot Air-Microwave Vacuum Drying of Sea Cucumbers

Generally, the drying efficiency is low during the later stage of hot air drying, in which it would take two-thirds of the whole process to remove the remaining one-third of the moisture. Compared to hot air drying, microwave drying has so many advantages in the final drying stage that could be applied to materials at the falling rate drying period or with low moisture, eventually reducing the drying time. The authors' research team evaluated the drying process of sea cucumber with the combination of hot air and vacuum microwave and figured out that although it was still difficult for interior water to escape at the final stage of drying, most of the moisture had evaporated through the microwave, accelerating the overall drying efficiency (Xue 2015). In addition, it was concluded that the combined drying conditions for the sea cucumbers were that the sea cucumbers were first dried by hot air at 90 °C until the moisture content (on a dry basis) reached approximately 80% and then dried through microwave vacuum drying at 1.3 kPa until the moisture content reached below (13 ± 1) %. This drying process could be completed within 142 min.

In hot air drying, the rate of internal water in sea cucumber transferring to the surface was lower than the evaporating rate of surface water, causing a falling rate, as well as a moisture gradient between the inside and the outside of the sea cucumber.

However, microwave drying led to a huge "pumping effect," which would quickly transport the internal water to the outside, significantly improving the drying rate. These results indicated that microwave vacuum drying was efficient enough to remove the remaining one-third of the moisture left by hot air drying and was responsible for the high drying rate even at the drying endpoint. The drying rate had achieved a peak value near the drying endpoint when the moisture content was about 45%. From then on, a slight "puffing effect" had also occurred, but to a lower extent than that in the case of vacuum microwave drying.

4.3 Combined Heat Pump-Microwave Drying of Sea Cucumbers

The microwave could also be applied to heat pump drying to enhance drying efficiency and product quality. Song et al. (2009) compared the drying efficiency of the combined heat pump-microwave drying with that of the heat pump drying. Firstly, the sea cucumber was dried to different moisture content by heat pump (temperature 30 °C, humidity 30%, wind speed 1 m/s) and then dried to 13% moisture content by microwave vacuum (microwave power 230 W, vacuum degree 0.060 MPa). The drying conditions of different drying methods were as follows:

1. Hpd60% + mv13%: heat pump drying to 60% moisture content, microwave vacuum drying to 13% moisture content.
2. Hpd50% + mv13%: heat pump drying to 50% moisture content, microwave vacuum drying to 13% moisture content.
3. Hpd40% + mv13%: heat pump drying to 40% moisture content, microwave vacuum drying to 13% moisture content.
4. Hpd30% + mv13%: heat pump drying to 30% moisture content, microwave vacuum drying to 13% moisture content.
5. Hpd13%: heat pump drying to 13% moisture content.

The higher the moisture content of the material at the drying transition point, the shorter the total drying time. Obviously, this is because the higher the moisture content of the material at the conversion point, the shorter the heat pump drying time required, thus shortening the total drying time of the material. When the moisture content of materials at the drying transition point is 60%, 50%, 40%, and 30% respectively, the total drying time under the same microwave drying conditions is 134 min, 250 min, 400 min, and 558 min respectively.

Based on the comparison of the rehydration rate, shrinkage, and appearance quality, it was found that the sea cucumbers only dried by the heat pump method had the smallest shrinkage and a large rehydration ratio, but poor morphological characteristics. The sea cucumbers experienced a similar quality after being dried by the combined drying methods including (1), (2), and (4), and the shrinkage and sensory quality had been improved when compared to those from the heat pump drying. Furthermore, the rehydration ratio of the dried sea cucumbers from the combined drying method (3) (HPD 40% + MV 13%) increased by 14.3%, and the overall drying time was shortened by 400 min. Meanwhile, the sensory quality of the dried sea cucumbers was the best.

4.4 Combined Freeze-Microwave Drying of Sea Cucumbers

Although the vacuum freeze-drying method could obtain dried sea cucumber products with the best quality, it needs enormous energy consumption. Microwave vacuum drying, characteristic of fast heating, and low evaporation point could not only remove moisture quickly but also maintain the high-quality of dried sea cucumber. The combined method, which first dries by freeze-drying and then uses microwave drying, could not only shorten the freeze-drying process but also reduce the dry load of the microwave equipment. Furthermore, the porous structure formed in freeze-drying features a slighter deformation, which is conducive to maintaining the appearance of the product.

For example, Duan et al. (2009) subjected sea cucumbers to the combined drying with both vacuum freezing and vacuum microwaving and compared this combined method with other single drying methods. It was found that the combined drying could effectively shorten the freeze-drying time, and there was no significant difference between its products and conventional freeze-dried products.

In that study, a self-designed and independently manufactured multifunctional drying machine was used, which featured two drying chambers sharing a vacuum system and a refrigeration system. The combined drying was compared with four single drying methods including vacuum microwave drying, hot air drying, freeze-drying, and microwave drying. The drying conditions of the different drying methods are presented as follows:

FD-VMD samples were first dried to specific moisture contents (20%, 30%, 40%, 50%, 60%, and 70%), respectively, in the FD drying chamber. The temperature of the heating plate was set to 60 °C, the cold trap temperature was −40 °C, and the air pressure was controlled at 50 Pa. Then the samples were dried until the moisture content reached 7% in the microwave drying chamber with a power of 500 W and air pressure of 20 kPa.

AD: The sea cucumbers were dried by hot air (60 °C) with an air speed of 2 m/s and relative humidity of 20%, and the final moisture content was 7%.

FD samples were dried by freeze-drying with a heating plate temperature of 60 ° C, a cold trap temperature of −40 °C, and air pressure of 50 Pa. The final moisture content was also 7%.

MD samples were dried by microwave with a power of 500 W at atmospheric pressure. The final moisture content was 7%.

VMD samples were dried by microwave with a power of 500 W, a cold trap temperature of −40 °C, and an air pressure of 20 kPa. The final moisture content was 7%.

The results showed that the drying rate of FD was the lowest, the falling rate of the drying stage began from 35% moisture content, and this stage took such a long time that it almost accounted for half of the entire FD process. Consistent with FD, AD also had a falling rate during the drying stage. This was because, during hot air drying, the tight binding of water and collagen bundles in the sea cucumber caused hardening crusts on the surface, which could hinder water from evaporating. Furthermore, AD and FD would sustain 8 h and 18 h, respectively. However, the drying rate of MD and VMD was much higher, which only took 150 and 120 min, respectively. Those drying methods did not experience the falling rate drying stage, mainly due to the high efficiency of microwave heating. If the microwave was used before the falling rate drying stage of FD, it would efficiently shorten the drying time. That was also the key to the combined drying.

In terms of nutritional quality, the acidic mucopolysaccharide in the sea cucumbers had dramatically decreased in AD and MD, while it experienced no change in FD-VMD when compared to FD. For energy consumption, FD had the highest, while AD with poorer product quality had the lowest. Although the energy consumption of the FD-VMD combined drying was much higher than that of AD or MD, it was still about 40% lower than that of FD. Considering the slight changes in the rehydration rate, acidic mucopolysaccharide content, and appearance, such a low energy consumption implied that the FD-VDM combined drying could replace FD to a certain extent.

In another study, Duan et al. (2007) used a three-factor central composite experiment design to optimize FD-VMD conditions. The results showed that the optimal parameters were as follows: moisture content of 40% at shift point, air pressure of 20–30 kPa, and microwave power of 350–450 W. These conditions would shorten half of the FD process, while the rehydration rate was close to that of FD products.

5 Conclusion

This chapter has discussed in detail the preprocessing of dried sea cucumbers and the influence of the drying process on the quality of sea cucumbers. The preprocessing methods such as water boiling, salt water boiling, pickling, ultrasonic treatment, enzyme treatment, and soaking could change the internal structural state of the sea cucumber body wall, which are conducive to the formation of special product structure and rehydration characteristics. In addition, compared with single drying methods such as hot air drying, freeze-drying, microwave drying, and heat pump drying, combined drying technologies such as heat pump-hot air drying, heat pump-microwave drying, and freeze-microwave drying could obtain better quality dried sea cucumbers at lower processing cost.

References

Chen Z, Wang M, Zhang J, Hou Z, Jin Y, Wang G, Zhang Y, Liu F (2017) Optimization of cold-air drying of *Apostichopus japonicus* and comparison with other drying methods. Food Sci 38:196–203

Duan X, Zhang M, Arun S, Mujumdar (2007) Study on a combination drying technique of sea cucumber. Dry Technol 25:2011–2019

Duan X, Zhang M, Li XL, Arun S, Mujumdar (2008) Ultrasonically enhanced osmotic pretreatment of sea cucumber prior to microwave freeze-drying. Dry Technol 26:420–426

Duan X, Zhang M, Zhu W (2009) Study on freeze-drying-microwave hybrid drying technology of sea cucumbers. Pack Food Mach 27:36–41

Duan X, Zhang M, Arun S, Mujumdar, Wang S (2010) Microwave freeze-drying of sea cucumber (*Stichopus japonicus*). J Food Eng 96:491–497

Li ZJ, Xue CH, Dong P (2005) A processing method of dried sea cucumber. China Patent. Appl. 1559296

Ma X, Zhang S, Hong B, Li X, Wu L, Xie Z (2015) Comparison of sea cucumber dried by cold air and combination of cold and hot air. J Dalian Ocean Univ 30:536–539

Mu G, Zhang G, Shao L (2007) Preliminary study on drying sea cucumber by the heat pump. Fish Modern 34:47–50

Song Y, Zhang G, Wang C, Wang N, Wang J, Wang L (2009) Preliminary study on combined heat pump and microwave vacuum drying of sea cucumber. Fish Modern 36:47–51

Sui SS, Yin YS (2016) Optimization of processing technology of dried instant sea cucumber. Storage Process 16:41–44

Xue C (2015) Intensive processing of sea cucumber theory and technology. Science Press, China

Yanara T, José MB, Luis AS, Rodrigo ED, Marco FG, Roberto AQ (2020) Effect of electrohydrodynamic pretreatment on drying rate and rehydration properties of Chilean sea cucumber (*Athyonidium chilensis*). Food Bioprod Process 123:284–295

Yuan W, Liu H, Wang X, Zhang M, Meng X, Xia X, Liu X, Zhang Y, Hu W (2010) Research on vacuum freeze-drying technology for sea cucumber *Apostichopus japonicus*. Shandong Sci 23: 67–70

Yun X, Yang H, Zhang J, Ge H (2006) Study of freeze-dried sea cucumber pretreatment process. Sci Technol Food Ind 27:103–105

Zhang G, Mu G, Wang J, Zhang Q (2012) Microwave vacuum drying process of sea cucumber. J Dalian Ocean Univ 27:186–189

Zhang W, Yang B, Cao Y (2017) Optimizing of *Stichopus japonicas* vacuum freeze-drying craft through response surface method. Food Technol 42:165–169

Zhang F, Zhang X, Li S, Li X, Han Y, Cao Y (2018) Effect of different drying methods on the quality of *Stichopus japonicas*. Food Mach 34:209–212

Zhu C, Zhang G, Zhang Q, Huo J, Mu G, Deng W (2015) Microwave vacuum drying process of semi-dry salted sea cucumber. Agric Eng 5:56–59

Ready-to-Eat Sea Cucumber Products and Collagen Stabilization Technology

Airong Jia, Miansong Zhang, Xue Liu, Tingting Cui, Hu Hou, and Changheng Liu

Abstract The ready-to-eat sea cucumber, prepared by advanced processing technologies such as high-temperature short-time sterilization and vacuum cooling, modified atmosphere packaging of high barrier materials, and stepwise secondary sterilization, has become a trend in sea cucumber processing, thanks to its convenience of consumption. In addition, it not only retains the original appearance and flavor of sea cucumber, but this process also results in a relatively small loss of nutrients and bioactive compounds. However, the body wall of ready-to-eat sea cucumber is still self-degraded during storage at room temperature due to the inactivation of endogenous enzymes and microorganisms, which would limit the development of the ready-to-eat sea cucumber sector. Cross-linking against non-enzymatic degradation can improve the storage instability of ready-to-eat sea cucumber. Thus, it is of great economic significance for the sea cucumber industry to explore the stabilization mechanism of cross-linking on ready-to-eat sea cucumber collagen. This chapter mainly summarizes the definition, classification, and processing technologies pertaining to ready-to-eat sea cucumber, the changes in ready-to-eat sea cucumber during processing and storage, as well as the effect of cross-linking on the storage stability of ready-to-eat sea cucumber and its mechanism.

A. Jia · X. Liu · T. Cui · C. Liu (✉)
Biology Institute, Qilu University of Technology (Shandong Academy of Sciences), Jinan, Shandong, P. R. China
e-mail: jiaar@sdas.org; liuxue@sdas.org; tingtingcui@sdas.org; liuchh@sdas.org

M. Zhang
Advanced Marine Biomanufacturing Laboratory, Centre for Marine Bioproducts Development, College of Medicine and Public Health, Flinders University, Adelaide, SA, Australia
e-mail: zhangms@sdas.org

H. Hou
College of Food Science and Engineering, Ocean University of China, Qingdao, Shandong, P. R. China
e-mail: houhu@ouc.edu.cn

C. Xue (ed.), *Advances in Sea Cucumber Processing Technology and Product Development*, Advances in Marine Bioprocesses and Bioproducts,
https://doi.org/10.1007/978-3-031-16512-2_7

Keywords Ready-to-eat sea cucumber · Processing technology · Non-enzymatic degradation · Stabilization mechanism · Cross-linking · Collagen · Structure · Endogenous and exogenous factors

1 Introduction

The ready-to-eat (RTE) sea cucumber is a convenient product that has become a trend in the research and development of sea cucumber products. Compared with dried sea cucumber, it can retain the original appearance and flavor of sea cucumber with negligible loss of nutrients and bioactive substances (Peng et al. 2015a, b). In the processing of RTE sea cucumber, the endogenous enzymes and microorganisms are inactivated, but will still deteriorate during storage because of significant non-enzymatic self-degradation, constraining the development of the RTE sea cucumber sector (Sun 2021). This chapter mainly summarizes the changes in RTE sea cucumber during processing and storage and the effect of cross-linking on the storage stability of RTE sea cucumber.

2 Processing of RTE Sea Cucumber

The sea cucumber is one of the "Eight Treasures of the Sea" in Chinese history, as famous as abalone and shark fin. Due to their biological characteristics, sea cucumbers contain a variety of enzymes, which are prone to result in "self-melting" after harvesting. Since ancient times, sea cucumbers have been stored and sold in the form of dried products after being boiled, salted, and dried. In addition to dried sea cucumbers, sea cucumber products on the Chinese market also include a small amount of rehydrated fresh, frozen, and RTE sea cucumbers.

The dried sea cucumber is currently the most recognized type of sea cucumber product on the Chinese market and the most representative sea cucumber product in international trade. Due to the low moisture content (<13%), dried sea cucumber products are very convenient for transportation and long-term preservation, making them free from regional and seasonal constraints. Although the products retain their good shape and have long shelf life, they have been processed under severe conditions, such as long-time high-temperature drying, and need complex reprocessing before being edible. Those processes will cause a partial loss of nutrients in the sea cucumber. As a dried sea cucumber product, frozen sea cucumber has attracted increasing attention in recent years. It can effectively overcome problems such as damage to water-soluble and heat-sensitive nutrients caused by traditional processing methods. However, there are still certain problems with freeze-dried sea cucumber, such as the need for high lump-sum investment in equipment for production and consumers' frequent questions about cost performance. For these reasons, the RTE cucumber has come into being. It is a kind of RTE food that can be eaten after opening the package with simple processing or no processing at all. This avoids cumbersome secondary processing such as soak, desalination, seasoning, and nutritional loss caused by improper processing. At the same time, the RTE sea cucumber retains the original form of the sea cucumber to the greatest extent, so it is more readily accepted by consumers (Hou et al. 2015; Hong 2005).

2.1 Categories of RTE Sea Cucumber

The RTE sea cucumber is a convenient food that is made of fresh sea cucumbers, dried sea cucumbers, or salted sea cucumbers. It can be eaten directly or after simple cooking. At present, RTE sea cucumbers sold in the Chinese market are mainly divided into the following categories: ready-to-eat sea cucumber, high-pressure cooked frozen sea cucumber, rehydrated sea cucumber, canned sea cucumber, and fast-to-rehydrate dried sea cucumber. Specific features are presented as follows:

2.1.1 RTE Sea Cucumber

RTE fresh sea cucumber is a snack food that is easy to store and transport and ready to eat after opening. It is produced through the use of fresh sea cucumber as raw material and the adoption of advanced processing technologies such as high-temperature short-time sterilization and vacuum cooling, modified atmosphere packaging of high barrier materials, and stepwise secondary sterilization. Since there is no need to go through salting, repeated rinsing, and boiling during production, the loss of nutrients, especially water-soluble nutrients and biologically active substances, is relatively small. The technical progress developed by Professor Liu Changheng's research team from the Shandong Academy of Sciences won the first prize of the Shandong Science and Technology Progress Award in 2010, representing the pinnacle of sea cucumber processing in China at that time (Liu et al. 2007, 2008; Shandong Homey Aquatic Development Co., Ltd. 2010).

2.1.2 High-Pressure Cooked Frozen Sea Cucumber

The high-pressure cooked frozen sea cucumber is made by processing fresh sea cucumber by gutting, blanching, lime ring removing, cleaning, autoclaving at 110–115 °C for 10–15 min, rehydrating for 1–2 days, quick-freezing, and vacuum packaging. The product features simple processing technologies, low cost, and convenient consumption. The high-pressure cooking process ultimately maintains the original form of sea cucumber. Moreover, under high pressure, the covalent cross-linking between collagen molecules and the covalent bond in the non-helical region is destroyed, resulting in the rapid disintegration of the triple helix structure and the degradation of the main chain into peptides, which solves the problem of digestion and absorption (Zheng and Wu 2014; Yao 2013). Unfortunately, the product still needs to be stored and transported in a low-temperature environment at −18 °C, and secondary processing including thawing and seasoning is also required.

2.1.3 Rehydrated Sea Cucumber

The rehydrated sea cucumber is a product obtained from dried sea cucumber after soaking, desalinating, lime ring removing, repeated pre-cooking, and rehydrating. It needs to be frozen, and the soak time is usually 5–7 days. Due to repeated boiling and rehydration, the nutrition loss of sea cucumber is significant, and the content of water-soluble protein in rehydration solution is 40.40–59.87 mg g^{-1} (Hong et al. 2014). The loss rate of acidic mucopolysaccharide of sea cucumber was also more than 30% under the best cold rehydration conditions (Yuan et al. 2016). Moreover, it needs secondary processing before eating, including thawing and seasoning, which is inconvenient to eat.

2.1.4 Canned Sea Cucumber

The canned sea cucumber is a product using blanched and cleaned sea cucumber as raw material. The sea cucumber and cooking liquor are canned, sterilized at high temperature and high pressure, and refrigerated. The nutrients and flavor change slightly during processing. It maintains the original nutritional value and flavor of sea cucumber to a large extent and can be eaten without secondary processing, so it is relatively convenient. However, during storage, the sea cucumber can easily absorb water, resulting in flaccid or even broken sea cucumber tissue in cloudy liquor. With the extension of sterilization time, the mass loss rate of canned sea cucumber changed gently at first and then increased. When the sterilization time exceeded 14 min, the mass loss rate of sea cucumber changed obviously. When the sterilization time is 12 min, the mass loss rate is 3.9%; when the sterilization time is 16 min, the mass loss rate is more than 8.9%. This is because of the extension of sterilization time, resulting in the main components in the body wall of sea cucumber, such as water-soluble albumin, under the condition of heat, easy to degrade, resulting in a serious loss of quality and quantity (Sang et al. 2013).

2.1.5 Fast-to-Rehydrate Dried Sea Cucumber

The fast-to-rehydrate sea cucumber is a kind of dried sea cucumber product subjected to a low-temperature drying process (Sui and Yin 2016).

Process flow: raw material selecting → eviscerating → first boiling → salting → cleaning → second boiling → heat preserving → low-temperature drying → packaging → finished product. The first boiling temperature is 80 °C, the salting time is 22 h, the soak time is 14 h, and the drying time is 72 h. Fast-to-rehydrate sea cucumber is rehydrated in 60–80 °C hot water for 5 h and then seasoned before eating. Compared with the rehydration of regular mild-salted dried sea cucumber subjected to traditional processes, the fast-to-rehydrate sea cucumber has the advantages of simple rehydration, reduced soak time, larger rehydration ratio, and less nutrient loss. Average families or business professionals only need to put the sea cucumber in a vacuum cup for rehydration, and then convenient, nutritious, and safe sea cucumber can be served anywhere and anytime.

2.2 *Processing Technology of RTE Sea Cucumber*

The "RTE sea cucumber" that can be eaten directly without secondary processing such as soaking, desalting, and seasoning and won the first prize of the Shandong Province Science and Technology Progress Award yielded from the program Key Technology and Industrialization of Ready-to-Eat Seafood Processing in 2010 will be taken as an example in what follows for a detailed explanation.

According to the biological characteristics of sea cucumbers, the technical process of RTE sea cucumber processing is as follows: fresh or live sea cucumber → temporary cultivating → raw material selecting → cleaning → laparotomy and eviscerating → cleaning and finishing → blanching → flavor soaking → high-temperature short-time sterilizing → vacuum cooling → packaging → stepwise secondary sterilizing → finished product (Liu et al. 2007). The key points of operation are introduced as follows:

2.2.1 Requirements for Raw Materials

Fresh sea cucumbers that are not contaminated and display normal development, with a growth cycle of more than 2 and a half years and with a piece weight of over 150 g, can be selected. Sea cucumbers that are deformed or inconsistent in size are eliminated (Liu et al. 2007; Li et al. 2010).

2.2.2 Temporary Cultivation

The sea cucumber is a "scavenger" of the seabed. Sea cucumbers mainly feed on seabed sediments and plankton and have a special fishy smell. Therefore, excreting the dirt from the sea cucumber before processing can help to improve product quality. The temperature and salinity of temporary cultivation are two important factors. Studies have shown that the salinity of seawater at 20 Baume degrees and seawater temperature at about 20 °C is preferred. In addition, using appropriate oxygenation equipment according to the size of the holding pond can facilitate sea cucumbers to expel the waste matter in their abdomen (Liu et al. 2007).

2.2.3 Selection

Selection is to exclude sea cucumbers that are incomplete, with rotting skin or weighing less than 150 g. During picking, sea cucumbers should be handled with care and as quickly as possible. Selected sea cucumbers are respectively placed in seawater according to the sorting results and then sent to the next process. If frozen sea cucumbers are to be processed, they need to be thawed first, and there must be fixed equipment and site for thawing. Thawing methods mainly include water shower and water soaking. The water temperature is generally controlled below 18 °C, and running water is the best option. During the thawing process, hooking, knocking, and smashing must be avoided. Moreover, the sea cucumbers must be processed while thawing to strictly prevent backlog (Liu et al. 2007; Li et al. 2010).

2.2.4 Cleaning

Sediment and lichen are first removed with clean, running seawater. The sea cucumbers are then soaked in seawater and put into clean, running freshwater below 10 °C to remove the residual sediment, lichen, sea salt, and other dirt (Liu et al. 2007; Li et al. 2010; Zhang et al. 2012).

2.2.5 Laparotomy and Evisceration

The cleaned fresh sea cucumbers are cut in the abdomen, and the guts and the lime ring at the mouth of the sea cucumbers are then removed (Liu et al. 2007).

Mechanized impurity removal and screening equipment in automated production is used for sea cucumbers. Circular tubes are arranged in parallel at a specific interval to form a grid sieve plate, which is inclined at a certain angle as a slideway. During operation, sea cucumbers enter the upper end of the slideway through the lifting mechanism and slide freely under gravity. Impurities and visceral tissues fall directly into the lower layer through the gaps between the circular tubes and are then separated. Rotten and broken sea cucumbers will fall out of the gaps during the sliding process due to the short residence time, while fresh sea cucumbers are not affected and quickly pass through the slideway. To prevent normal sea cucumbers from falling, a low-temperature seawater spray is added above the slideway, which stimulates the sea cucumbers to undergo a stress response and promotes hardening of the sea cucumbers, while rotten and broken sea cucumbers will not undergo such stress changes, ensuring that they can still be removed. In actual production, the contact surface of the screening slideway composed of circular tubes is arc-shaped with no edges and corners, so as not to damage the surface of the sea cucumbers and to facilitate ease of sliding. This design not only ensures sea cucumber quality but also provides high screening efficiency without manual operation (Lin et al. 2010).

2.2.6 Cleaning and Finishing

After laparotomy, the sea cucumbers are washed with cold water and drained.

In actual production, flexible bubble cleaning equipment is used. It adopts a water tank structure, with multiple gas pipelines laid at the bottom and 1 mm thick perforated mesh plates mounted above. After compressed air enters the water body, it does not directly act on the surface of sea cucumbers, but generates large air bubbles that then form a large amount of flexible fine air bubbles via the thick perforated mesh plates above. The flexible fine bubbles have a gentle cleansing effect without damaging the sea cucumber surface. To achieve continuous cleaning, a water jet propulsion system is added at the feed end of the cleaning tank, and a material conveying mechanism is mounted at the discharge end. During cleaning, the water jet propulsion system is used to generate a horizontal propulsive water flow to push the water body forward. Sea cucumbers are subjected to the upward force of the bubbles and the horizontal thrust at the same time. Under the action of the joint force, the sea cucumbers roll up and down in the water while moving forward. Meanwhile, any unseen dead parts of sea cucumbers are also thoroughly cleansed through being surrounded by dense fine bubbles. When they reach the front of the water tank, the sea cucumbers leave the cleaning water tank through the conveying mechanism to complete the cleaning process (Yang and Zhang 2018).

2.2.7 Blanching

Sea cucumbers are put into water at a specific temperature to achieve enzyme inactivation, dehydration, and fixation. Due to the individual differences in sea cucumbers, blanching all to a single standard presents a significant challenge. The double-blanching method is therefore adopted: in the first blanching, fresh sea cucumbers are blanched in hot water of 80–85 °C for 3 min and cooled; in the second blanching, "the sea cucumbers" are then soaked in water at 95–100 °C for 4 min (Liu et al. 2007).

In actual production, a baffle-type material hoist is set up at the front end of the continuous cooking equipment as the feeding port. A circulating conveyor belt for material conveying is placed inside the cooking tank, using a stainless steel mesh chain structure equipped with equal-spaced dividers for the conveyor belt. To prevent sea cucumber materials from slipping off the conveyor belt during the blanching process, protective side plates are installed on both sides of the conveyor belt, and there are detachable pressing mesh plates on the top to prevent sea cucumbers from floating on the water surface. During cooking, it is only necessary to pre-fill the water tank (the water level must exceed the pressing mesh plates) and preheat to the specified blanching temperature. At the same time, the corresponding running speed of the conveyor belt must be set according to the cooking time required for the sea cucumber. During cooking, sea cucumber materials enter the cooking tank through the feeding mechanism and undergo the blanching process as they pass through the water on the conveyor belt. To remove the floating foam, a compressed air pump with timed activation is used. The airflow enters the tank from a pipeline through an air pipe installed on the inner sidewall of the tank, with the air then being expelled onto the water surface through nozzles at different positions on the pipe to guide the floating foam to the overflow port on the opposite side, thus discharging the foam out of the tank uniformly. The application of continuous cooking equipment can fully realize the automatic cooking process (Xu et al. 2011).

2.2.8 Flavor Soaking

Blanched fresh sea cucumbers are placed in seasoning (or seasoning liquid) for flavoring treatment. Since sea cucumber is a high-end nutritious food, and considering consumer groups and taste, seasoning should be mild, fresh, and delicate, without any significant change to the original flavor of sea cucumber (Liu et al. 2007).

2.2.9 High-Temperature Short-Time Sterilization

Sterilization is carried out in the cooking tank through a two-step steaming process: the first steaming is at 105 °C, for 4 min; the second steaming is at 135 °C and stops when the F value reaches 2.9. During this process, the sea cucumber is rapidly dehydrated, from 92% of the raw material moisture content to 78% after cooking. At the same time, the total number of bacteria in the material drops sharply, achieving a sound sterilization effect. The temperature and pressure curves in the tank and in the material center during high-temperature short-time sterilization are shown in Fig. 1 (Liu et al. 2007; Shandong Homey Aquatic development 2010).

2.2.10 Vacuum Cooling

To reduce product temperature as soon as possible, vacuum cooling technology is used after high-temperature sterilization. When the vacuum degree in the container is close to -0.9 kg/cm^2, the material temperature drops sharply. Thus, the material immediately drops from a high temperature. The temperature and pressure curves in the tank and material in the center are shown in Fig. 2 (Liu et al. 2007; Shandong Homey Aquatic development 2010).

2.2.11 Packaging

Packing materials relate to not only product quality but also to its appearance. When selecting packaging materials, the shelf life of products, soluble substances of packaging materials during sterilization, and the appearance of products are

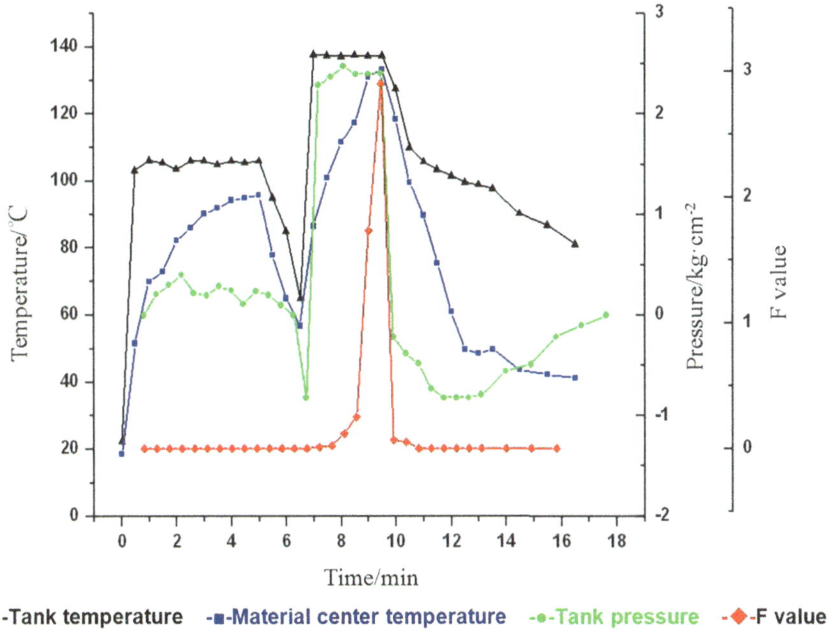

Fig. 1 Temperature and pressure curves of RTE sea cucumber during high-temperature short-time sterilization

considered. Therefore, nylon/aluminum oxide/high-temperature resistant polypropylene materials have been selected. They exhibit a high barrier property (the oxygen permeability is 0.7 mL/24 h m^2 at 20 °C), which not only meets the requirements for the shelf life of the product, high-temperature sterilization, and other related aspects but also has the advantage of visual transparency. The product will be visible through openings in the package. During packaging, the air in the bag is removed and the bag is then filled with 99.97% pure nitrogen to ensure vacuum and nitrogen-filled packaging of RTE sea cucumber.

2.2.12 Stepwise Secondary Sterilization

Stepwise sterilization technology is a kind of new technology that is suitable for processing all types of fresh instant foods or semi-finished products and is designed to overcome the shortcomings of conventional methods such as vacuum packaging and high-temperature and high-pressure sterilization. The purpose of stepwise sterilization is to kill the microorganisms introduced in the packaging process after high-temperature short-term sterilization. The method of multistage temperature step-by-step heating and cooling is adopted to keep the central temperature of solid materials basically close to the tank temperature at a lower temperature and reduce the high-temperature sterilization time. Sterilization treatments and mild sterilization

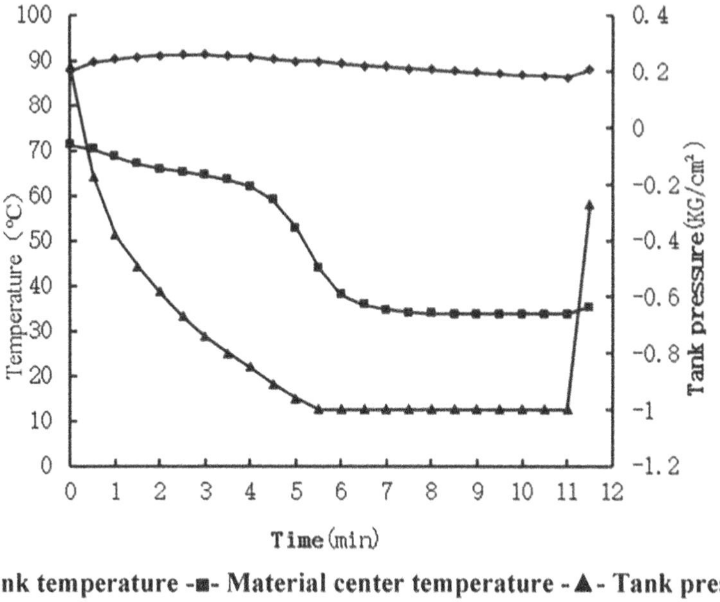

Fig. 2 Curve of vacuum cooling temperature and pressure of RTE sea cucumber. -●- Tank temperature -■- Material center temperature -▲- Tank pressure

Table 1 Quality comparison of RTE sea cucumber under stepwise sterilization and high-temperature and high-pressure sterilization

Processing method	60 days	120 days	180 days
Stepwise sterilization	The sea cucumber is in complete and full shape, with complete and upright thorns on the back and in shiny brown color	Normal	Normal
High-temperature and high-pressure sterilization	The sea cucumber is in complete but not full shape, with soft thorns on the back, in brown color, and with typical taste	Typical flavor, soft thorns and limited chewiness	Typical flavor, loose thorns and limited chewiness

processes such as nitrogen-filled packaging and stepwise sterilization can essentially preserve the quality and nutritional composition of cooked food. The original color, flavor, taste, and appearance of foods change only minimally. This not only solves the problem of quality deterioration for vacuum-packed foods or foods processed at a high temperature and a high pressure but also overcomes the disadvantages of short shelf life and high cost in the circulation of refrigerated and frozen foods.

Stepwise sterilization technology has been used in the processing of sea cucumber products and has been compared with high-temperature and high-pressure sterilization conditions (Table 1). Moreover, a stepwise sterilization formula suitable for sea cucumber has been developed (Table 2, Fig. 3). Stepwise-sterilized products

Table 2 Formula of RTE sea cucumber stepwise sterilization

Item	Temperature (°C)	Time (min)
Temperature rise (Step 1)	30–100	3.5
Sterilization (Step 1)	100	5
Temperature rise (Step 2)	100–110	2.5
Sterilization (Step 2)	110	10
Temperature rise (Step 3)	110–117	1.5
Sterilization (Step 3)	117	13
Cooling (Step 1)	117–80	3
Cooling (Step 2)	80–36.5	7

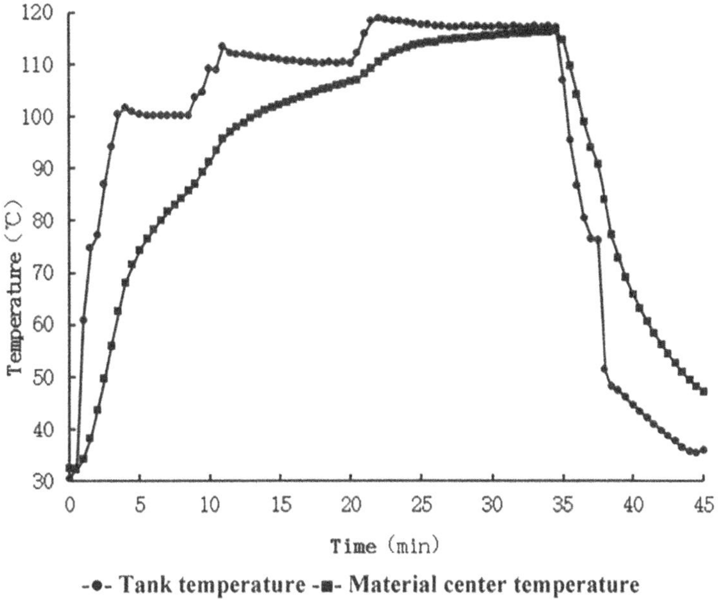

Fig. 3 Comparison of tank temperature and material center temperature in stepwise sterilization. -●- Tank temperature -■- Material center temperature

can be stored effectively and safely for 6 months, while products from high-temperature and high-pressure sterilization experience a change in quality after only 3 months (Liu et al. 2007).

2.3 *Effects of Processing on Nutrient Content and Sensory Quality of RTE Sea Cucumber*

In Sect. 2.2.12, the sensory quality of RTE sea cucumber under stepwise sterilization and under high-temperature and high-pressure sterilization has been compared. This section focuses on comparing the differences in basic nutritional composition, amino acid content, fatty acid composition, and texture between RTE sea cucumbers yielded from advanced processing technologies. These include high-temperature short-time sterilization and vacuum cooling, modified atmosphere packaging of high barrier materials, and stepwise secondary sterilization and RTE sea cucumbers produced under traditional high-temperature and high-pressure sterilization (Yuan et al. 2008; Zheng and Wu 2014; Tang et al. 2015).

2.3.1 Effects on Some Nutritional Components of RTE Sea Cucumber

Temperature has a significant effect on vitamins. The vitamins B_1 and B_2 obtained by high-temperature short-time stepwise sterilization were 1.35 and 1.96 times higher than those obtained by high-temperature and high-pressure sterilization, respectively. High-temperature and high-pressure sterilization conditions had little effect on polysaccharide and saponin content, the primary physiological active substances of sea cucumber. Temperature-insensitive nutrients (such as crude protein) also showed no significant difference under different sterilization methods. In terms of the level of vitamin changes, high-temperature short-time sterilization and stepwise sterilization caused minor damage to the nutritional components of RTE sea cucumber, which exhibited higher nutritional value (Zu 2017) (Table 3).

2.3.2 Effects on Amino Acid Content of RTE Sea Cucumber

In terms of total amino acid content, the amino acid content of RTE sea cucumber treated with high-temperature short-time stepwise sterilization was slightly higher than that treated with high-temperature and high-pressure sterilization by about 1922.23 mg/100 g. Due to the longer time required for mild sterilization, the content of essential amino acids decreased, but the difference was not significant. The actual amino acid contents of the products obtained by high-temperature short-time stepwise sterilization and high-temperature and high-pressure sterilization were 12,651.74 mg/100 g (24.3% of total amino acids) and 12,926.22 mg/100 g (25.78% of total amino acids), respectively. The contents of umami amino acids in products that underwent high-temperature short-time stepwise sterilization were 3.42% higher than those with high-temperature and high-pressure sterilization (Table 4).

In addition, it is significant that hydroxyproline, which is the primary amino acid in collagen, experienced a relatively significant loss at high temperature. The hydroxyproline content in products treated with high-temperature short-time stepwise sterilization was 988.09 mg/100 g. In comparison, the content of hydroxyproline in products treated with high-temperature and high-pressure sterilization was 656.83 mg/100 g. The proline content in products treated with high-temperature short-time stepwise sterilization was relatively low. Under different processing methods, the proportion of hydroxyproline and proline differed. Under high-

Table 3 Table of nutritional components of RTE sea cucumber with different sterilization methods

Item	Moisture /%	Crude protein /%	Polysaccharide /%	Saponin (mg/g)	Vitamin B_1 (mg/100 g)	Vitamin B_2 (ug/100 g)	Vitamin A (ug/100 g)
High-temperature short-time step-wise sterilization	79.72	69.32	3.16	1.97	0.44	2.92	Not detected
High-temperature and high-pressure sterilization	78.27	68.14	3.18	1.89	0.32	1.49	Not detected

Note: Except for water content, all other nutrients are measured on a dry basis

Table 4 Amino acid content of RTE sea cucumber with different processing methods

Symbol	Designation	High-temperature short-time stepwise sterilization mg/100 g (dry basis)	High-temperature and high-pressure sterilization mg/100 g (dry basis)
TAU	Taurine	0.00	0.00
HYPRO	Hydroxyproline	988.09	656.83
ASP	Aspartate[a]	5325.11	4936.56
THR	Threonine	2715.35	2544.33
SER	Serine	2541.87	2302.35
GLU	Glutamic acid[a]	10,024.18	10,039.06
PRO	Proline	1531.16	3394.75
GLY	Glycine[a]	8822.49	7022.19
ALA	Alanine[a]	4408.90	3687.84
CYS	Cystine	245.85	233.94
VAL	Valine[b]	2210.00	2170.98
MET	Methionine[b]	588.33	725.96
ILE	Isoleucine[b]	1764.98	1783.80
LEU	Leucine[b]	2303.22	2361.70
TYR	Tyrosine	1214.37	1182.29
PHE	Phenylalanine[b]	1289.79	1348.22
ORN	Ornithine	0.00	0.00
LYS	Lysine[b]	1780.07	1991.22
HIS	Histidine	221.25	253.51
TRP	Tryptophan[b]	–	–
ARG	Arginine[a]	4088.12	3505.37
Total amino acids		52,063.13	50,140.90
Essential amino acid content		12,651.74	12,926.22
Proportion of essential amino acids in total		24.30%	25.78%
Umami amino acid content		28,259.90	25,503.18
Proportion of umami amino acids in total		54.28%	50.86%

[a]Umami amino acid
[b]Essential amino acid

temperature short-time stepwise sterilization conditions, it was 0.6, while under high-temperature and high-pressure situations, it was only 0.2. The body wall of sea cucumber is mainly composed of collagen and acidic mucopolysaccharide. The content of proline and hydroxyproline in collagen was the highest among all kinds of proteins. These amino acids are cyclic amino acids, which lock the whole collagen molecule and make it difficult to open, giving collagen slight elasticity and muscular tensile strength. Therefore, the change in the ratio of the two is bound to affect the elasticity and tensile strength of collagen, thus affecting product taste. Combined with sensory evaluation, products with high hydroxyproline/proline value had better elasticity and chewiness.

2.3.3 Effects on Fatty Acid Composition of RTE Sea Cucumber

The saturated fatty acid content of fresh sea cucumber under high-temperature and high-pressure sterilization was higher than the content of monounsaturated fatty acid and polyunsaturated fatty acid. The ratio of *n*-6 fatty acids/*n*-3 fatty acids was decreased by high-temperature and high-pressure sterilization (Table 5).

2.3.4 Effects on Texture of RTE Sea Cucumber

The TPA texture analysis method was established in 1967 and is suitable for general texture measuring instruments. The TPA texture test, also known as the two-bite test (TBT), mainly compresses solid and semi-solid samples twice by simulating the chewing movement of the human mouth. The test is connected to a computer, and the texture test curve is output via the interface. It can be seen from the texture data that the hardness, cohesiveness, elasticity, and chewiness parameters of sea cucumber produced by high-temperature short-time stepwise sterilization were slightly higher than those produced by high-temperature and high-pressure sterilization. Combined with the sensory evaluation results, the overall average score of the products from high-temperature short-time stepwise sterilization was higher than that of the products from high-temperature and high-pressure sterilization, and the RTE sea cucumber was tougher, with better chewiness, taste, and quality (Table 6).

Thanks to the high-temperature short-time sterilization technology for solid raw materials, modified atmosphere packaging of high barrier materials, stepwise secondary sterilization, and other advanced processing technologies, the "RTE sea cucumber" has come into being. The product has a complete and full appearance, with less nutritional loss than previous processing methods, good taste, and fine flavor, and is convenient to eat. It is obviously superior to comparable marine foods on the market.

3 Quality Changes of RTE Sea Cucumber During Storage

High-temperature steam can inactivate the endogenous enzymes and microorganisms in sea cucumbers. However, previous studies reported that ready-to-eat sea cucumber can also degrade after a period of storage (Peng et al. 2015a). Moreover, the non-enzymatic degradation of RTE sea cucumber might lead to a change of textural properties, microstructure, ammonia nitrogen content, free hydroxyproline

Table 5 Fatty acid composition of RTE sea cucumber with different processing methods

Fatty acid	High-temperature short-time stepwise sterilization g/100 g	High-temperature and high-pressure sterilization g/100 g
Caproic acid[a]	0.1048	0.1639
Myristic acid[a]	1.4171	1.5581
Myristoleic acid[b]	0.7701	0.7480
Pentadecanoic acid[a]	0.1942	0.4169
Palmitic acid[a]	11.7123	13.4843
Hypogaeic acid[b]	10.4152	13.6768
Margaric acid[a]	0.8955	1.2406
Heptadecenoic acid[b]	–	–
Stearic acid[a]	6.9208	7.4772
Oleic acid[b]	14.7149	7.1460
Linoleic acid[c, d]	4.8717	3.0514
Linolenic acid[c,e]	1.1836	1.4073
Twenty carbonate[a]	2.0190	1.6279
Eicosenoic acid[b]	13.8764	8.4013
Eicosadienoic acid[c, d]	2.8304	1.6852
Heneicosanoic acid[a]	1.1420	1.5192
Eicosatrienoic acid[c, e]	13.7689	12.8494
Arachidonic acid[c, d]	–	–
Dodicarbonate[a]	6.9823	14.9368
Tetracosanoic acid[a]	0.5471	1.0998
Tetracosenoic acid[b]	5.6335	7.5100
Saturated fatty acids	31.9351	43.5247
Monounsaturated fatty acids	45.4101	37.4821
Polyunsaturated fatty acids	22.6546	18.9933
n-6 fatty acids/n-3 fatty acids	0.5151	0.3322
Monounsaturated fatty acids %	45.41%	37.48%
Polyunsaturated fatty acids %	22.65%	18.99%

[a]Saturated fatty acid
[b]Monounsaturated fatty acid
[c]Polyunsaturated fatty acid
[d]*n*-6 fatty acids
[e]*n*-3 fatty acids

content, and water distribution during storage, which could be related to the fracture of collagen fiber and the degradation of collagen. Non-enzymatic degradation of RTE sea cucumber would seriously limit the development of the sea cucumber

Table 6 Texture parameters and sensory scores of RTE sea cucumbers with different processing methods

Sterilization method	Hardness	Springiness	Cohesiveness	Chewiness	General sensory evaluation
High-temperature short-time stepwise sterilization	8.369	0.604	0.508	2.219	
Sensory score	6	7	7	7	27
High-temperature and high-pressure sterilization	7.855	0.325	0.483	1.644	
Sensory score	5	5	4	5	21

processing industry. Therefore, it is important to explore the non-enzymatic softening mechanism of RTE sea cucumber during storage (Chen et al. 2016).

3.1 Changes of Texture Characteristics of RTE Sea Cucumber During Storage

Texture profile analysis (TPA) was used to evaluate the textural properties and sensory acceptability of sea cucumber body wall (Zhang et al. 2017). Hardness, chewiness, springiness, cohesiveness, and restorability could be used to evaluate the textural characteristics of sea cucumber (Lin et al. 2019). It was found that sea cucumbers treated with high temperature and high pressure would degrade and soften during storage at 4 °C (Chen et al. 2016). After storage for 160 days, the hardness of sea cucumber decreased from 295.85 gf to 15.67 gf, and the chewiness decreased from 227.08 gf^2 s to 9.18 gf^2 s (Chen et al. 2016). The cohesiveness initially increased but then decreased sharply during storage and reached the maximum value (93.86 gf s) at 75 days (Chen et al. 2016). Ready-to-eat sea cucumber was stored at 5 °C, 10 °C, and 15 °C, respectively (Hou et al. 2014). The results of TPA showed that the hardness of sea cucumber decreased from 2744.064 g to 1350.976 g after storage at 15 °C for 42 days. After storage at 5 °C for 74 days, the hardness decreased to 1064.175 g (Hou et al. 2014). This indicated that the hardness of RTE sea cucumber decreased with the increase of storage time. The rate of decrease in hardness slowed down when storage temperature decreased (Hou et al. 2014). In addition, a negative correlation was found between the changes of textural properties (springiness, chewiness, and cohesiveness) and the storage time (Hou et al. 2014). Compared with storage at 5 °C, the changes of springiness for RTE sea cucumber were significant during storage at 10 °C and 15 °C (Hou et al. 2014).

Zhao et al. (2015b) reported that the hardness and chewiness of ready-to-eat sea cucumber decreased sharply after storage at 37 °C for 10 days, which might be attributed to the protein structure of sea cucumber body wall being destroyed and collagen degrading during storage. It was reported that the hardness of RTE sea cucumber decreased from 374.60 gf to 35.00 gf after storage at 37 °C for 33 days, and the cohesiveness, chewiness, and restorability were also significantly reduced (Peng et al. 2015a, b). In the first 24 days, the springiness of sea cucumber decreased from 1.01 to 0.70 (Peng et al. 2015a, b). It was demonstrated that the body wall of RTE sea cucumber deteriorated significantly during storage at 37 °C, and the hardness decreased from 6.23 ± 0.21 N to 0.83 ± 0.17 N after storage for 30 days (Sun 2021). The hardness of sea cucumber decreased by 53.4% on the fifth d of storage, but the rate of decrease slowed with the increase of storage time (Sun 2021). In order to explore the effect of storage temperature on the body wall of RTE sea cucumber, sea cucumbers were treated with high temperature and high pressure and then stored at 4 °C, 20 °C, 37 °C, 50 °C, and 60 °C, respectively (Lin et al. 2019). The results of TPA showed that with the increase of storage time, the springiness of sea cucumber body wall decreased during storage at 20–60 °C, and the rate of decrease was positively correlated with the storage temperature, while there was no significant correlation between the storage temperature and the changes of cohesiveness and restorability (Lin et al. 2019).

In addition, it was reported that moisture content could affect the textural properties of the body wall of ready-to-eat sea cucumber (Lin et al. 2021). RTE sea cucumbers with different moisture contents were prepared by vacuum drying method (Lin et al. 2021). The hardness and chewiness of RTE sea cucumber gradually decreased during storage at 37 °C for 30 days, indicating that the difference of moisture content did not change the non-enzymatic degradation of the body wall of RTE sea cucumber (Lin et al. 2021). Half-life (T_{50}) was calculated, and it was found that the T_{50} values of hardness and chewiness decreased significantly with the increase of moisture content, which proved that the moisture content was significantly negatively correlated with the hardness and chewiness (Lin et al. 2021).

In conclusion, storage temperature and moisture content could affect the storage stability for the body wall of ready-to-eat sea cucumber, and TPA parameters decreased significantly with the increase of storage temperature and moisture content. Thus, moisture content and storage temperature could be decreased to reduce the degree of damage to the tissue structure of RTE sea cucumber body wall, slowing down the deterioration rate of texture and prolonging its shelf life.

3.2 Microstructure Changes of RTE Sea Cucumber During Storage

Van-Gieson (VG) staining can be used to stain muscle and connective tissues by dying muscle fibers into yellow and collagen fibers into red. Chen et al. (2016) explored the changes of collagen fibers for ready-to-eat sea cucumber during storage at 4 °C using Van-Gieson staining. It was found that the collagen fibers of RTE sea cucumber were slender, clear, and closely arranged at 0 day (Chen et al. 2016). After storage for 30 days, the collagen fibers became disordered, the long collagen fibers fractured at the end of storage, and the structure became loose and disarranged (Chen et al. 2016). Collagen fibers seriously degraded from 90 to 160 days, and the results of VG staining showed no significant difference (Chen et al. 2016). It was demonstrated that the fracture of collagen fiber and degradation of gel structure destroyed the bond between moisture and collagen, resulting in moisture migration, which led to severe deterioration of RTE sea cucumber (Chen et al. 2016).

In addition, sea cucumbers were prepared with high-pressure steam and stored at 37 °C (Peng et al. 2015a, b). The results of VG staining showed that the red long fibers gradually faded and became shorter during storage (Peng et al. 2015a, b). The collagen fibers became loose and disordered and showed a porous structure (Peng et al. 2015a, b). It was indicated that the collagen fibers were destroyed gradually during storage, and the self-degradation of ready-to-eat sea cucumber occurred (Peng et al. 2015a, b). Sun (2021) observed the changes in the microstructure of RTE sea cucumber during storage at 37 °C by scanning electron microscopy (SEM). The collagen fibers of RTE sea cucumber were dense at 0 day (Sun 2021). After storage for 30 days, the collagen fibers were broken and disorganized, showing a foam-like structure (Sun 2021). Moreover, SEM analysis showed that the body wall tissues were ruptured on a large scale after storage at 37 °C for 15 days with obvious observable areas of agglutination and foam-like structures, while the collagen fibers were destroyed, leading to degradation of the body wall of RTE sea cucumber (Zhu et al. 2022a).

SEM analysis also revealed that storage temperature had a significant effect on ready-to-eat sea cucumber at the same storage time and that the degree of degradation was positively correlated with storage temperature (Lin et al. 2019). In addition, moisture condition is one of the important factors affecting the storage stability of RTE sea cucumbers. It was found that the collagen fibers in the body wall of RTE sea cucumber with low moisture content showed a network structure (Zhang et al. 2016). The degradation degree of the gel structure of RTE sea cucumber decreased gradually during storage with the decrease of moisture content, indicating that non-enzymatic degradation was closely related to moisture content (Zhang et al. 2016). Lin et al. (2021) studied the effect of moisture content on the microstructure of the body wall of RTE sea cucumber using SEM. It was found that the diameter of collagen fiber bundles declined significantly with the decrease of moisture content and that these gradually formed a dense network structure (Lin et al. 2021). After storage at 37 °C for 20 days, the RTE sea cucumbers with high moisture content showed a porous structure (Lin et al. 2021). Moreover, results of TEM showed that the collagen fiber bundles were destroyed and the triple helix structure of the collagen molecule degraded after treatment at 121 °C and 0.2 MPa (Lin et al. 2021). This further degraded during storage at 37 °C (Lin et al. 2021). The RTE sea cucumbers with high moisture content had obvious fragmented structure, while those with low moisture content still showed a banded structure (Lin et al. 2021). Therefore, non-enzymatic self-degradation of RTE sea cucumber could be effectively controlled by the reduction of storage temperature and moisture content to prolong its shelf life (Lin et al. 2019; Zhang et al. 2016).

3.3 Changes of Ammonia Nitrogen and Free Hydroxyproline Content in RTE Sea Cucumber During Storage

Free hydroxyproline and ammonia nitrogen content could be used to evaluate the quality of sea cucumber to reflect the collagen degradation in sea cucumber body wall (Zhang et al. 2017). It was demonstrated that the free hydroxyproline content of ready-to-eat sea cucumber experienced little change in the first 60 days at 4 °C, but increased significantly from 1.91 μg/mL to 3.52 μg/mL during storage from 60 days to 160 days (Chen et al. 2016). The total hydroxyproline content (0.706–0.823%) was not significant during storage (Chen et al. 2016). The ammonia nitrogen content of RTE cucumber was 7.65 μmol/g at 0 days and did not change significantly during storage from 15 days to 75 days, but increased from 13.73 μmol/g to 18.46 μmol/g during storage from 75 days to 90 days, indicating that the body wall of RTE sea cucumber degraded during storage (Chen et al. 2016). It was revealed that the free hydroxyproline content and ammonia nitrogen content were 24.12 μg/g and 7.37 μmol/mL at 30 days, respectively, which were significantly higher than those at 0 days (8.33 μg/g and 1.91 μmol/mL, respectively) (Peng et al. 2015a, b). Sun (2021) found that the free hydroxyproline content and ammonia nitrogen content increased by 13.79 μg/g and 6.24 μg/g, after storage at 37 °C for 30 days, respectively. The changes in free hydroxyproline content and ammonia nitrogen content for RTE sea cucumber were fitted with the degradation model $\ln A = kt + \ln A_0$ (Sun 2021). The first-order kinetic models were $\ln A = 0.0249\ t + \ln 11.416$ and $\ln A = 0.0354\ t + \ln 3.589$, respectively, and the correlation coefficients R^2 were 0.9906 and 0.8288, respectively (Sun 2021). It was demonstrated that the degradation degree of collagen gradually increased with the increase of storage time (Sun 2021). It was reported that the free hydroxyproline content was 13.58 μg/g and 46.63 μg/g during 37 °C storage at 0 d and 30 days, respectively (Zhu et al. 2022a). The content of free ammonia nitrogen increased by 13.28 μg/g at 30 days, which was consistent with the change of free hydroxyproline content (Zhu et al. 2022a). The results of correlation analysis indicated that the free ammonia nitrogen content was significantly correlated with the free hydroxyproline content ($P < 0.01$), and the correlation coefficient was 0.969 (Zhu et al. 2022a).

It was found that the free hydroxyproline and ammonia nitrogen content of ready-to-eat sea cucumbers with different moisture contents increased significantly during storage at 37 °C, suggesting that the collagen degraded (Zhang et al. 2016). The free hydroxyproline content of RTE sea cucumber with low moisture content increased from 1.08 μg/mL to 5.60 μg/mL at 30 days, and the change was lower than that of RTE sea cucumber with high moisture content (Zhang et al. 2016). When the moisture content decreased to about 62.49%, the ammonia nitrogen content of the RTE sea cucumber with low moisture content (4.16 μmol/mL) was significantly lower than that of the RTE sea cucumber with high moisture content after storage for 15 days (Zhang et al. 2016). It was suggested that the free hydroxyproline and ammonia nitrogen content of the RTE sea cucumber with low moisture content increased at a slow rate during storage, and the low moisture content interfered with the degradation of the collagen in sea cucumber body wall (Zhang et al. 2016).

It showed that non-enzymatic gel deterioration of ready-to-eat sea cucumber occurred during storage, and the collagen was broken and degraded (Chen et al. 2016). The free hydroxyproline and free ammonia nitrogen content for the body wall of RTE sea cucumber showed an uptrend with the increase of storage time (Chen et al. 2016). In addition, moisture content could affect the change rate of free hydroxyproline content and free ammonia nitrogen content (Zhang et al. 2016). Low moisture content (<75%) was beneficial in protecting the gel structure of RTE sea cucumber body wall and reducing the degree of degradation of collagen, thereby prolonging the storage time (Zhang et al. 2016).

3.4 *Changes of Water State of RTE Sea Cucumber During Storage*

Water activity could reflect the ratio of free water to bound water in sea cucumber tissue, which is an important factor affecting the quality and storage stability of ready-to-eat sea cucumbers (Zhu et al. 2022a). Zhu et al. (2015) determined the water activity of RTE sea cucumbers during storage at 0 °C. The water activity of the RTE sea cucumbers did not change significantly at the beginning of storage and increased gradually after storage for 25 days (Zhu et al. 2015). It was reported that the water activity of RTE sea cucumber was 0.963 after storage at 4 °C for 90 days (Chen et al. 2016). It was found that the water activity of RTE sea cucumber gradually increased with the increase of storage time at 37 °C (Zhu et al. 2022a). This might be related to the destruction of tissue structure during storage and the increase of free water content, leading to an increase in water activity (Zhu et al. 2022a). Lin et al. (2019) investigated the changes of water activity for the body wall of RTE sea cucumbers stored at 4, 20, 37, 50, and 60 °C. The water activity showed an uptrend during storage, and its change rate was proportional to the storage temperature (Lin et al. 2019). It was revealed that the water activity of RTE sea cucumbers with different moisture contents (62.49–87.49%) showed an uptrend during storage. Moreover, the water activity of RTE sea cucumber with low moisture content was significantly lower than that of RTE sea cucumber with high moisture content, which indicated that low moisture content (<75%) was beneficial in improving the storage stability of sea cucumber (Zhang et al. 2016).

Low field nuclear magnetic resonance (LF-NMR), an effective, highly sensitive, and non-invasive detection method, was used to analyze the water distribution in food products (Zhu et al. 2022a). Transverse relaxation time T_2 was closely related to the freedom of water (Zhang et al. 2017). The longer the relaxation time, the weaker the combination of water molecules and the substrate, and the greater the freedom of water molecules (Zhang et al. 2017). Thus, T_2 relaxation spectra could be divided into three parts according to T_2 relaxation time, these being bound water, immobile water, and free water. It was revealed that the water state for the body wall of ready-to-eat sea cucumber changed during storage at 4 °C, resulting in changes of the relative proportion of T_2 (Chen et al. 2016). The relative proportion of T_{21} decreased during storage, the relative proportion of T_{22} decreased from 0.151 to 0.072 at 45 days, while the relative proportion of T_{23} increased significantly (Chen et al. 2016). The relaxation time of T_{23} was 328.119 ms at 160 days, which was higher than that at 0 day (Chen et al. 2016). It was suggested that the interaction was weakened and that part of the immobile water was converted into free water during storage (Chen et al. 2016). Sun (2021) found that the relaxation time of T_{21} increased from 0.43 ms to 0.76 ms after storage at 37 °C for 30 days, and the relative proportion of T_{21} decreased from 2.92% to 2.61%. At the same time, the relative proportion of T_{24} increased from 75.98% to 88.69%, indicating that the relative content of free water in RTE sea cucumber increased significantly during storage at 37 °C (Sun 2021).

In addition, T_1-weighted imaging and T_2-weighted imaging can directly reflect water distribution (Zhu et al. 2022b). It was found that the free water content in the body wall of ready-to-eat sea cucumber increased sharply after storage at 4 °C for 90 days through T_1-weighted imaging (Chen et al. 2016). It was reported that the proportion of red area in the body wall of RTE sea cucumber gradually decreased with the increase of storage time, suggesting that the content of bound water gradually decreased during storage (Sun 2021). It was demonstrated that the RTE sea cucumber with moisture content lower than 75% had a stable gel structure during storage (Zhang et al. 2016). The bound water in the sea cucumber body wall was uniformly distributed, and the change of water state was not obvious (Zhang et al. 2016). When the moisture content was higher than 80%, the gel structure of RTE sea cucumber degraded severely, resulting in a significant decrease in bound water content during storage (Zhang et al. 2016).

4 Effects of Cross-Linking on the Stability of RTE Sea Cucumber During Storage

Non-enzymatic degradation of ready-to-eat sea cucumbers occurs during storage, seriously affecting the quality of RTE cucumbers, hindering development of the sea cucumber processing industry, and causing serious economic losses (Peng et al. 2015a, b). Preventing the degradation of RTE sea cucumbers and improving the stability during storage has become an urgent problem for the industry. Currently, commonly used methods include the transaminase cross-linking method and the plant polyphenol cross-linking method (Liu 2022; Velickovic and Stanic-Vucinic 2017). In addition, the proteins could be modified through oxidative oligosaccharide cross-linking to enhance the gel properties of sea cucumber, stabilize the collagen fiber structure, and reduce the degree of fracture of collagen molecules in the body wall of sea cucumber, significantly improving the storage stability of RTE sea cucumbers. This has important economic and practical significance for the development of the sea cucumber industry (Wang et al. 2020; Zhang et al. 2017).

4.1 *Effects of Glutamine Aminotransferase on the Stability of RTE Sea Cucumber*

Transglutaminase (TG) is a protein cross-linking agent, which has been widely used in the food industry to enhance gel stability and mechanical strength and to improve the quality and nutritional value for food products such as surimi (Li et al. 2017), sausage (Choi et al. 2016), and so on. Through the catalysis of the acyl transfer reaction between the γ-carboxyamine group of glutamine residue and the primary amino group, TG cross-linking modifies the protein through three means: amine import, cross-linking, and deamination (Liu 2022). The ε-amino of lysine in protein acts as acyl receptor to realize intramolecular and intermolecular cross-linking of ε-(γ-glutamyl)-Lys isopeptides (Kieliszek and Misiewicz 2014). In reaction systems without lysine residue, water acts as the acceptor of acyl group, and the carboxamide group of glutamine residue is deamidated to form glutamic acid and ammonia residue, thus changing the charge and stability of protein (Liu 2022). TG cross-linking can improve the sensory properties of protein foods. This modification will not reduce the nutritional value of food, but can improve food texture and taste. Any protein containing glutamine and lysine residues, whether natural or synthetic, may constitute the substrates of TG (Hu et al. 2010).

Liu (2022) found that when the concentration of TG was 2%, the cross-linking degree was 72.44%. When the concentration of TG decreased, the substrates in the body wall of sea cucumber could not be completely cross-linked, and the sea cucumber body wall appeared to be "melting" after high-temperature sterilization. The cross-linking degree increased with the increase of TG concentration, but when TG reached a certain concentration, substrate saturation and steric hindrance increased with the increase of TG, which was not conducive to the cross-linking of enzymes into the body wall (Liu 2022). Therefore, it could be seen that the cross-linking effect decreased when TG concentration was greater than 2%. Compared with uncross-linked sea cucumber, 2% TG cross-linking could improve the sensory quality and thermal stability of sea cucumber after high-temperature sterilization (121 °C, 10 min) (Liu 2022). TPA analysis showed that sea cucumbers cross-linked by TG performed better in terms of texture indexes than did uncross-linked sea cucumbers. The contents of free ammonia nitrogen and free hydroxyproline in the cross-linking group were 0.17 mg/g and 10.15 ug/g, respectively, while those in the uncross-linked group were 0.68 mg/g and 17.33 ug/g, respectively. After cross-linking by TG, the thermal denaturation temperature of sea cucumber increased from 92.8 °C to 104.8 °C. SEM images showed that the microstructure of ready-to-eat sea cucumbers cross-linked by TG was better than that of uncross-linked sea cucumbers, and the sensory quality of cross-linked sea cucumbers was improved (Liu 2022). TG cross-linking could increase the thermal denaturation temperature of collagen fibers in the body wall of sea cucumber. The reason might be that cross-linking formed intramolecular and intermolecular covalent bonds, thus maintaining the stable spatial network structure of collagen and supporting the organizational structure and conformation transformation of collagen during heating (Liu 2022). Zhang et al. (2009) stabilized collagen in the body wall of RTE sea cucumber by curing it in transglutaminase solution, which prolonged the shelf life of RTE sea cucumber by up to 3 months at room temperature.

Liu studied the effect of TG cross-linking on the digestion and absorption of sea cucumber body wall through an in vitro simulation experiment in the gastrointestinal tract and found that the final digestibility of TG cross-linked sea cucumbers and uncross-linked sea cucumbers was above 80% (Liu 2022). At the initial digestion stage, the contents of TCA soluble polypeptide and free amino acid in the uncross-linked sea cucumber were higher than those in the cross-linked group, but at the end of the digestion treatment, the contents of TCA soluble polypeptide and free amino acid in the cross-linked group were higher (Liu 2022). Polyacrylamide gel electrophoresis (SDS-PAGE) analysis showed that the molecular weight of the digested products of sea cucumber body wall in the two groups was less than 10 kDa after digestion for 4 h. There was no significant difference in the molecular weight of the final digested products (Liu 2022). The results of $ABTS^+$ scavenging rate and iron reducing ability showed that the antioxidant activity of sea cucumber digested products was not affected by TG cross-linking (Liu 2022).

4.2 *Effects of Plant Polyphenols on the Stability of RTE Sea Cucumber*

Plant polyphenols are an important class of substance, mainly existing in grains, fruits, coffee, and tea (Balange and Benjakul 2009) and which are widely used in medicine and food processing. The phenolic hydroxyl group in plant polyphenols could combine with $-NH_2$ on the side chain of polar amino acids in collagen, thus promoting intermolecular cross-linking (Velickovic and Stanic-Vucinic 2017). In addition, phenolic hydroxyl groups in plant polyphenols could also react with collagen via secondary bonds (Poungchawanwong et al. 2020).

In order to obtain a sound level of sensory acceptability in ready-to-eat sea cucumber in terms of hardness, elasticity, taste and flavor, combined with prolonged storage life, Sun (2021) and Zhu et al. (2022b) investigated a large number of plants. The effective plants (celery and Yunnan green tea) were screened out, and the phenols in celery and Yunnan green tea were identified. An accelerated shelf life test showed that sea cucumbers treated with chlorogenic acid and gallic acid still maintained good sensory acceptability (Sun 2021). When RTE sea cucumbers were treated by 12 mg/mL of gallic acid and 10 mg/mL of chlorogenic acid, the ε-amino group level decreased by 50.2% and 47.9%, and hardness increased by 74.8% and

80.0%, respectively. The proportion of free water decreased by 10.9% and 8.8%, and the relaxation time decreased by 52.9 and 57.8 ms, respectively (Zhu et al. 2022b). Zhu et al. (2022a) reported that the hardness was significantly increased by 108% or 254% at 30 days, the relaxation time decreased by 31.90 or 39.89 ms, while the proportion of T_{23} reduced by 0.40% or 1.15% after RTE sea cucumbers were cross-linked by celery or chlorogenic acid, respectively. In addition, the moisture content of sea cucumbers treated with high-pressure steam after cross-linking was obviously lower than that of the sea cucumbers in the uncross-linked group during storage, indicating that chlorogenic acid and celery cross-linking had a definite influence on the water content of RTE sea cucumber (Zhu et al. 2022b). The reason might be that the free hydroxyl groups in the sample would be reduced after the cross-linking of polyphenols and collagen, which reduced the hydrophilicity of the sample, thus resulting in decreased moisture content of the sample (He et al. 2011). This effectively enhanced the stability of sea cucumber collagen and extended the shelf life of sea cucumber products.

The ready-to-eat sea cucumbers cross-linked by chlorogenic acid exhibited smaller pore size with finer collagen fibrils. In addition, the aggregation and density of collagen fibers increased and their porosity decreased, as observed by Van Gieson (VG) staining and scanning electron microscope images. After cross-linking by 12 mg/mL gallic acid and 10 mg/mL chlorogenic acid, the peak temperatures increased to 137.5 °C and 130.0 °C, respectively, similar to the changes of activation energy, indicating that the thermal stability of RTE sea cucumbers was improved (Zhu et al. 2022b). Moreover, chlorogenic acid cross-linking also inhibited the breaking of peptide bonds in RTE sea cucumber collagen, achieving a decrease in the free hydroxyproline level from 46.63 to 34.53 μg/g and a reduction in free ammonia nitrogen from 20.96 to 15.30 μmol/g (Zhu et al. 2022a). The resistance of molecular cross-linked RTE sea cucumbers to non-enzymatic degradation might be attributed to ameliorated moisture migration, breakage of collagen fibers, and degradation of collagen during storage (Zhu et al. 2022a). Zhou et al. (2019) cross-linked sea cucumbers with polyphenolic extracts (tea polyphenols, bamboo leaf extracts, rosemary extracts, and theaflavins) to improve the water holding capacity of RTE sea cucumber, reduce the release of collagen hydrolysates in the body wall of sea cucumber, and prolong the shelf life of RTE sea cucumber. Hou et al. (2021) cross-linked sea cucumbers in 0.01–5% tea extract or tea polyphenols or grape seed extract, which greatly shortened the production cycle of RTE sea cucumber, retained the nutritional components and unique flavor, and significantly extended sea cucumber's storage period.

Zhao et al. (2015a) studied the effects of *Fructus chebulae* and *Galla chinensis* extracts on ready-to-eat sea cucumbers and compared these samples with the body wall of uncross-linked RTE sea cucumbers. The elasticity of the body wall of RTE sea cucumber cross-linked by *Galla chinensis* and *Fructus chebulae* extracts significantly increased by 1.58 and 1.16 times, respectively. Moreover, the degradation degree of protein in RTE sea cucumber body wall cross-linked by *Galla chinensis* and *Fructus chebulae* extracts was reduced by 20–30% (Zhao et al. 2015a). This study showed that the water contents of RTE sea cucumbers cross-linked by *Galla chinensis* and *Fructus chebulae* extracts were 13.12% and 16.78%, respectively, lower than that of uncross-linked sea cucumbers (Zhao et al. 2015a). In addition, the sensory scores of *Galla chinensis* and *Fructus chebulae* extracts cross-linked RTE sea cucumbers were twice as high as those of uncross-linked RTE sea cucumbers during storage at 37 °C, making them more acceptable to consumers. The maximum cross-linking degrees of collagen treated with *Fructus chebulae* and *Galla chinensis* extracts were 76.9% and 78.3%, respectively. It was noteworthy that the cross-linking of the two extracts did not affect the triplex structure of collagen (Zhao et al. 2015a).

4.3 Effects of Other Cross-Linkers on the Stability of RTE Sea Cucumber

Zhang et al. (2009) transferred vacuum impregnated sea cucumbers in neutral chitosan solution into transglutaminase solution for cross-linking, which stabilized the collagen in the body wall of ready-to-eat sea cucumbers and significantly extended the shelf life of RTE sea cucumbers at room temperature (to 3 months). Oxidized oligosaccharides are non-toxic small polyaldehydes. A previous study managed to diffuse various oxidized oligosaccharides into the protein wall of rehydrated sea cucumbers, and the texture profile analysis, total soluble substance assay, and scanning electron microscope (SEM) images suggested the treated sea cucumbers featured significantly enhanced thermal stability (Liu et al. 2020). Oxidized oligosaccharides, especially sucrose oxide, showed great potential for use in food-related industries as a small molecule stabilizer that could be diffused into protein hydrophilic colloid without compromising its appearance and structural integrity (Liu 2022).

5 Study on the Stabilization Mechanism of Cross-Linking on Collagen of RTE Sea Cucumber

As the major edible part of sea cucumber, the body wall is mainly composed of collagen, accounting for about 70% of the total protein (Tian et al. 2020). Therefore, collagen plays a crucial role in the quality (especially textural properties) of sea cucumber and its processed products (Zhu et al. 2012). In order to improve the stability and mechanical strength of collagen, attention has increasingly been paid to ways of modifying it.

5.1 Study on the Structure and Stability of Sea Cucumber Collagen

The collagen in sea cucumber is a highly cross-linked polymer of collagen molecules. The collagen fibers are 30 μm–1 mm in length, in the shape of a symmetrical fusiform. The ratio of average length to fiber diameter is 2000–2500, and the collagen only dissolves to a small degree before the peptide bonds break. The collagen can be assembled to fibrils (Wilkie 2005), and the collagen fibrils can further form into collagen fibers through proteoglycan connections (Ribeiro et al. 2011), which play an important role in the mechanical stability of sea cucumber body wall (Thurmond et al. 1997; Wilkie 2005). Due to its insolubility, sea cucumber collagen is mainly extracted by the pepsin-soluble method (Dong et al. 2011; Matsumura 1974; Cui et al. 2006; Liu et al. 2010; Zhu et al. 2012). It has been found that 70% of the body wall protein of *Stichopus japonicus* is collagen, which is rich in glycine, alanine, and hydroxyproline. Cui et al. extracted pepsin-soluble collagen from *Apostichopus japonicus* and preliminarily analyzed its amino acid composition and physicochemical properties (Cui et al. 2006). It was found that the collagen fibers of sea cucumber are heteromorphic (Tian et al. 2020). The collagens were identified as two clade A collagens, one clade B collagen, and two fibril-associated collagens with interrupted triple helices (FACIT) collagens (Tian et al. 2020). Liu et al. prepared collagen from the body wall of giant red sea cucumber (*Parastichopus californicus*) by the pepsin digestion method, and its T_d and T_m values were 18.5 °C and 33.2 °C, respectively (Liu et al. 2010).

The collagen fibers in the body wall of *Stichopus japonicus* had periodic light and dark bands with a period of 61 nm (Si 2018). The pepsin-soluble collagen (PSC) of *Stichopus japonicus* was prepared by the enzymatic assistant extraction method and consisted of three α chains, each of which had a molecular weight of 135 kDa and the subunit composition of $(\alpha_1)_3$ (Si 2018). The contents of protein and carbohydrate in the PSC of *Stichopus japonicus* were 95.1% and 2.5%, respectively. The PSC and CCF in the body wall of *Stichopus japonicus* contained about 1/3 of glycine and were rich in alanine, proline, and hydroxyproline (Si 2018). The maximum ultraviolet absorption wavelength of the PSC was around 234 nm. There was a positive peak at 221 nm and a negative peak at 197 nm in circular dichroism spectrum (CD) (Si 2018). Fourier transform infrared spectroscopy (FTIR) and X-ray diffraction results showed that the structures of PSC and CCF were slightly different (Si et al. 2018; Hou et al. 2013).

Si et al. (2018) found that when the temperature increased from 20 °C to 100 °C, the amide A band of PSC shifted from 3322 cm^{-1} to 3315 cm^{-1}, which was related to N-H stretching vibration. The amide B band fluctuated between 3076 cm^{-1} and 3080 cm^{-1}, which was related to the asymmetric stretching of CH_2 (Si et al. 2018). The amide I band changed from 1654 cm^{-1} to 1660 cm^{-1}, which was related to the stretching vibration of C=O or the hydrogen bond coupled with COO– (Li et al. 2004). The amide II band fluctuated slightly between 1543 cm^{-1} and 1546 cm^{-1}, which was usually a combination of N–H in-plane bending and C–N stretching vibrations (Matmaroh et al. 2011). The amide band III represented N–H bending vibration, C–N stretching vibration, and C–H stretching (Veeruraj et al. 2015), which fluctuated slightly between 1237 and 1241 cm^{-1} (Si et al. 2018). As the temperature increased, the amide A band of PSC on *Stichopus japonicus* shifted to a lower frequency and the amide I band shifted to a higher frequency (Si et al. 2018). This indicated that some N–H groups in the peptide chain were involved in hydrogen bonding, but the partial hydrogen bond coupling with COO– was broken during heat treatment (Si et al. 2018). However, amide B, II, and III bands were not sensitive to temperature and basically remained unchanged during the temperature change, indicating that the corresponding groups of these amide bands had higher stability (Si et al. 2018).

When the temperature of heat treatment was between 110 °C and 120 °C, the content of free hydroxyproline in *Stichopus japonicus* increased significantly with the extended heat treatment time (10–160 min), indicating that collagen was degraded (Lin 2019). Zhang et al. showed that when the heat treatment temperature of sea cucumber was 80–100 °C, the content of free hydroxyproline did not change significantly (Zhang et al. 2017). When the temperature was 110–120 °C, the content of free hydroxyproline increased with heating time (Zhang et al. 2017). When the sea cucumber was treated at 80 °C for 40 min, the D period decreased to 21.43 ± 0.28 nm and the average diameter increased to 139.29 ± 8.58 nm, indicating that the collagen fiber bundle experienced thermal shrinkage (Lin 2019). The D-periodic structure could not be observed after 160 min of heat treatment, indicating that the structure of collagen fiber was significantly damaged (Lin 2019). When sea cucumber was heated at 110 °C and 120 °C, collagen was greatly degraded, and any observable collagen fiber or network structure disappeared in SEM images (Lin 2019). The average pore diameter decreased from 1.57 ± 0.65 μm (40 min) to 1.16 ± 0.34 μm (160 min), indicating that the layered structure was gradually destroyed (Wang et al. 2020). During heat treatment, sea cucumber collagen was prone to break at the sites of G, P, A, L, Y, K, and R (Wang et al. 2020).

When the temperature increased from 20 °C to 100 °C, the first peak of the X-ray diffraction pattern for sea cucumber collagen shifted from 7.45° to 7.98°, while the second peak decreased from 22.79° to 20.88° (20–80 °C). It then rose to 22.06° at 100 °C (Si et al. 2018). According to the Bragg equation, the d value corresponding to the first peak gradually decreased from 1.18 nm to 1.11 nm, indicating that the distance between molecular chains gradually narrowed, which might be due to the thermal shrinkage of the PSC of *Stichopus japonicus* (Si et al. 2018). Meanwhile, heat treatment caused changes of diffuse scattering in the amorphous region, leading to slight fluctuations in the position of the second peak (Si et al. 2018). For the thermal degradation process of PSC of *Stichopus japonicus*, the initial temperature of degradation and the maximum weight loss temperature corresponding to the two stages showed a linear uptrend with the increase of heating rate (Si et al. 2018). The Flynn–Wall–Ozawa method and Satava–Sestak method were used to analyze the thermal degradation kinetics of *Stichopus japonicus*. The activation energy of PSC of *Stichopus japonicus* was in the range of 163–173 kJ/mol. The thermal degradation mechanism function could be expressed as $G(\alpha) = [-\ln(1\text{-}\alpha)]^{2/3}$ (Si et al. 2018).

5.2 *Effects of Endogenous and Exogenous Factors on Collagen Stability of Sea Cucumber*

NaCl can induce the degradation of PSC at a certain concentration, destroying the electrostatic interaction between the side chains of PSC molecules (Wang et al. 2020). In addition, the degradation of collagen is influenced by N- and C- terminal peptide, increasing its stability (Wang et al. 2020). Adding polysaccharide sulfate to PSC can also prevent its degradation. The reason might be that the positive charge of PSC could attract the negative charge of polysaccharide sulfate to slow down its degradation rate and enhance the structural stability of PSC (Wang et al. 2020).

Adding gallic acid could retard the degradation of PSC, probably because gallic acid could cross-link with the ε-NH_2 of PSC to improve the thermal stability of PSC and thus inhibit degradation (Wang et al. 2020). The free ammonia nitrogen (FAN) contents of both PSC cross-linked by gallic acid and uncross-linked PSC increased with the increase of heating temperature (60–120 °C) when heated for 20 min, and the FAN content of cross-linked PSC was lower than that of uncross-linked PSC (Wang et al. 2020). This indicated that gallic acid could cross-link with FAN, resulting in a decreased degree of breakage in PSC, which proved that gallic acid could retard the degradation of PSC (Wang et al. 2020). In addition, the free hydroxyproline (FHYP) content decreased significantly after the addition of gallic acid at the same heating time and temperature, indicating that gallic acid made the structure of PSC more stable, which further proved that gallic acid could slow down the degradation of PSC (Wang et al. 2020).

5.3 *Mechanism of Cross-Linking and Stabilizing Sea Cucumber Collagen*

Polyphenols are widely distributed in plant tissues with important functions. Common polyphenols include caffeic acid, chlorogenic acid, gallic acid, quercetin, rutin, and so on (Zhu et al. 2022a). Phenolic compounds can interact with collagen through covalent cross-linking or secondary bonds (Velickovic and Stanic-Vucinic 2017; Poungchawanwong et al. 2020).

The changes in collagen fiber in ready-to-eat sea cucumbers treated with gallic acid or chlorogenic acid at different concentrations have been studied. It was found that RTE sea cucumbers cross-linked by gallic acid or chlorogenic acid exhibited small pore diameter and fine collagen fibril compared with an uncross-linked sea cucumber group (Zhu et al. 2022b). In addition, Zhu et al. (2022b) found that when the concentration of gallic acid and chlorogenic acid was increased to 12 and 10 mg/mL, respectively, the degradation of collagen fiber was reduced, which confirmed the interaction between cross-linkers (gallic acid and chlorogenic acid) and the collagen in the body wall of RTE sea cucumber body wall. When the collagen fiber was cross-linked by chlorogenic acid, the degradation of collagen fiber was inhibited and its stability was enhanced (Zhu et al. 2022b). The collagen fiber structure of sea cucumber was arranged neatly and clearly after being treated with *Fructus chebulae* and *Galla chinensis* extracts (Zhao et al. 2015a). Sun (2021) revealed that the collagen fibers of uncross-linked RTE sea cucumbers were damaged to certain extent after storage for more than 1 day. The network structure was loosened and the pore size was significantly increased at 30 days. Under the same storage conditions, the damage of collagen fibers was significantly reduced after cross-linking by green tea extract. The network structure of collagen fiber was dense and clear after storage for 15 days. Moreover, the collagen fibrils were still relatively dense, and the pore size was small after storage for 30 days (Sun 2021).

Compared with uncross-linked sea cucumber, the peak temperature of ready-to-eat sea cucumber treated with 12 mg/mL gallic acid increased from 125.0 °C to 137.5 °C, and the peak temperature of sea cucumber treated with 10 mg/mL chlorogenic acid increased to 130.0 °C (Zhu et al. 2022b). The reason might be that a part of gallic acid or chlorogenic acid was oxidized and connected with the collagen of sea cucumber body wall through hydrogen bond and hydrophobic interaction, improving the thermal stability of RTE sea cucumber (Tian et al. 2021). The thermal degradation activation energy of uncross-linked sea cucumber, of sea cucumber treated with gallic acid, and of sea cucumber treated with chlorogenic acid were 268.15–317.44 kJ/mol, 302.07–356.28 kJ/mol, and 254.99–382.13 kJ/mol, respectively. The activation energy of sea cucumber treated with 12 mg/mL gallic acid or 10 mg/mL chlorogenic acid was higher than that of uncross-linked sea cucumber, the reason for which might be that the phenolic hydroxyl group of cross-linkers could bind with sea cucumber collagen to form hydrogen bonds, thus improving the thermal stability (Sun 2021). Wang et al. (2020) found that as the heating temperature increased from 60 °C to 120 °C, α chain was observed at 135–190 kDa in sea cucumber treated with gallic acid without significant degradation. At the same temperature, small molecular bands of cross-linked sea cucumber collagen were fewer than those of uncross-linked collagen. α chains of uncross-linked collagen were significantly degraded at 110 °C and 120 °C, while those of the collagen cross-linked by gallic acid were basically not degraded (Wang et al. 2020). This indicated that adding cross-linkers such as gallic acid could effectively prevent against and delay the degradation of sea cucumber collagen (Wang et al. 2020).

Zhao et al. (2015a) indicated that the amide I and amide II bands of gelatin cross-linked by *Fructus chebulae* extract or genipin both shifted to low wavenumbers. The bands at 3300 cm^{-1} of cross-linked gelatin shifted to low wavenumbers, and the bands became wider than those of the uncross-linked gelatin. The reason might be that cross-linking bonds were generated between the cross-linkers and gelatins through hydrogen bonding interaction (Zhao et al. 2015a). An increase of temperature led to a red shift of absorption peak of PSC, suggesting the exposure of a large amount of Tyr and Phe residues (Wang et al. 2020). Compared with PSC, the absorption peak showed a red shift after the addition of gallic acid (Wang et al. 2020). The peak curve for the amide A band of gelatin became wider after the addition of gallnut extract. This was mainly due to the abundant hydroxyl groups and

C=O groups of gallnut extract, which could interact with the gelatin molecular chains to form a large number of hydrogen bonds (Zhao et al. 2015a). The d value of gelatin cross-linked by gallnut extract was lower than that of uncross-linked gelatin, which might be due to the interaction between the cross-linker and the gelatin molecular chains decreasing the distance between the gelatin molecular chains and enhancing the interaction between the molecular chains. It was suggested that gallnut extract could react with gelatin molecules, resulting in enhanced intermolecular hydrogen bonding and van der Waals forces as well as reduced intermolecular gaps (Zhao et al. 2015a).

Liu et al. (2021) reported that the free ammonia nitrogen content of uncross-linked sea cucumber was higher than that of sea cucumber cross-linked by microbial transglutaminase, suggesting that the peptide bonds of uncross-linked sea cucumber collagen were severely fractured and more amino groups were released. The content in the sea cucumber cross-linked by microbial transglutaminase was maintained at a low level without any obvious increase during storage, indicating that cross-linking inhibited the decomposition of peptides (Liu et al. 2021). Sun (2021) revealed that the free hydroxyproline content and ammonium nitrogen content of ready-to-eat sea cucumber cross-linked by gallic acid or chlorogenic acid were lower than those of uncross-linked sea cucumber. It was demonstrated that cross-linking treatment could inhibit the destruction of collagen network structure and the breakage of collagen fibers and that it had an effect on the degradation of RTE sea cucumber body wall during storage. In addition, compared with uncross-linked sea cucumber, the hydrolysis breakage rate of cross-linked dipeptides was reduced, indicating that cross-linking could effectively inhibit the degradation of peptide during 37 °C storage (Sun 2021). The collagen of sea cucumber body wall was cross-linked, improving the stability of collagen to overcome collagen degradation and extending the shelf life of RTE sea cucumber products (Xue et al. 2018).

6 Conclusion

The ready-to-eat sea cucumber prepared by advanced processing technologies such as high-temperature short-time sterilization and vacuum cooling, modified atmosphere packaging of high barrier materials, and stepwise secondary sterilization possesses high nutrition value and good sensory properties. However, non-enzymatic degradation of RTE sea cucumber can occur during storage, leading to changes of texture parameters, increased water activity, higher free water content, and the fracture and degradation of collagen. These can seriously affect the quality of RTE sea cucumber and caused serious economic losses. Transaminase, polyphenol, and oxo-oligosaccharide treatment can be used to modify sea cucumber proteins, which could enhance the gel properties of sea cucumber, stabilize the collagen fiber structure, and reduce the degree of breakage of collagen. This could significantly improve the storage stability of RTE sea cucumber. The body wall of sea cucumber is its main edible part, and this is mainly composed of collagen. The body wall is therefore essential for sea cucumber quality, and the modification of collagen could improve its stability and mechanical strength and thus improve the stability of RTE sea cucumber, which is of great significance to the development of the sea cucumber industry.

Conflict of Interest All authors declare they have no conflict of interest.

References

Balange A, Benjakul S (2009) Enhancement of gel strength of bigeye snapper (*Priacanthus tayenus*) surimi using oxidised phenolic compounds. Food Chem 113:61–70. https://doi.org/10.1016/j.foodchem.2008.07.039

Chen TJ, Peng Z, Lu IH, Li B, Hou H (2016) Self-degradation of sea cucumber body wall under 4c storage condition. J Food Process Preserv 40:715–723. https://doi.org/10.1111/jfpp.12652

Choi YS, Jeong TJ, Hwang KE, Song DH, Ham YK, Kim HW, Kim CJ (2016) Combined effects of *laminaria japonica* and transglutaminase on physicochemical and sensory characteristics of semi-dried chicken sausages. Poult Sci 95:1943–1949. https://doi.org/10.3382/ps/pew093

Cui FX, Xue CH, Li ZJ, Dong P, Xue DM (2006) Extraction and the physical and chemical characterization of collagen in *Stichopus japonicus*. J Fish China 30:549–553. https://doi.org/10.3724/SP.J.1231

Dong XP, Zhu BW, Sun LM et al (2011) Changes of collagen in sea cucumber (*Stichopus japonicas*) during cooking. Food Sci Biotechnol 20(1137):1141. https://doi.org/10.1007/s10068-011-0155-x

He LR, Mu CD, Shi JB, Zhang Q, Shi B, Lin W (2011) Modification of collagen with a natural cross-linker, procyanidin. Int J Biol Macromol 48:354–359. https://doi.org/10.1016/j.ijbiomac.2010.12.012

Hong Z (2005) Market potential and processing technology of ready-to-eat sea cucumber. Fish Modern 1:42. https://doi.org/CNKI:SUN:HDXY.0.2005-01-00J

Hong J, Cheng W, Zhang H, Dan-Chen L, Chen L (2014) Effects of rehydration conditions on water soluble protein content in the rehydration solution of dried sea cucumber. Subtrop Agric Res 10:51–55. https://doi.org/10.13321/j.cnki.subtrop.agric.res.2014.01.002

Hou H, Sun J, Mao JN, Li QQ, Xue CH (2013) Collagen fibers morphology and physical and chemical properties of collagen of sea cucumber. Mod Food Sci Technol 29:1491–1495. https://doi.org/10.13982/j.mfst.1673-9078.2013.07.001

Hou HM, Cui YN, Tang L, Zhang GL, Sun LM (2014) The sensory quality and textural property of ready-to-eat sea cucumber in storage period. Adv Mater Res 887–888:619–622. https://doi.org/10.4028/www.scientific.net/AMR.887-888.619

Hou Z, Wang M, Zhang J, Zhao Y, Wang Y, Wang G (2015) The research progress of sea cucumber processing technology. Food Ind 36:248–252

Hou H, Xue CH, Fan Y, Li BF, Zhang ZH, Zhao X et al (2021) A high-temperature High-Pressure Sea cucumber product with stable storage and its preparation. CN Patent. Appl. 105265940B

Hu S, Tao H, Huang Y, Kong L (2010) Research advance of transglutaminase China Brewing 6:9–12

Kieliszek M, Misiewicz A (2014) Microbial transglutaminase and its application in the food. Folia Microbiol 59:241–250. https://doi.org/10.1007/s12223-013-0287-x

Li FG, Li JS, Yan LJ, Huang GX (2017) Optimization of modification process parameters of soybean 11S protein by glutamine transaminase using response surface methodology. China Oils Fats 42:93–97. https://doi.org/10.3969/j.issn.1003-7969.2017.02.022

Li H, Liu B, Gao L, Chen H (2004) Studies on bullfrog skin collagen. Food Chem 84:65–69. https://doi.org/10.1016/S0308-8146(03)00167-5

Li YT, Li YJ, Chen YX, Shao RD (2010) Production process of instant fresh sea cucumber. Meet Res 6:78–81

Lin L (2019) Study on the formation of thermal gel and the process and mechanism of non-enzymatic gel degradation of sea cucumber (*Stichopus japonicus*) body wall. Dissertation, Ocean University of China

Lin W, Shen J, Xu H, Xue CH, Zheng XW, Xu WQ, Zhang JF, Gu JH, Ou-Yang J, Cai SJ (2010) A sea cucumber pretreatment machining process and its corresponding equipment. CN Patent. Appl. 101617837A

Lin L, Sun X, Hou H (2019) Effects of storage temperature on the structural changes of high temperature and high pressure sea cucumber body wall. Food Ferment Ind 45:194–199. https://doi.org/10.13995/j.cnki.11-1802/ts.021189

Lin L, Sun X, Qi X, Wang WX, Hou H (2021) Variation of body wall of sea cucumber with different humidity ratio in microscale. J Chin Instit Food Sci Technol 21(247):253. https://doi.org/10.16429/j.1009-7848.2021.02.030

Liu JY (2022) Microbial glutamine aminotransferase promotes the stability of instant sea cucumber at room temperature. Dissertation, Ocean University of China

Liu C, Liu K, Dang L, Wang X (2007) Processing method of ready-to-eat sea cucumber. CN Patent. Appl. 1663463A

Liu C, Yuan W, Liu J, Meng X, Wang X, Tang J, Hu W, Sun Y (2008) Study on processing technology of ready-to-eat sea cucumber. China Food Ind 2:56–58

Liu Z, Oliveira ACM, Su YC (2010) Purification and characterization of pepsin-solubilized collagen from skin and connective tissue of giant red sea cucumber (*Parastichopus californicus*). J Agric Food Chem 58:1270–1274. https://doi.org/10.1021/jf9032415

Liu J, Xu Y, Xia T, Xue C, Sun X (2020) Oxidized oligosaccharides stabilize rehydrated sea cucumbers against high-temperature impact. Int J Mol Sci 21:5204. https://doi.org/10.3390/ijms21155204

Liu J, Chen S, Jiang X, Xue C, Cao H, Liu D, Sun X (2021) Microbial transglutaminase inhibits the deterioration of high-temperature-treated sea cucumber. J Food Process Preserv 45:e15985. https://doi.org/10.1111/jfpp.15985

Matmaroh K, Benjakul S, Prodpran T, Encarnacion AB, Kishimura H (2011) Characteristics of acid soluble collagen and pepsin soluble collagen from scale of spotted golden goatfish (*Parupeneus heptacanthus*). Food Chem 129:1179–1186. https://doi.org/10.1016/j.foodchem.2011.05.099

Matsumura T (1974) Collagen fibrils of the sea cucumber, *Stichopus japonicus*: purification and morphological study. Connect Tissue Res 2:117–125. https://doi.org/10.3109/03008207409152097

Peng Z, Hou H, Feng YL, Cai SX, Huang M, Zhang ZH, Xue CH, Zhao X, Li BF (2015a) Effects of phosphorylation and calcium cross-linking on the gel properties of sea cucumber collagen aggregates. Mod Food Sci Technol 31:190–195. https://doi.org/10.13982/j.mfst.1673-9078.2015.10.032

Peng Z, Hou H, Bu L, Li BF, Zhang ZH, Xue CH (2015b) Nonenzymatic softening mechanism of collagen gel of sea cucumber (*Apostichopus japonicus*). J Food Process Preserv 39:2322–2331. https://doi.org/10.1111/jfpp.12479

Poungchawanwong S, Klaypradit W, Li QQ, Wang J, Hou H (2020) Interaction effect of phenolic compounds on Alaska Pollock skin gelatin and associated changes – science direct. LWT 133:110018. https://doi.org/10.1016/j.lwt.2020.110018

Ribeiro AR, Barbaglio A, Benedetto CD, Ribeiro CC, Wilkie IC, Carnevali MDC, Barbosa MA (2011) New insights into mutable collagenous tissue: correlations between the microstructure and mechanical state of a sea-urchin ligament. PLoS One 6:1–10. https://doi.org/10.1371/journal.pone.0024822

Sang L, Weng QD, Pan CR (2013) Study on sterilization technology of canned sea cucumber. Sci Technol Food Ind 34:257–261. https://doi.org/10.13386/j.issn1002-0306.2013.19.087

Shandong Homey Aquatic Development Co., Ltd. (2010) Ready-to-eat fresh sea cucumber. In (Vol. Q/SHD0031S)
Si LL (2018) Study on the mechanism of thermal denaturation and degradation of collagen from sea cucumber (*Stichopus japonicus*). Master dissertation. Ocean University of China
Si L, Fan Y, Wang Y, Sun L, Li B, Xue C, Hou H (2018) Thermal degradation behavior of collagen from sea cucumber (*Stichopus japonicus*) using tg-ftir analysis. Thermochim Acta 659:166–171. https://doi.org/10.1016/j.tca.2017.12.004
Sui S, Yin Y (2016) Study on processing technology optimization of dry ready-to-eat sea cucumber. Storage Process 16:41–44
Sun X (2021) Study on the mechanism of non-enzymatic degradation and interventional pathway of sea cucumber treated by high temperature and high pressure during storage. Dissertation, Ocean University of China
Tang L, Kong X, Zhang G, Sun L, Hou H (2015) The quality changes of water soaked sea cucumber during storage. J Chin Instit Food Sci Technol 15:254–259. https://doi.org/10.16429/j.1009-7848.2015.08.036
Thurmond FA, Koob TJ, Bowness JM, Trotter JA (1997) Partial biochemical and immunologic characterization of fibrillin microfibrils from sea cucumber dermis. Connect Tissue Res 36:211–222. https://doi.org/10.3109/03008209709160221
Tian M, Xue CH, Chang YG, Shen JJ, Zhang YY, Li ZJ, Wang YC (2020) Collagen fibrils of sea cucumber (*Apostichopus japonicus*) are heterotypic. Food Chem 316:126272. https://doi.org/10.1016/j.foodchem.2020.126272
Tian X, Wang Y, Duan S, Hao Y, Wang W (2021) Evaluation of a novel nano-size collagenous matrix film cross-linked with gallotannins catalyzed by laccase. Food Chem 351:129335. https://doi.org/10.1016/j.foodchem.2021.129335
Veeruraj A, Arumugam M, Ajithkumar T, Balasubramanian T (2015) Isolation and characterization of collagen from the outer skin of squid (*Doryteuthis singhalensis*). Food Hydrocoll 43:708–716. https://doi.org/10.1016/j.foodhyd.2014.07.025
Velickovic TDC, Stanic-Vucinic DJ (2017) The role of dietary phenolic compounds in protein digestion and processing technologies to improve their antinutritive properties. Compr Rev Food Sci Food Saf 17:82–103. https://doi.org/10.1111/1541-4337.12320
Wang J, Lin L, Sun X, Hou H (2020) Mechanism of sea cucumbers (*Apostichopus japonicus*) body wall changes under different thermal treatment at micro-scale. LWT-Food Sci Technol 130: 109461. https://doi.org/10.1016/j.lwt.2020.109461
Wilkie IC (2005) Mutable collagenous tissue: overview and biotechnological perspective. Prog Mol Subcell Biol 39:221–250. https://doi.org/10.1007/3-540-27683-1_10
Xu W, Cai S, Shen J (2011) Continuous cooking technology and equipment of fresh sea cucumber. Food Sci Technol 36:108–111. https://doi.org/10.13684/j.cnki.spkj.2011.01.046
Xue Y, Zhao YY, Xue CH et al (2018) A method of extending the storage of ready-to-eat sea cucumber. CN Patent. Appl. 104522836A
Yang T, Zhang T (2018) A new bubble agitated sea cucumber cleaning device. CN Patent. Appl. 209573046U
Yao HH (2013) Research on the impact of cooking methods for the nutrition quality and characteristic in sea cucumber. Master dissertation, Yantai University
Yuan WP, Liu CH, Meng XM, Wang XJ, Xia XQ, Zhang MS (2008) Effects of different sterilization methods on quality of ready-to-eat sea cucumber. China Food Ind 12:54–55
Yuan L, Li HJ, Ni DD, He ZF, Xie YJ (2016) Effects of different rehydration conditions on short dry sea cucumber quality. J Southwest Univ (Nat Sci Edition) 38:25–31. https://doi.org/10.13718/j.cnki.xdzk.2016.02.004
Zhang YQ, Xue CH, Xue Y et al (2009) A preparation method of ready-to-eat sea cucumber with collagen stabilization. CN Patent. Appl.101773260A
Zhang S, Ma MY, Dai XL, Zhang CX (2012) Processing technology of instant fresh sea cucumber by response surface method designs. Guangdong Agric Sci 39:106–107. https://doi.org/10.16768/j.issn.1004-874x.2012.12.011

Zhang Y, Hou H, Fan Y, Zhang FF, Li BF, Xue CH (2016) Effect of moisture status on the stability of thermal gels from the body wall of sea cucumbers (*Apostichopus japonicus*). LWT Food Sci Technol 74:294–302. https://doi.org/10.1016/j.lwt.2016.07.058

Zhang K, Hou H, Bu L, Li B, Xue C, Peng Z, Su S (2017) Effects of heat treatment on the gel properties of the body wall of sea cucumber (*Apostichopus japonicus*). J Food Sci Technol 54(707–717). https://doi.org/10.1007/s13197-017-2509-8

Zhao YY, Xue Y, Dong J, Li ZJ, Xue CH (2015a) Effect of two plant extracts on the stability of the body wall collagen in instant sea cucumber. Mod Food Sci Technol 31:113–119. https://doi.org/10.13982/j.mfst.1673-9078.2015.11.019

Zhao YY, Xue Y, Li ZJ, Xue CH, Huang WC (2015b) Preparation of fresh-eaten sea cucumber and changes in its quality during room temperature storage. Sci Technol Food Ind 36:305–308. https://doi.org/10.13386/j.issn1002-0306.2015.11.053

Zheng LH, Wu RN (2014) Comparison of different processing methods of sea cucumber and their effects on its nutritional components. Agric Technol Service 31:202. https://doi.org/10.3969/j.issn.1004-8421.2014.06.152

Zhou DY, Liu ZQ, Zhu BW, Liu YX, Dong XP, Huang WC (2019) Preparation method for a ready-to-eat sea cucumber product with stable storage. CN Patent. Appl.110338368A

Zhu BW, Dong XP, Zhou DY et al (2012) Physicochemical properties and radical scavenging capacities of pepsin-solubilized collagen from sea cucumber *Stichopus japonicus*. Food Hydrocoll 28:182–188. https://doi.org/10.1016/j.foodhyd.2011.12.010

Zhu JN, Wang L, Lu HX, Chen Q (2015) The storage stability of instant sea cucumber *Acaudina molpadioidea*. Food Sci Technol 5:42–45. https://doi.org/10.13684/j.cnki.spkj.2015.05.008

Zhu LL, Qi X, Bai J, Sun X, Hou H (2022a) The mechanism of molecular cross-linking against nonenzymatic degradation in the body wall of ready-to-eat sea cucumber. Food Chem 373: 131359. https://doi.org/10.1016/j.foodchem.2021.131359

Zhu LL, Sun X, Fan Y, Wang YF, Qi X, Hou H (2022b) Effect of gallic acid and chlorogenic acid on physicochemical, microstructural and thermal degradation properties of ready-to-eat sea cucumber. Food Chem 380:132186. https://doi.org/10.1016/j.foodchem.2022.132186

Zu Y (2017) Effects of repeated freezing and thawing on water dynamics of ready-to-eat sea cucumbers. Dissertation, Dalian Polytechnic University

The Extraction, Separation Technology, and New Product Development of Sulfated Polysaccharides from Sea Cucumber

Zihao Wei and Yaoguang Chang

Abstract The polysaccharide in sea cucumbers possesses a wide range of biological activities and health benefits. Sea cucumbers are rich in sea cucumber polysaccharides, accounting for more than 6% of the dry weight of sea cucumber. Polysaccharides in sea cucumbers have antioxidant, antitumor, and anti-inflammatory activities, enjoying broad application prospects. At present, a series of progress has been achieved in the research and application of sea cucumber polysaccharides. This chapter mainly introduces the technologies for sea cucumber polysaccharide processing by focusing on the extraction and separation of sea cucumber polysaccharides and discusses the application progress of sea cucumber polysaccharides in the development of healthy foods and cosmetics.

Keywords Sea cucumber · Fucosylated chondroitin sulfate · Sulfated fucan · Extraction · Separation

1 Introduction

Two kinds of sulfated polysaccharide are present in sea cucumbers. One is sea cucumber fucosylated chondroitin sulfate (fCS), which is a branched heteropolysaccharide with small relative molecular weight. fCS is composed of glucuronic acid, *N*-acetylaminogalactose, and fucose. The fucose branched chain of chondroitin sulfate from sea cucumber is a remarkable feature, which is different from other sources of chondroitin sulfate. Another is sea cucumber sulfate fucan

Z. Wei
College of Food Science and Engineering, Ocean University of China, Qingdao, R.P. China

Y. Chang (✉)
College of Food Science and Engineering, Ocean University of China, Qingdao, R.P. China

Laboratory for Marine Drugs and Bioproducts, Pilot National Laboratory for Marine Science and Technology (Qingdao), Qingdao, China
e-mail: changyg@ouc.edu.cn

C. Xue (ed.), *Advances in Sea Cucumber Processing Technology and Product Development*, Advances in Marine Bioprocesses and Bioproducts,
https://doi.org/10.1007/978-3-031-16512-2_8

(SF), which is composed of L-fucose connected by α-1,3 glycosidic bonds. It is a linear polysaccharide with a relatively large relative molecular mass. The content and ratio of the two polysaccharides are different in sea cucumbers by species, origin, and harvest season. Sulfated polysaccharides extracted from sea cucumbers have various biological activities, such as antibacterial, antiviral, antioxidant, anticoagulant, anti-inflammatory, antitumor, antithrombotic, antifibrotic, and immunomodulatory activities (Zhu et al. 2021). Sea cucumber polysaccharide can be used as a promising functional ingredient in food and medicine.

The separation and purification of polysaccharides significantly affected the purity and yield of polysaccharides. Polysaccharide preparation can be divided into extraction, refining, and purification. The extraction of polysaccharide refers to the preparation process of crude polysaccharides, which mainly consists of two steps. The first step is to release the polysaccharide from the existing tissue into an aqueous solution. Secondly, polysaccharides dispersed in an aqueous solution are separated to obtain crude polysaccharides. Sea cucumber polysaccharides are generally connected with proteins by covalent bonds, and the two also form macromolecular substances in spatial conformations such as coiled, curled, helical, and folded through hydrophobic interactions and non-covalent bonds (such as hydrogen bonds), which exist in the free state. The key issue for the extraction and separation lies in how to remove bound proteins under conditions where the polysaccharides are not significantly degraded, which means how to not only break the secondary bonds between the proteoglycan aggregates but also degrade the core protein chain to destroy the covalent binding of the polysaccharide chain to the protein, thereby releasing the polysaccharide chain for extraction and separation. At present, sea cucumber polysaccharides are mainly extracted from the body wall of sea cucumber, and some reports have also indicated that sea cucumber polysaccharides could be extracted from other tissues of sea cucumber. Enzymatic extraction, alkaline extraction, hot water extraction, ultrasonic-assisted extraction, and microwave-assisted extraction are often used to release polysaccharides from sea cucumbers. The released polysaccharides are usually separated by membrane separation or precipitation to obtain crude polysaccharides.

Polysaccharide refining is to remove impurities and obtain high purity polysaccharides. The common methods for protein removal include Sevage method, trichloroacetic acid (TCA) method, and isoelectric point (IP) method. The commonly used decolorization methods are activated carbon decolorization, hydrogen peroxide decolorization, resin adsorption, and so on.

The purification of polysaccharide refers to the separation of specific polysaccharide components from the crude polysaccharide of sea cucumber. The methods are mainly ion exchange column chromatography and gel column chromatography. The two methods are often used in combination to obtain samples with higher purity. The purified polysaccharide usually shows a single symmetrical peak in chromatographic detection, which proves a good purity. The preparation flow chart of polysaccharide from sea cucumber is shown in Fig. 1. This chapter will review the extraction, refining, and purification method of sea cucumber polysaccharides.

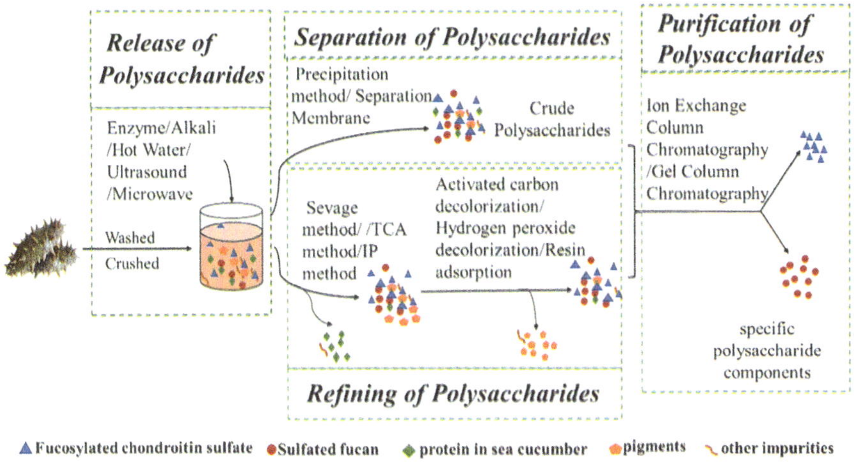

Fig. 1 Flowchart of preparation of polysaccharide from sea cucumber

2 Extraction of Sea Cucumber Polysaccharides

2.1 Release of Sea Cucumber Polysaccharides

2.1.1 Enzymatic Extraction

Enzymatic extraction is a common method for releasing polysaccharides from sea cucumber. Based on the specificity of enzymatic reaction that can effectively hydrolyze or degrade cell membrane components, proteins linked to polysaccharides are hydrolyzed by specific proteases, thus releasing polysaccharides (Guan and Wang 1996). Enzymatic extraction, as a mild and efficient method, has been widely utilized. It can be categorized into the single enzymatic extraction method and the compound enzymatic extraction method. Proteases with low specificity for peptide bond breaking are generally selected, of which pepsin, trypsin, and papain are commonly used. Enzymatic hydrolysis has many advantages, such as mild conditions, easy removal of impurities, high recovery rate, and energy saving. Some literature on the preparation of polysaccharides by enzymatic extraction is listed in Table 1.

In order to ensure sufficient contact between the sea cucumbers raw material and enzymes, pretreatment is carried out before enzymatic extraction. Pretreatment steps include crushing, homogenizing, and defatting. In some studies, ethanol (Liu et al. 2016; Feng and Zhao 2017), acetone (Li et al. 2018), chloroform, and methanol (Chen et al. 2011) were used to defat sea cucumber powder or sea cucumber homogenate. Papain is the most commonly used enzyme. The operating temperature is usually 50–60 °C, and the pH is 6–7. The buffer is usually sodium acetate buffer (Shi et al. 2019), and EDTA and cysteine are often added (Chen et al. 2011). Yang et al. (2020) prepared polysaccharides from the sea cucumber's gonad under the

Table 1 Related studies on the preparation of sea cucumber polysaccharides by enzymatic extraction

Sea cucumber species	Processing method	Precipitation method	Polysaccharide yield	References
Holothuria scabra jaeger	Washed, crushed, 0.4% papain, 55 °C, pH 6.0–6.5, 12 h	40% ethanol, treated by acetone and ethanol	–	Chen et al. (2006)
Holothuria mexicana	Chopped, defatted by acetone, 0.5% papain (containing 5 mM EDTA and 5 mM cysteine, pH 5.7), 60 °C, 12 h	4 V ethanol	–	Li et al. (2018)
Holothuria leucospilota	Fresh sea cucumber homogenated, serine protease, 50 °C, 6 h	Ethanol	–	Qiu et al. (2020)
Stichopus japonicus	Fresh sea cucumber homogenated, 0.4% pepsin and 0.45% trypsin, 50 °C, 5 h	2 V ethanol	–	Zhang et al. (2010)
Thelenota anax	(1) Homogenated, 10 V 95% ethanol, 45 °C, 4 h, repeated twice (2) pH 1.5, pepsin (400 U/g), 37 °C, 3 h (3) pH 7.8, trypsin (400 U/g), 37 °C, 3 h	3 V ethanol	–	Liu et al. (2016)
Acaudina molpadioides *Holothuria atra* *Apostichopus japonica* *Bohadschia marmorata*	Papain	CPC	3.8% 3.0% 2.4% 3.7%	Chang et al. (2010)
Holothuria tubulosa	Papain	CPC	2.5%	Chang et al. (2016)
Holothuria floridana	(1) Dried, chopped, treated with acetone (2) Papain, sodium acetate buffer solution (15 mM EDTA and 15 mM cysteine, pH 6.0), 60 °C, 24 h	(1) CPC (2) 85% ethanol	–	Shi et al. (2019)
Pearsonothuria graeffei *Isostichopus badionotus* *Stichopus tremulus*	(1) Defatted by $CHCl_3$/MeOH (4:1, v/v) (2) 60 °C, papain (in 5 mM EDTA and 5 mM cysteine), 10 h	(1) CPC (2) 95% ethanol	11.0% 6.3% 7.0% 9.9%	Chen et al. (2011)

(continued)

Table 1 (continued)

Sea cucumber species	Processing method	Precipitation method	Polysaccharide yield	References
Holothuria vagabunda				
Cucumaria syracusana	(1) Ground, treated by $CHCl_3$/MeOH 2:1 (v/v) (2) 5 mM EDTA and 5 mM cysteine, papain at 60 °C, 24 h	(1) 0.4% CPC (2) 95% ethanol	–	Chahed et al. (2020)
Stichopus naso *Holothuria floridana* *Holothuria tubulosa*	(1) Ground into powder (2) 0.1 mol/L, pH 6.0, sodium acetate buffer solution (5 mM EDTA and 5 mM cysteine hydrochloride), 10% papain, 60 °C, 12 h	(1) CPC (2) 2 V 95% ethanol	7.37% 8.06% 6.94%	Hu et al. (2015a)
Pearsonothuria graeffei *Isostichopus badionotus*	5 mM EDTA and 5 mM cysteine, papain, 60 °C, 10 h	(1) CPC (2) Stepwise ethanol precipitation	–	Li et al. (2017)
Sea cucumber viscera	(1) Washed, freeze-dried, ground into powder (2) 0.1 M sodium acetate buffer, 5 mM EDTA and 5 mM cysteine, pH 6.0, 5% papain, 60 °C, 24 h	(1) CPC (2) Ethanol precipitation	4.9%	Yang et al. (2020)
Pearsonothuria graeffei	Ground into powder, 5 mM EDTA, 5 mM cysteine, papain, 60 °C, 10 h	(1) CPC (2) Stepwise ethanol precipitation	–	Hu et al. (2015b)
Cucumaria japonica	(1) Homogenated (2) Papain, EDTA, 0.1 M sodium acetate buffer (pH 6.0) (1.46 g) and cysteine (0.78 g), 45–50 °C, 24 h	CTAB (10%, 300 mL)	7.75/1054	Ustyuzhanina et al. (2016)
Thelenota ananas *Stichopus variegatus* *Stichopus chloronotus* *Holothuria nobilis* *Holothuria albiventer* *Holothuria scabra*	(1) Salt-pretreated dried sea cucumber, washed in deionized water, 50 ° C, 12 h, ground twice with meat grinder (2) Papain (enzyme: substrate, 4:1000 w/w), 55 °C, 4 h	(1) Membrane filtration (10 kDa) (2) 4 V ethanol	9.60% 7.01% 11.44% 6.31% 4.14% 6.81% 4.17% 7.24% 3.90% 7.41%	Zhu et al. (2020)

(continued)

Table 1 (continued)

Sea cucumber species	Processing method	Precipitation method	Polysaccharide yield	References
Holothuria tubulosa *Holothuria impatiens* *Ypsilothuria bitentaculata* *Cucumaria frondosa*				
Acaudina molpadioides	(1) Sea cucumber ground powder (2) 7% neutral protease, 45 °C, 4 h (3) 8% papain, 60 °C, 4 h, pH 7.0	Ultrafiltration	14.51%	Chen et al. (2018)
Apostichopus japonicus	(1) 0.1% papain, 50 °C, 6 h (2) 0.5 M NaOH, 60 °C, 2 h	–	–	Guan et al. (2019)
Stichopus variegatus *Holothuria fuscopunctata*	(1) 0.1% papain, 50 °C, 6 h (2) 0.5 M NaOH, 60 °C, 2 h	Ethanol	–	Liu et al. (2020)
Pattalus mollis	(1) 0.1% papain, 50 °C, 6 h (2) 60 °C, 0.5 M NaOH, 2 h	40%, 60% ethanol	–	Ma et al. (2021)
Holothuria coluber	(1) Dried sea cucumber powder (2) 1.0% papain, 50 °C, 6 h (3) 50 °C, 0.25 M NaOH, 2 h	40%, 60% ethanol	5.1%	Yang et al. (2018)
Pattalus mollis	(1) Papain (2) 0.5 M NaOH	40%, 60% ethanol	–	Zheng et al. (2019)
Holothuria fuscopunctata	(1) Sea cucumber powder, NaOH (2) Papain	–	11%	Gao et al. (2020)
Acaudina leucoprocta	(1) Sea cucumber powder, 1% NaOH (1:50, m/v), 60 °C, 3 h (2) 4% papain (m/v), 60 °C, 2.5 h, pH 6.7	–	4%	He et al. (2020)
Acaudina molpadioides *Holothuria nobilis*	(1) Sea cucumber powder, 6 °C, 2 V 6% NaOH, 24 h (2) pH 6.5, 55 °C, 1% papain, 5 h	3 V ethanol	7.60% 8.29%	Dong et al. (2014)

following conditions: 0.1 M sodium acetate buffer, pH 6.0, 5 mM EDTA and 5 mM cysteine contained, 60 °C, and hydrolysis of raw material with 5% papain solution for 24 h. The polysaccharide yield was 4.9%.

In addition to papain, serine protease, pepsin, and trypsin have also been used in the extraction of sea cucumber polysaccharides. Qiu et al. (2020) used serine protease to prepare crude polysaccharides from *H. leucospilota*. Zhang et al. (2010) extracted crude polysaccharides from *Stichopus japonicas* using pepsin and trypsin.

In addition to single enzymatic extraction, compound enzymatic extraction is also a good method to extract sea cucumber polysaccharides. Chen et al. (2018) prepared polysaccharides from *Acaudina molpadioides* by secondary enzymatic hydrolysis. First, 7% neutral protease was used to hydrolyze sea cucumber at 45 °C for 4 h, and then they hydrolyzed sea cucumber with 8% papain at 60 °C for 4 h. The pH value of the whole process was maintained at 7.0, and the extraction rate of polysaccharides reached 14.511%. For the method of enzymatic extraction using papain alone, the yield of crude polysaccharides from *Acaudina molpadioides* was only 3.8%, which indicated that the compound enzymatic extraction method is more suitable for the polysaccharide preparation from *Acaudina molpadioides* (Chang et al. 2010).

Wang et al. (2011) reported a comprehensive extraction process for polypeptides and polysaccharides from sea cucumber. The process used fresh sea cucumbers as the raw material. Sea cucumbers were colloidized first and then hydrolyzed by the compound trypsin. Then, A.S1398 purified neutral protease was added to continue the hydrolysis. Finally, ethanol precipitation was repeated several times to obtain crude polysaccharides. The prepared polysaccharides from sea cucumbers featured a high yield and high purity. Moreover, the process ensured conditions of a high safety standard with mild reactions.

As an efficient and green extraction method, enzymatic hydrolysis has been applied to the preparation of polysaccharides from dozens of sea cucumber species such as *Thelenota ananas*, *Stichopus variegatus*, *Stichopus chloronotus*, *Holothuria nobilis*, *Holothuria albiventer*, *Holothuria scabra*, *Ypsilothuria bitentaculata*, *Cucumaria frondosa*, etc. Sea cucumber polysaccharides are mainly extracted from the body wall of sea cucumbers, which are partially listed in Table 1, and there are also a few studies on the extraction of polysaccharides from sea cucumber viscera (Yang et al. 2020).

2.1.2 Alkaline Extraction

The alkaline extraction method has long been applied to the extraction of polysaccharides. Based on β-elimination reaction, the use of alkaline extraction can break the glycopeptide bonds, so polysaccharides are released from proteoglycan for a more complete extraction. The alkaline extraction method is suitable for the extraction of polysaccharides which are either acidic or have a large molecular weight. This method can not only destroy the glycosidic bonds of glycoproteins, but the alkaline solution used can also form salts with acidic polysaccharides to increase polysaccharide solubility (Xiong et al. 2020). A high concentration of alkaline liquor may break the glycosidic bonds and thus disrupt the polysaccharide structure, so the alkaline liquor used is generally a diluted base such as diluted NaOH or KOH. The advantage of the alkaline extraction method lies in its high yield of crude polysaccharides, but the polysaccharide structure is susceptible to destruction when the alkaline extraction method is used. Desulfation occurs due to the susceptibility to Walden conversion or the formation of 3,6-endoether derivatives after base treatment, thus limiting the application of this method. Therefore, the temperature and time of the reaction and the concentration of the base needs to be carefully screened. Compared with enzymatic extraction, alkaline extraction may cause the shedding and structural change of glycosyls, resulting in the degradation of polysaccharide molecules due to the violent reaction conditions (Fang 2003). However, the reaction conditions of enzymatic hydrolysis are relatively mild, and the integrity of polysaccharide molecular structure can be maintained in the extraction process.

Han and Ma (2012) used alkaline extraction and ethanol precipitation methods to obtain sea cucumber polysaccharides. The process chose 3% KOH with a solid-liquid ratio of 1:50, extracted polysaccharides at 60 ° C for 4 h, and precipitated the polysaccharides by 2 volumes of ethanol to obtain crude polysaccharides. The yield of final polysaccharide was 20.69%, though product impurities were prominent.

Among the preparation methods for sea cucumber polysaccharide, the method of extracting sea cucumber polysaccharide by alkaline extraction alone was rarely used. More researches focused on the combination of alkaline extraction and enzymatic extraction. There are two sequences for extraction: first, enzymatic extraction followed by alkaline extraction and, second, alkaline extraction followed by enzymatic extraction. Ma et al. (2021) extracted crude polysaccharides from sea cucumber *Pattalus mollis* by enzymatic extraction and then by alkaline extraction. After purification, five different sulfated polysaccharide fractions were obtained. This was the first report on obtaining such diverse sulfated polysaccharides from the same sea cucumber species.

Luo et al. (2013) extracted polysaccharides from the body wall of sea cucumbers by alkaline extraction and then enzymatic extraction. The dried body wall tissue was digested with 0.5 M sodium hydroxide. Then, the core proteins bound to polysaccharides were released by papain. The dry weight yields of crude polysaccharides isolated from three sea cucumbers *H. edulis*, *A. japonicas*, and *H. nobilis* were 6.3%, 2.6%, and 6.3%, respectively. Dong et al. (2014) treated sea cucumbers *Acaudina molpadioides* and *Holothuria nobilis* with 6% NaOH at low temperatures, hydrolyzed them with papain, removed proteins with 15% trichloroacetic acid, precipitated polysaccharides with 3 times the volume of ethanol, and used vacuum filtration to obtain sea cucumber crude polysaccharides after freeze-drying. The dry weight yields were 7.60% and 8.29%, respectively. Cai et al. (2009) hydrolyzed *Acaudina molpadioides* with 4% NaOH at 6 °C for 24 h and then used papain for enzymolysis for 4 h. The yield of crude polysaccharides from *Acaudina molpadioides* was 12.08%.

2.1.3 Hot Water Extraction

Hot water extraction is a method that uses high temperatures to increase the diffusion rate of polysaccharides and eventually improve extraction efficiency. This method has many advantages, such as being a simple process, easy to control, and low cost. However, this method cannot destroy the cell membrane, so it can only be used to extract extracellular polysaccharides, and the yield of polysaccharides is relatively low (Xiong et al. 2020).

Cao et al. (2017) extracted crude polysaccharides from *Stichopus japonicus* by using the hot water extraction method. The process defatted the sea cucumber by soaking sea cucumber in 2 V ethanol for 2 h, removed the ethanol, extracted crude polysaccharides in water at 80 °C for 2 h, precipitated the polysaccharides with 70% ethanol, and removed free proteins by the Sevage method. The yield of crude sea cucumber polysaccharides was 9.1%, which contained 56.6% proteins. It was inferred that incomplete hydrolysis of glycopeptide bond led to the high protein content in the crude polysaccharides, indicating that the efficiency of hot water extraction was relatively low, which might explain why this method was rarely used.

2.1.4 Ultrasonic-Assisted Extraction

The ultrasonic-assisted extraction method mainly uses the cavitation, mechanical, and thermal effects of ultrasonic waves to rupture cells and accelerate the release, diffusion, and dissolution of intracellular polysaccharides, which significantly improves the extraction efficiency and rate. It is a modern method for polysaccharide extraction. The disadvantage is that ultrasound can cause the degradation of sea cucumber polysaccharides, which reduces the yield of sea cucumber polysaccharides.

The ultrasound-assisted extraction method is often combined with enzymatic extraction to extract polysaccharides. The yield of polysaccharides by enzymatic extraction is limited, which may be due to the limited enzymatic action sites exposed on the surface of glycoproteins. Ultrasound can enhance the molecular collision rate between enzymes and substrates, destroy the intermolecular force, increase enzymatic action sites, and improve the efficiency of enzymatic hydrolysis (Hou et al. 2020). Wang et al. (2022) extracted polysaccharides from the gonads of the sea cucumber (*Stichopus japonicas*) by using the ultrasonic method, the neutral protease enzymolysis method, and the ultrasonic enzymolysis method separately. The polysaccharide yields were 4.35%, 5.47%, and 6.09%, respectively. The extraction efficiency of the ultrasonic method was higher than that of the enzymatic hydrolysis method. However, an even higher efficiency was obtained through a combination of the two methods. It has been reported that negative pressure could be generated by compression and rarefaction when ultrasonic waves are propagated in the solvent, thereby forming tiny bubbles. The ultrasonic bubbles kept growing until collapsing during the ultrasonic process, which could provide strong shearing force and generate a huge amount of energy, destroying the aggregation of glycoproteins and increasing the restriction enzyme cutting sites of glycoproteins (Ojha et al. 2020). In addition, Wang et al. (2005) combined the alkaline extraction method with the ultrasonic extraction method and used an orthogonal experimental design to study the extraction process of polysaccharides in *Thelenota ananas*. The optimal extraction condition was 3% KOH solution, a solid-liquid ratio of 1:40, and the ultrasonic at 40–45 °C for 2 h. The product was purified by ethanol twice, and the yield of crude polysaccharides was 20.25%.

2.1.5 Microwave-Assisted Extraction

Microwave technology is combined with traditional water extraction method. It mainly uses microwave energy and electromagnetic energy to directly heat extracted raw materials, which can rapidly break down cells and transfer the mass to increase extraction efficiency, improve the extraction speed and shorten extraction time.

Wang et al. (2016) used *Stichopus japonicus* from Xiapu, Fujian, China, as the raw material and optimized the ultrasonic-microwave synergistic extraction process for sea cucumber polysaccharide by orthogonal experimental design: Ultrasonic microwave synergistic extraction was carried out at 80 °C for 30 min and repeated for 4 times. The yield of crude polysaccharides was 12.387%.

2.1.6 Other Methods

In addition to the above conventional methods, there are also some other extraction methods. Kariya et al. (2004) twice added the volume of chloroform/methanol (2:1, v/v) to the homogenate of *Stichopus japonicas*. The mixture was filtered to collect the extract, and this process was repeated twice. The combined filtrate was evaporated to dryness. The product was extracted in distilled water for 16 h at 4 °C, precipitated with 3 volumes of ethanol, and then dried under reduced pressure to obtain crude polysaccharides.

Fang (2005) invented a preparation method for sea cucumber polysaccharide. With the help of autolytic enzymes in fresh sea cucumbers, supplemented by an appropriate number of natural organisms as the degradant and combined with physical catalysis, without concentration or addition of any toxic or harmful chemical solvents, the biological activity of sea cucumber polysaccharides was relatively completely maintained. This preparation method has the advantages of simple production process, low production cost, large output, and high quality of sea cucumber polysaccharides. At the same time, protein-based residual materials could also be obtained, which could be made into various nutrient-rich oral liquids or food additives.

2.2 Separation of Sea Cucumber Polysaccharides

After the degradation of sea cucumbers, the next step is to separate polysaccharides from the polysaccharide extract (a mixture of proteins, polysaccharides, and pigments). At present, the two commonly used methods are the precipitation method and the membrane separation method.

2.2.1 Precipitation

The precipitation method is an approach to transfer polysaccharides from solution to precipitation. The polysaccharides are transferred to the solid phase by precipitant in the polysaccharide extract, so they are separated from most impurities. By precipitant, there are mainly two methods: ethanol precipitation and quaternary ammonium salt precipitation.

Ethanol is one of the most commonly used precipitants. Ethanol can destroy the hydrated shell on the surface of polysaccharide, reduce the ionization constant of the solution, and precipitate polysaccharides in the form of salts. The polysaccharide precipitate is separated from alcohol-soluble impurities (including some pigments and small molecular impurities) via centrifugation. An important parameter to consider when ethanol is used to precipitate polysaccharides is the final concentration of ethanol in the system. The commonly used ethanol concentration range is 40–85%, and different final concentrations of ethanol will achieve different precipitation effects.

Chen et al. (2006) extracted crude polysaccharides from *Holothuria scabra* Jaeger and used 40% ethanol for precipitation. After detection, the main component of the polysaccharides obtained was fucosylated chondroitin sulfate. Shi et al. (2019) precipitated the enzymatic hydrolysis solution of sea cucumber (*Holothuria floridana*) by using ethanol with a final concentration of 85% and finally obtained crude polysaccharides which contained two main components including sulfated fucan and fucosylated chondroitin sulfate. Yang et al. (2018) used the method of step-by-step alcohol precipitation to prepare crude polysaccharides in sea cucumber (*Holothuria coluber*). First, 40% alcohol precipitation was performed to obtain crude polysaccharides. The main component was fucosylated chondroitin sulfate. Then, the ethanol content was increased to 60% to obtain the second batch of crude polysaccharides, mainly consisting of sulfated fucan.

There are two major types of polysaccharide in sea cucumbers. When ethanol concentration is 40–60%, the precipitated polysaccharide is mainly fucosylated chondroitin sulfate. When ethanol concentration exceeds 60%, sulfated fucan is gradually transferred from the solution to the precipitate. The composition of the obtained polysaccharides depends on the final concentration of ethanol. Therefore, during the alcohol precipitation operation, it is necessary to add ethanol while stirring to prevent excessive local ethanol concentration.

Quaternary ammonium salt is a cationic surfactant which can form water-insoluble salts with polyanionic glycosaminoglycan. Commonly used quaternary ammonium salt precipitants are cetylpyridinium chloride (CPC) and cetyltrimethyl ammonium bromide (CTAB). In some cases, the concentration of polysaccharides in enzymatic hydrolysis solution is low, so quaternary ammonium salt precipitation could be a good choice.

The CPC and CTAB precipitation methods are widely used in the preparation of polysaccharides, as shown in Table 1. After polysaccharides react with CPC, they precipitate in the form of insoluble salts. The proteins and some of the pigments which do not react with CPC are still dispersed in the solution. The supernatant is removed through centrifugation, which removes most proteins and pigments. The salts formed by polysaccharides and CPC are insoluble in water, so they need to be dissolved in NaCl-ethanol (100:15 v/v) solution. The concentration of the NaCl solution is relatively high, generally 2–4 mol/L. During this time, the polysaccharides in the solution will be separated from CPC and obtained via ethanol precipitation. The principle of the CTAB method is the same, but NaI is usually used instead of NaCl. Ustyuzhanina et al. (2016) used the CTAB method to precipitate sulfated polysaccharides in *Cucumaria japonica*. Based on the mechanism of quaternary ammonium salts precipitation, there is no way to use quaternary ammonium salt precipitation alone to complete the preparation of polysaccharides, so the ethanol precipitation method is often required to transfer polysaccharides to a solid phase (Guan et al. 2019; Shi et al. 2019; Yang et al. 2020).

Hu et al. (2015a) used papain enzymatic hydrolysis, CPC precipitation, ethanol precipitation, and other steps to obtain sea cucumber polysaccharides from three low value sea cucumbers including *Stichopus naso*, *Holothuria floridana*, and *Holothuria tubulosa*. The yields of sea cucumber polysaccharides were 7.37%, 8.06%, and 6.94%, respectively.

2.2.2 Separation Membrane

The separation membrane method utilizes the characteristics of small molecules such as inorganic salts, monosaccharides, and disaccharides which are able to pass through semipermeable membranes (dialysis membranes, ultrafiltration membranes, reverse osmosis membranes, etc.) in the solution, while macromolecules cannot pass through semipermeable membranes.

The molecular weights of sea cucumber polysaccharides are relatively large, which vary from tens to hundreds of kDa depending on the type of sea cucumber and the preparation process. Some small molecular impurities can be effectively removed by separation of the membrane.

Chen et al. (2014) used a hollow fiber membrane with a cut-off molecular weight of 6000 Da to ultrafilter the enzymatic hydrolyzed solution of *Pearsonothria graeffei*. The crude polysaccharide solution was desalted and lyophilized to obtain the crude product of fucosylated chondroitin sulfate. Li (2009) applied ultrafiltration to purify the polysaccharide components in sea cucumber (*Stichopus japonicas*), and the final concentration of sea cucumber polysaccharides could reach 93.4%, with a significant effect. Chen et al. (2018) used neutral protease and papain to hydrolyze *Acaudina molpadioides* and optimized the ultrafiltration process through a single factor experiment. The operating pressure was 0.20 MPa, the mass fraction of the feed liquid was 6%, and the operating temperature was 35 °C. Feng (2016) proposed a method for extracting sea cucumber polysaccharides by enzymatic hydrolysis combined with ultrafiltration technology. The sea cucumber was crushed and enzymatically hydrolyzed with compound enzyme. The enzymatic hydrolysis product was concentrated and filtered twice with ultrafiltration membrane featuring a cut-off molecular weight of 6–10 kDa. After freeze-drying, sea cucumber polysaccharides were obtained. This extraction process required less water resources, and the extraction rate was significant. Liu (2018) used compound enzymes (a mixture of papain, lipase, and trypsin) to hydrolyze sea cucumbers and filtered them by using ultrafiltration membranes with a cut-off molecular weight of 200,000 to 500,000 Da. The filtrate was treated with a magnetic field and then spray-dried to obtain sea cucumber polysaccharides.

The separation membrane is also a good approach to prepare sea cucumber polysaccharides, which can effectively remove some small molecular impurities. The membrane usually uses 6–10 kDa.

3 Refining of Sea Cucumber Polysaccharides

The refining of polysaccharides is the process of transforming crude polysaccharides into high-purity of polysaccharides. Some impurities such as protein and pigment are difficult to remove completely when sea cucumber polysaccharides are being extracted. Through the refining method, these residual impurities can be further removed to improve the purity of polysaccharides. There are two main processes for refining sea cucumber polysaccharides: one is to remove protein, and the other is to remove small molecular substances such as pigment. Some studies related to polysaccharide refining are listed in Table 2.

3.1 Protein Removal

The protein in sea cucumber is one of the major impurities of crude sea cucumber polysaccharides, which mainly has two forms. One is the protein existing in free state and not combining with polysaccharides, which is easy to remove. The other is the protein covalently binding polysaccharides (Li et al. 2021).

When protein impurities exist in a free state, they are mostly removed with chemical methods, mainly including the Sevage method and the trichloroacetic acid (TCA) method. The mechanism of protein removal by using the Sevage method is based on the denaturation of protein in organic solvents such as chloroform. Two

Table 2 Purification methods of polysaccharides

Sea cucumber species	Polysaccharide type	Deproteinization	Decolorization	References
Apostichopus japonicus	–	IP, pH = 2.8	–	Guan et al. (2019)
Stichopus variegatus Holothuria fuscopunctata	–	IP, pH = 2.8	–	Liu et al. (2020)
Holothuria mexicana	fCS	IP, pH = 2.5	–	Li et al. (2018)
Pattalus mollis	–	IP, pH = 2.8	50 °C, 3% H_2O_2, 2 h	Ma et al. (2021)
Holothuria leucospilota	fCS	–	50 °C, 3% H_2O_2, 1 h	Qiu et al. (2020)
Holothuria coluber	SF fCS	–	3% H_2O_2, 50 °C, 2 h	Yang et al. (2018)
Pattalus mollis	SF fCS	–	3% H_2O_2,45 °C, 2 h (pH 10)	Zheng et al. (2019)
Holothuria scabra jaeger	fCS	TCA method	Decolorization with 30% H_2O_2 solution	Chen et al. (2006)
Acaudina molpadioides Holothuria nobilis	SF fCS	15% trichloroacetic acid	–	Dong et al. (2014)
Stichopus japonicus	–	2 M potassium acetate, 4 °C, 24 h	Decolorization with D101 macroporous adsorption resin and precipitation with 0.15 M potassium acetate for 24 h	Zhang et al. (2010)
Thelenota ananas Stichopus variegatus Stichopus chloronotus Holothuria nobilis Holothuria albiventer Holothuria scabra Holothuria tubulosa Holothuria impatiens Ypsilothuria bitentaculata Cucumaria frondosa		Sevage reagent ($CHCl_3$: *n*-BuOH, 4:1 v/v) deproteinization for six times	–	Zhu et al. (2020)

layers are formed in the protein, and the denatured protein layer is easy to remove. This method has been used to remove protein from various crude sea cucumber polysaccharides, but it usually needs to be repeated over and over again due to low efficiency. Zhu et al. (2020) prepared 10 kinds of crude sea cucumber polysaccharides by enzymatic hydrolysis and treated them with Sevage reagent 6 times to remove protein.

The mechanism of the TCA method is that TCA can form insoluble salts with protein under acidic conditions and lead to the exposure of lipophilic groups through changes in conformation, resulting in protein aggregation. Chen et al. (2006) used 4% trichloroacetic acid to remove the protein of *Stichopus naso* crude polysaccharides. Cai et al. (2009) used the TCA method to remove protein from crude polysaccharides with 50.86% deproteinizing rate and only 3.92% polysaccharide loss rate, achieving an ideal effect.

However, chemical methods also have some disadvantages. The two methods can only remove free proteins (Li et al. 2021). The Sevage method can cause the precipitation of polysaccharides, which form compounds with proteins, reducing the yield of polysaccharides. The reaction conditions of the TCA method are violent, which easily cause polysaccharide degradation and reduce molecular weight (Ge and Yang 2010).

Apart from the previous two methods, another relatively common method for protein removal is the isoelectric point (IP) method. The isoelectric point (IP) method achieves a separation effect by using the lowest solubility of protein at the isoelectric point, at which proteins precipitate easily. In this method, low temperature standing is usually used to promote precipitation, and the supernatant polysaccharide solution can be obtained after centrifugation. The operation is simple. Moreover, it has been found that by adjusting the pH to 2.8 and standing overnight at 4 °C, it can remove protein impurities in the crude polysaccharides of many sea cucumbers, such as *P. mollis*, *A. japonicus*, *S. variegatus*, and *H. fuscopunctata* (Guan et al. 2019; Liu et al. 2020; Ma et al. 2021).

3.2 Depigmentation

There are abundant of pigments in sea cucumbers. In order to improve the purity of polysaccharides, the pigments need to be removed. Commonly used decolorization methods include activated carbon decolorization, hydrogen peroxide decolorization, resin adsorption, and so on.

The hydrogen peroxide decolorization method mainly converts pigment substances through oxidation. A high concentration of hydrogen peroxide will lead to polysaccharide degradation. When hydrogen peroxide is used for decolorization, low concentration is often chosen, generally 3%. H_2O_2 is useful for the decolorization of crude polysaccharides in *Pattalus mollis* and *Holothuria coluber* (Yang et al. 2018; Ma et al. 2021). Han and Ma (2012) refined sea cucumber polysaccharides by decolorization with hydrogen peroxide and removed the protein by using potassium acetate precipitation. The final yield of polysaccharides was 7.96%, and the purity was 82.39%.

The activated carbon decolorization method is mainly used to adsorb and remove pigment substances through the micropores of activated carbon. This method is generally suitable for removing water-soluble pigments and free small molecule pigments in polysaccharide extract (Niu et al. 2021). Yu et al. (2020) used ultrafiltration, activated carbon adsorption decolorization, nanofiltration, and other processes to remove small molecular substances such as pigment in sea cucumber crude polysaccharides.

Commonly used resins mainly include macroporous resin and ion exchange resin. Macroporous resin is characterized by its huge specific surface, which can produce adsorption through surface action and the formation of hydrogen bonds. At the same time, its own porous structure also has a screening effect, but at the same time, some polysaccharides will be adsorbed and removed (Xiong et al. 2020). Ion exchange resin can utilize the ion exchange capacity to treat charged pigment molecules, increasing the efficiency of decolorization significantly (Niu et al. 2021). Zhang et al. (2010) used D101 macroporous adsorption resin to remove the pigments in *Stichopus japonicus* (Table 2).

4 Purification of Sea Cucumber Polysaccharides

The purification of polysaccharides is often used to separate specific polysaccharide components from crude polysaccharides of sea cucumbers. The most common methods to achieve a high purity in polysaccharide components are ion exchange column chromatography and gel column chromatography. Purification also has a refining effect, but it does not aim at refining.

4.1 Ion Exchange Column Chromatography

The ion exchange column chromatography realized the separation of components in the sample based on the difference between the charges. The ion exchanger (in stationary phase) combined with charged ions in the sample through electrostatic interaction. The charged ions in the sample could be eluted by ions with the same charge in the mobile phase. Different components in the sample differed in affinity, so different components could be eluted by the mobile phases with different concentration gradients. The ionic strengths and pH values of those mobile phases also differed.

The two polysaccharides in sea cucumber are acidic polysaccharide and anionic polysaccharide. Containing charged groups, carboxylic acid groups, and sulfuric acid groups, they can be effectively purified by anion exchange column. Common anion exchange chromatographic columns include Amberlite FPA98 ion exchange resin, DEAE-cellulose 52 column, DEAE-sepharose fast flow column, DEAE-sepharose gel column, DEAE-sephacel column (in Cl-), Q-sepharose fast flow column, and Express-Ion D column, as shown in Table 3. During anion exchange, NaCl solutions with different concentrations are the most commonly used elution solutions. Some studies have added sodium acetate buffer solution (pH 5.0) (Dong et al. 2014) or NaH_2PO_4-NaOH (pH 6.3) solution (Shi et al. 2019) to adjust the pH value to weak acidity.

In anion exchange chromatography, fCS is eluted earlier compared to SF in the anion exchange chromatography, due to the charge difference of the two polysaccharides. Chen et al. (2011) purified the polysaccharides of four sea cucumbers *I. badionotus*, *S. tremulus*, *P. graeffei*, and *H. vagabunda* by DEAE-cellulose column. Anion exchange chromatography analysis showed that polysaccharides of the four sea cucumbers had two peaks, corresponding to fCS and SF, respectively. The elution time of SF and fCS may vary among different species of sea cucum-bers. The peak sequence of the four fCSs was fCS-Pg $<$ fCS-Hv $<$ fCS-Ib $\sim$ fCS-St, indicating that there were differences in sulfate content.

In order to enhance the purification effect, studies have used anion exchange chromatography twice in a row. Gao et al. (2020) purified the crude polysaccharides of *H. fuscopunctata* by anion exchange chromatography by utilizing FPA98 ion exchange resin column, eluting with distilled water, and then successively using 0.5 M NaCl, 1.0 M NaCl, and 2.0 M sodium chloride solutions to obtain fCS in 2.0 M NaCl eluent. The components were further purified by weak anion exchange chromatography with DEAE-cellulose 52 column, and SF was obtained with gradient NaCl solutions (0, 0.2, 0.5, 0.8, 1.0, 1.2 M) in 50 mM sodium acetate buffer (pH 5.0) and 1.2 M NaCl eluent.

4.2 Gel Column Chromatography

Gel column chromatography is defined as purifying components according to molecular weight difference, which represents the molecular sieve mechanism. This method is also called gel exclusion chromatography. Under the principle of molecular exclusion, the movement speed of solute in the stationary phase is separated due to the difference in molecular weight and the blocking effect on the stationary phase. Substances with large molecular weight have large volume and cannot enter into the pore size of the gel. They flow out via the gap, and the outflow speed is fast. On the contrary, substances with small molecular weight flow out slowly. Ion exchange chromatography and gel filtration chromatography provide an ideal purification effect, but chromatography resin is expensive, the operation is cumbersome, and the treatment scale is limited.

The columns used for purifying sea cucumber polysaccharides include Sepharose CL-6B (Guan et al. 2019), Sephadex G series columns (G-200 (Wang et al. 2022), G-100 (Ma et al. 2021), G-50 (Liu et al. 2016)), Superdex™ 200 (Zhang et al. 2010), Sephacryl S-200 (He et al. 2020), HiPrep 26/60 Sephacryl S-500 HR column

Table 3 Polysaccharide purification methods

Sea cucumber species	Polysaccharide type	Anion exchange chromatography		Gel permeation chromatography		References
		Chromatographic column	Mobile phase	Chromatographic column	Mobile phase	
Apostichopus japonicus	SF, fCS	Amberlite FPA98 ion exchange resin (8 × 50 cm)	–	Sepharose CL-6B (2.0 × 150 cm)	–	Guan et al. (2019)
Stichopus variegatus Holothuria fuscopunctata	SF	Amberlite FPA98 resin	NaCl solution	–	–	Liu et al. (2020)
Holothuria coluber	SF, fCS	Strong ion exchange chromatography on FPA98 column	H_2O, 0.5 M NaCl, 1.0 M NaCl, 2.0 M NaCl and 3.0 M NaCl	–	–	Yang et al. (2018)
Pattalus mollis	SF, fCS	Strong FPA98 ion exchange chromatography	Eluted with H_2O and 0.5 M, 1.0 M, 1.5 M, 2.0 M and 3.0 M NaCl aqueous solutions	–	–	Zheng et al. (2019)
Pattalus mollis	SF	FPA98 ion exchange resin column (11.5 × 50 cm)	Sodium chloride solution eluted with distilled water, 0.5 M, 1.0 M, 1.5 M, 2.0 M and 3.0 M.	Sephadex G-100	0.3 M NaCl	Ma et al. (2021)
Holothuria fuscopunctata	SF, fCS	(1) FPA98 ion exchange resin (2) DEAE-cellulose 52 column by weak anion exchange chromatography	(1) Eluted with distilled water, 0.5 M NaCl, 1.0 M NaCl and 2.0 M NaCl; fCS was obtained from the 2.0 M NaCl eluent. (2) NaCl solutions (0, 0.2, 0.5, 0.8, 1.0, 1.2 m) in 50 mM sodium acetate buffer (pH 5.0); SF obtained in 1.2 M NaCl eluent	–		Gao et al. (2020)

Acaudina molpadioides *Holothuria nobilis*	fCS	DEAE-sepharose fast flow column	0–2 M NaCl dissolved in 0.1 M sodium acetate, pH 5.0	Sephadex G-100 column	–	Dong et al. (2014)
Stichopus japonicus Selenka	fCS	DEAE-sepharose gel fast flow anion exchange column	0–1 M NaCl	Sephadex G-200 column (1.5 × 100 cm)	0.5 mol/L NaCl	Wang et al. (2022)
Cucumaria syracusana	fCS	DEAE-cellulose column (2.5 × 30 cm)	0–1.2 M NaCl, pH 5.5	–		Chahed et al. (2020)
Thelenota anax	SF, fCS	DEAE-52 further separated and purified	NaCl	Sephadex G-50 chromatography	H_2O	Liu et al. (2016)
Sea cucumber viscera	fCS	DEAE-52 cellulose chromatographic column (2.4 × 40 cm)	0–1.0 M NaCl	Sephadex G-200 column (1.5 × 100 cm)	0.5 M NaCl	Yang et al. (2020)
Stichopus japonicus	SF, fCS	DEAE anion exchange chromatography	0–2.0 M NaCl	Superdex™ 200	0.15 M NaCl	Zhang et al. (2010)
Cucumaria japonica	SF, fCS	DEAE-sephacel with Cl- form (about 380 mL)	NaCl (0.5, 0.75, 1.0 and 1.5 M)	–		Ustyuzhanina et al. (2016)
Pearsonothuria graeffei *Stichopus tremulus* *Holothuria vagabunda* *Isostichopus badionotus*	fCS	DEAE-cellulose column (2.6 × 40 cm)	0–1.2 M NaCl (in 0.1 M sodium acetate, pH 5.0)	–		Chen et al. (2011)
Holothuria scabra jaeger	fCS	DEAE-sepharose F·F·column	–	–	–	Chen et al. (2006)
Holothuria mexicana	fCS		0–2.0 M NaCl	Sephadex G-200 column	0.1 M NH_4HCO_3	Li et al. (2018)

(continued)

Table 3 (continued)

Sea cucumber species	Polysaccharide type	Anion exchange chromatography		Gel permeation chromatography		References
		Chromatographic column	Mobile phase	Chromatographic column	Mobile phase	
		Q sepharose fast flow anion exchange column separation				
Acaudina molpadioides	SF, fCS	Q sepharose-fast-flow anion exchange column	–	–		Chen et al. (2018)
Acaudina leucoprocta	SF, fCS	Q sepharose FF column (3 × 7 cm) anion exchange chromatography	–	Sephacryl S-200 (2 × 100 cm)	0.2 mol/L NH_4HCO_3	He et al. (2020)
Pearsonothuria graeffei	SF	Q-sepharose fast flow column (4.6 × 20 cm)	0–3 M NaCl	–		Hu et al. (2015b)
Holothuria floridana	SF, fCS	Q-sepharose FF (2.6 × 30 cm) (Amersham Pharmacia Biotech, USA)	fCS washed with 1.4 mol/L NaCl (25 mmol/L NaH_2PO_4-NaOH, pH 6.3), SF was washed with 1.8 mol/L NaCl (25 mol/L NaH_2PO_4-NaOH, pH 6.3)	–		Shi et al. (2019)
Holothuria tubulosa	SF	Express-Ion D (Whatman, USA) column	0–2.0 M NaCl	Gel filter column (HiPrep 26/60 Sephacryl S-500 HR column, GE Healthcare, USA)	0.2 M NH_4HCO_3	Chang et al. (2016)

(He et al. 2020), etc. The commonly used mobile phase is NaCl solution. In gel exclusion chromatography, the concentration of the mobile phase is constant without gradient change.

Since gel exclusion chromatography is based on the molecular weight of substances during component purification, it is difficult to purify two polysaccharides with similar molecular weights by gel exclusion chromatography alone. Therefore, in the separation and purification of sea cucumber polysaccharides, gel exclusion chromatography is often used in conjunction with ion exchange chromatography, with the aim to further purify the components purified by ion exchange chromatography. Guan et al. (2019) prepared crude polysaccharides of *Apostichopus japonicus*. The crude polysaccharides were purified by ion exchange chromatography with Amberlite FPA98 ion exchange resin, and then the obtained components were further purified by Sepharose CL-6B gel column. 4.2 g pure sea cucumber polysaccharide was finally obtained. The purity of the purified polysaccharide was over 99% by using high performance gel permeation chromatography (HPGPC) with Shodex OH-pak SB-804 HQ column.

Based on the function of gel exclusion chromatography, when a polysaccharide is being purified, ion exchange chromatography is often performed first, followed by gel exclusion chromatography. However, this approach was used in the opposite order. Liu et al. (2016) purified sea cucumber (*Thelenota anax*) polysaccharides by Sephadex G-50 chromatography. With water as the eluent and a flow rate of 1 mL/min, crude sea cucumber polysaccharides II-I, II-II, and II-III were obtained. II-I was further separated and purified by DEAE-52 and eluted with different concentrations of sodium chloride solution at a flow rate of 1 mL/min. Six refined polysaccharide components were obtained with good purification effect. Some cases of purifying sea cucumber polysaccharides have been summarized in Table 3.

4.3 Other Methods

Aqueous two-phase extraction (ATPE) is also a method for purifying sea cucumber polysaccharides. ATPE is a new separation technology, which has a series of advantages such as mild conditions, fast mass transfer, short phase separation time, convenience in operation, and easy amplification. A system of ATPE was provided for the purification of sea cucumber polysaccharides (Xu et al. 2012). The two-phase extraction system was composed of polyethylene glycol, ammonium sulfate, and water. The polyethylene glycol in the two-phase system had a mass percentage of 5–25%, with an average molecular weight of 600–10,000, while the mass percentage of the ammonium sulfate was 10–30%. After being added to the extraction system, the crude sea cucumber polysaccharides were let stand for phase splitting. The top phase was then desalinated and dried to obtain sea cucumber sulfated fucan; the bottom phase was also desalinated and dried to obtain sea cucumber fucosylated chondroitin sulfate, completing the separation and purification of sea cucumber polysaccharides.

Separation membrane has also been applied to sea cucumber purification in some studies. Chen et al. (2018) used the ultrafiltration process, in which the polysaccharides of *Acaudina molpadioides* were graded in order from small molecular mass to macromolecular mass, and four different molecular mass segments of *Acaudina molpadioides* crude polysaccharides were obtained: G1 (<10 ku, 63.09%), G2 (10–100 ku, 7.24%), G3 (100–200 ku, 4.67%), and G4 (>200 ku, 25.00%).

5 Sea Cucumber Polysaccharide Products

5.1 *Sea Cucumber Polysaccharide Products on the Market*

Some industrialized products of sea cucumber polysaccharides have already appeared on the market. A kind of solid beverage granules of *Stichopus japonicus* polysaccharide is produced in Shandong Province, China. A sea cucumber extract, prepared by double enzymatic hydrolysis process, is produced in Shanxi Province, China. Moreover, the product description mentions that it contains sea cucumber polysaccharides. The product can be used to make nutritional "sea cucumber oral liquid," "sea cucumber milk," "sea cucumber soft capsule," and other products.

5.2 *Patents*

For the development of sea cucumber polysaccharide products, scholars have also carried out some regular researches in China and abroad. There have been studies on sea cucumber polysaccharide-related foods, such as candy (Song et al. 2020) and snakehead soup (Dong 2020). More researches have applied sea cucumber polysaccharides in health foods, such as sea cucumber liquor, sea cucumber polysaccharide oral liquid, sea cucumber polysaccharide capsule, sea cucumber polysaccharide granules (Zhu 2006), etc. Sea cucumber polysaccharide products mainly focus on lowering blood sugar, reducing blood lipids, improving immunity, and so on. There have also been studies on the application of sea cucumber polysaccharides in cosmetics for alleviating skin aging. Some patents related to sea cucumber polysaccharide products are listed in Table 4.

6 Conclusion

As an important component in the nutritional value chain of sea cucumbers, the sea cucumber polysaccharide has attracted more and more attentions in recent years, and products with sea cucumber polysaccharide as the selling point have gradually entered the market. From preparation to refinement, to purification of sea cucumber

Table 4 Sea cucumber polysaccharide product development

Category		Preparation or Formulation	Function	References
Food	Confectionery snack food	Add one or more than one of these ingredients: sulfated fucan, sea cucumber polysaccharides, and carrageenan	During preparation of this confectionery snack food, the activity of sulfated fucan, sea cucumber polysaccharide, and carrageenan does not disappear	Song et al. (2020)
	Snakehead soup product containing polysaccharides	The snakehead soup product is fortified with sulfated fucan and/or sea cucumber polysaccharide, which is mixed with other ingredients of the snakehead soup product prior to packaging	During preparation of the snakehead soup product, the activity of sulfated fucan and sea cucumber polysaccharide does not disappear	Dong (2020)
	Sea cucumber compound nutritious food	The active ingredient sea cucumber polysaccharide is extracted via fresh sea cucumber enzymatic extraction, and the extracted sea cucumber polysaccharides are mixed with the water extracts of black fungus, jujube, goji berries, hawthorn, chrysanthemum, wild jujube, dark plum, and cordyceps as well as black sesame seeds to make food supplements in the form of paste or soup	The present invention not only enhances the nutritional value of the product functionally, but also improves the medicinal value of the product from a medical point of view	Zhu et al. (2013)
Health food	Sea cucumber liquor Sea cucumber polysaccharide granules Sea cucumber polysaccharide oral liquid Sea cucumber polysaccharide capsule	Sea cucumbers are selected as the raw material. Alkali hydrolysis, single and double exogenous enzymatic hydrolysis, or combined autolytic enzyme and exogenous enzyme hydrolysis is applied. Then the isolated mother liquor is dissolved in ethanol and purified to	The present invention extracts the limited amount of mucopolysaccharides from sea cucumber and makes them into functional foods with a certain content, so as to give full play to the role of anticancer and immunity enhancement	Zhu (2006)

(continued)

Table 4 (continued)

Category		Preparation or Formulation	Function	References
		obtain pure mucopolysaccharide product. Add the mucopolysaccharide to the liquor to get sea cucumber liquor The mucopolysaccharide is mixed with the extract of traditional Chinese medicine, and the granules can be obtained through concentrating, granulating, and drying The mucopolysaccharide is mixed with the extract of traditional Chinese medicine, and the oral liquid is obtained through clarifying, flavoring, filtering, and sterilizing The mucopolysaccharide is mixed with the extract of traditional Chinese medicine, and starch is added for tableting to obtain tablets; the mucopolysaccharide is also mixed with β-cyclic dextrin and boiled, and the capsules are obtained after vacuum drying and packaging		
	Sea cucumber polysaccharide oral liquid	With sea cucumber as the raw material, special compound protease, flavor enzyme, and thermal insulation enzyme for the hydrolysis of seafood are utilized; the sea cucumber is then made into an oral liquid by enzymatic hydrolysis	The original flavor of sea cucumber is maintained, without the loss of original ingredients. The prepared sea cucumber polysaccharide oral liquid is conducive to human digestion, with the sea cucumber polysaccharide content of 5 mg/mL	Chi (2016)
	Sea cucumber health liquor	Preparation: Low molecular weight sea	The functional ingredients of the health	Shao (2011)

(continued)

Table 4 (continued)

Category		Preparation or Formulation	Function	References
		cucumber polysaccharide is added to the base liquor of 38–65 degrees according to the concentration of 0.01–5.00 wt%, and sea cucumber saponin is also added to the base liquor by 0.001–0.100 wt%, so as to obtain the final product	liquor are clear and distinguishable, the contents are stable and easy to absorb for human body, and the health-care effect is obvious	
	Sea cucumber nutritious health liquor	After cleaning sea cucumbers, put them in a closed container, add water, and heat at 50–115 °C for 10–180 min. Remove the sea cucumbers and soak them in water for 1 to 5 days by changing the water 1–5 times a day until the water is clear and colorless. After drying, the sea cucumbers are nano-powdered to an average particle size of 100–300 nm. Then, enzymatic hydrolysis treatment is performed: add 0.1–10.0% soluble calcium salt to the enzymatic hydrolysis supernatant and fully dissolve the salt; precipitate with ethanol, and then dry the ethanol supernatant to form a kind of solid sea cucumber health liquor additive. Add 0.1–10 wt% sea cucumber health liquor additives to liquor of 38–65 degrees (v/v). The product is golden		Shao and Jiao (2012)

(continued)

Table 4 (continued)

Category		Preparation or Formulation	Function	References
		yellow or reddish brown, containing 0.06–8.00% sea cucumber protein, 0.01–1.00% sea cucumber polysaccharides, and 0.01–1.00% calcium, with no fishy smell of sea cucumber, but with a unique umami taste of the sea cucumber protein		
	Sea cucumber polysaccharide products that assist in hypoglycemic effects	A powder with hypoglycemic health-care function developed with propolis extract and whole *Lycium ruthenicum* powder as compound ingredients	The ingredients are sea cucumber polysaccharides, flavonoids, and anthocyanins. Through the synergistic effect between the above raw materials, the prepared sea cucumber polysaccharide product has significant hypoglycemic activity	Zhou (2020)
	Hypoglycemic composition containing sea cucumber compound polysaccharide protein powder	The functional ingredients consist of sea cucumber compound polysaccharide protein powder, yam, lily, goji berries, mulberry leaves, sea buckthorn, and kudzu	It has hypoglycemic and hypolipidemic functions and enhances immunity	Yan et al. (2021)
Cosmetics	Anti-aging skincare matrix	The skin matrix contains 40–50% sea cucumber polysaccharides, 29.7–53.7% water, 5–15% propylene glycol, 1–5% vitamin C sodium phosphate, 0.2% methylparaben, and 0.1% hydroxyphenylpropyl vinegar	It can effectively improve the proliferative activity of skin cells and reduce the symptoms of skin aging	Bo et al. (2017)

polysaccharides, there are multiple strategies at each step, providing a wealth of options for the extraction of sea cucumber polysaccharides. However, compared with marine polysaccharides such as sulfated fucan, carrageenan, and agar gum, the research on sea cucumber polysaccharides started relatively late, and most of the extraction protocols are still in the laboratory stage, so their feasibility in actual production needs to be further verified.

References

Bo W, Huang JS, Zhou ZH et al (2017) A preparation method for an anti-aging skin care matrix containing sea cucumber polysaccharides (in Chinese). CN Patent. Appl. 107375009 A

Cai BX, Wu CY, Liu SJ (2009) Optimization of extraction conditions and deproteinization of polysaccharide from *Acaudina molpadioides*. Sci Technol Food Ind 30:194–196+199

Cao RA, Surayot U, You SG et al (2017) Structural characterization of immunostimulating protein-sulfated fucan complex extracted from the body wall of a sea cucumber, *Stichopus japonicus*. Int J Biol Macromol 99:539–548

Chahed L, Balti R, Elhiss S et al (2020) Anticoagulant activity of fucosylated chondroitin sulfate isolated from *Cucumaria syracusana*. Process Biochem 9:147–159

Chang YG, Xue CH, Tang QJ et al (2010) Isolation and characterization of a sea cucumber fucoidan-utilizing marine bacterium. Lett Appl Microbiol 50:301–307

Chang YG, Hu YF, Yu L et al (2016) Primary structure and chain conformation of fucoidan extracted from sea cucumber *Holothuria tubulosa*. Carbohyd Polym 136:1091–1097

Chen J, Zheng AC, Xiao KJ, Lin FL (2006) Isolation and characterization of acidic mucopolysaccharide from sea cucumber. Food Ferment Ind 32:123–126

Chen SG, Xue CH, Yin LA et al (2011) Comparison of structures and anticoagulant activities of fucosylated chondroitin sulfates from different sea cucumbers. Carbohyd Polym 83:688–696

Chen SG, Xue CH, Ye XQ et al (2014) Extraction methods and us-es of chondroitin sulfate polysaccharides from Pearsonothuria graeffei(in Chinese). CN Patent. Appl. 102367285 B

Chen FH, Li C, Wu GB (2018) Study on the extraction and antioxidant activities of polysaccharides and polypeptides from *Acaudina molpadioides* (semper). J Jimei Univ (Nat Sci) 23:105–118

Chi GL (2016) An oral solution of sea cucumber polysaccharide. CN Patent. Appl. 106539090 A

Dong XP (2020) Preparation of a black fish broth product containing polysaccharides. CN Patent. Appl. 111743128 A

Dong XD, Pan RJ, Deng XY et al (2014) Separation, purification, anticoagulant activity and preliminary structural characterization of two sulfated polysaccharides from sea cucumber *Acaudina molpadioides* and *Holothuria nobilis*. Process Biochem 49:1352–1361

Fang CB (2003) Study on extraction and pharmacological action of polysaccharides from *Acaudina molpadioides*. Lishizhen Med Mater Med Res 2:76–77

Fang H (2005) Preparation method of sea cucumber polysaccharide (in Chi-nese). CN Patent. Appl. 1194018 C

Feng QL (2016) An extraction procedure for polysaccharides from sea cucumbers. CN Patent. Appl. 104130334 B

Feng WS, Zhao B (2017) A combined extraction method for multiple nutrients in sea cucumbers. CN Patent. Appl. 107955081 A

Gao N, Chen R, Mou RR et al (2020) Purification, structural characterization and anticoagulant activities of four sulfated polysaccharides from sea cucumber *Holothuria fuscopunctata*. Int J Biol Macromol 164:3421–3428

Ge BY, Yang B (2010) Comparative study on three methods of removing proteins from polysaccharide extract from *Undaria pinnatifida*. Capital Food Med 12:96–98

Guan HS, Wang CY (1996) Advances in extraction, isolation, and analytical methods for aminopolysaccharide determination. Marine Drugs China 1:24–32

Guan RW, Peng Y, Zhou LT et al (2019) Precise structure and anticoagulant activity of fucosylated glycosaminoglycan from *Apostichopus japonicus*: analysis of its depolymerized fragments. Mar Drugs 17:195

Han QJ, Ma HF (2012) Extraction and purification of polysaccharides in sea cucumber (in Chinese). J Anhui Agri 40(14):8071–8072+8074

He WT, Sun HH, Su LJ et al (2020) Structure and anticoagulant activity of a sulfated fucan from the sea cucumber *Acaudina leucoprocta*. Int J Biol Macromol 164:87–94

Hou FR, Ma XB, Fan LH et al (2020) Enhancement of chitin suspension hydrolysis by a combination of ultrasound and chitinase. Carbohyd Polym 231:115669

Hu YF, Dong SJ, Xu XQ et al (2015a) Extraction, isolation and chemical composition of polysaccharides from sea cucumbers of three low value species. Food Ferment Ind 41:227–232

Hu YQ, Li S, Li JH et al (2015b) Identification of a highly sulfated fucoidan from sea cucumber *Pearsonothuria graeffei* with well-repeated tetrasaccharides units. Carbohyd Polym 134:808–816

Kariya Y, Mulloy B, Imai K et al (2004) Isolation and partial characterization of fucan sulfates from the body wall of sea cucumber *Stichopus japonicus* and their ability to inhibit osteoclastogenesis. Carbohydr Res 339:1339–1346

Li YN (2009) Separation and purification of acidic mucopolysaccharide from *Stichopus japonicus* by ultrafiltration. Chem Bioeng 26:39–41

Li S, Li JH, Zhi ZJ et al (2017) Macromolecular properties and hypolipidemic effects of four sulfated polysaccharides from sea cucumbers. Carbohyd Polym 173:330–337

Li QY, Cai C, Chang YG et al (2018) A novel structural fucosylated chondroitin sulfate from *Holothuria mexicana* and its effects on growth factors binding and anticoagulation. Carbohyd Polym 181:1160–1168

Li Y, Li M, Xu BJ et al (2021) The current status and future perspective in combination of the processing technologies of sulfated polysaccharides from sea cucumbers: a comprehensive review. J Funct Foods 87:104744

Liu CJ (2018) A method for the extraction of polysaccharides from sea cucumbers. CN Patent. Appl. 107857829 A

Liu HM, Zhou XQ, Yao YN et al (2016) Purification and antitumor activity of polysaccharides from *Thelenota anax*. J Univ Jinan (Sci & Tech) 30:403–408

Liu XX, Zhang ZX, Mao H et al (2020) Characterization of the hydrolysis kinetics of fucosylated glycosaminoglycan in mild acid and structures of the resulting oligosaccharides. Mar Drugs 18:286

Luo L, Wu MY, Xu L et al (2013) Comparison of physicochemical characteristics and anticoagulant activities of polysaccharides from three sea cucumbers. Mar Drugs 11:399–417

Ma Y, Gao N, Zuo ZC et al (2021) Five distinct fucan sulfates from sea cucumber pattalus mollis: purification, structural characterization and anticoagulant activities. Int J Biol Macromol 186:535–543

Niu JJ, Sun YP, Li X et al (2021) Research progress of microporous resin in decolorization of polysaccharides from traditional Chinese medicine. Chem Eng 35:48–51

Ojha KS, Aznar R, O'Donnell C, Tiwari BK (2020) Ultrasound technology for the extraction of biologically active molecules from plant, animal and marine sources. TrAC Trends Anal Chem 122:115663

Qiu PP, Wu FX, Yi L et al (2020) Structure characterization of a heavily fucosylated chondroitin sulfate from sea cucumber (*h. leucospilota*) with bottom-up strategies. Carbohyd Polym 240:116337

Shao JJ (2011) Preparation of a sea cucumber health care wine. CN Patent. Appl. 101935607 A

Shao JJ, Jiao J (2012) Preparation method of sea cucumber nutraceutical wine. CN Patent. Appl.101935608 B

Shi DL, Qi JH, Zhang HW et al (2019) Comparison of hydrothermal depolymerization and oligosaccharide profile of fucoidan and fucosylated chondroitin sulfate from *Holothuria floridana*. Int J Biol Macromol 132:738–747

Song S, Dong XP, Zhu BW et al (2020) Preparing sugar fruit leisure food comprises adding active substance in the process of preparing sugar fruit leisure food, where the active substance is fucoidan, sea cucumber polysaccharide or carrageenan. CN Patent Appl 111938005A

Ustyuzhanina NE, Bilan MI, Dmitrenok AS et al (2016) Structure and biological activity of a fucosylated chondroitin sulfate from the sea cucumber *Cucumaria japonica*. Glycobiology 26: 449–459

Wang YH, Lv ZH, Jiang TF et al (2005) Study on extraction technology and content determination of polysaccharides in *Thelenota ananas*. Periodical Ocean Univ China 6:109–112

Wang HT, Fu XJ, Jin HZ (2011) Comprehensive extraction technology of polypeptide and polysaccharide from sea cucumber. CN Patent. Appl. 101191139 B

Wang QF, Huang LY, Lin XJ et al (2016) Optimization of ultrasonic microwave synergistic extraction process of *Stichopus japonicus* polysaccharide. J Fujian Med Univ 50:78 81

Wang JJ, Shi S, Li FF et al (2022) Physicochemical properties and antioxidant activity of polysaccharides obtained from sea cucumber gonads via ultrasound-assisted enzymatic techniques. LWT 160:113307

Xiong QP, Song ZY, Hu WH et al (2020) Methods of extraction, separation, purification, structural characterization for polysaccharides from aquatic animals and their major pharmacological activities. Crit Rev Food Sci 60:48–63

Xu XQ, Chang YG, Su W et al (2012) Separation and purification of sea cucumber polysaccharide and application of aqueous two-phase extraction in this process. CN Patent. Appl. 102532340 B

Yang WJ, Cai Y, Yin RH (2018) Structural analysis and anticoagulant activities of two sulfated polysaccharides from the sea cucumber *Holothuria coluber*. Int J Biol Macromol 115:1055–1062

Yang DD, Lin FD, Huang YY et al (2020) Separation, purification, structural analysis and immune-enhancing activity of sulfated polysaccharide isolated from sea cucumber viscera. Int J Biol Macromol 155:1003–1018

Yan QX, Zeng Q, Chen PY et al (2021) Hypoglycemic composition used for hypoglycemic drug, comprises active ingredients comprising sea cucumber compounded polysaccharide protein powder, yam, lily, wolfberry, mulberry leaf, sea buckthorn and kudzu root, and additives. CN Patent. Appl. 112741873 A

Yu DJ, Wang CW, Wang MH et al (2020) An extraction method for bioactive substances from sea cucumbers (in Chinese). CN Patent. Appl. 111393537 A

Zhang YJ, Song SL, Song D et al (2010) Proliferative effects on neural stem/progenitor cells of a sulfated polysaccharide purified from the sea cucumber Stichopus japonicus. J Biosci Bioeng 109:67–72

Zheng WQ, Zhou LT, Lin LS et al (2019) Physicochemical characteristics and anticoagulant activities of the polysaccharides from sea cucumber *Pattalus mollis*. Mar Drugs 17:198

Zhou P (2020) Sea cucumber polysaccharide product used for reducing blood sugar, comprises sea cucumber polysaccharide, propolis extract and lycium ruthenicum whole powder, where propolis extract has high flavone content. CN Patent. Appl. 111543638 A

Zhu BW (2006) Sea cucumber mucopolysaccharide rich food and its preparation method. CN Patent. Appl.1281162C

Zhu BW, Liu XS, Zhao LL et al (2013) Preparation method of composite nutritional foods from sea cucumber(in Chinese). CN Patent. Appl. 101194733 B

Zhu QY, Lin LZ, Zhao MM (2020) Sulfated fucan/fucosylated chondroitin sulfate-dominated polysaccharide fraction from low-edible-value sea cucumber ameliorates type 2 diabetes in rats: new prospects for sea cucumber polysaccharide based-hypoglycemic functional food. Int J Biol Macromol 159:34–45

Zhu YL, Liu LB, Sun ZY et al (2021) Fucoidan as a marine-origin prebiotic modulates the growth and antibacterial ability of Lactobacillus rhamnosus. Int J Biol Macromol 180:599–607

The Extraction, Separation Technology, and New Product Development of Functional Lipids from Sea Cucumber

Jie Xu, Tatsuya Sugawara, Tiantian Zhang, and Takahashi Koretaro

Abstract Sea cucumber lipids with unique structures have various biological activities, such as metabolic syndrome improvement and nerve function enhancement, acting as essential active ingredients in sea cucumber. The extraction and separation methods of functional lipids from sea cucumber mainly include solvent extraction, supercritical fluid extraction (SFE), subcritical fluid extraction (SCFE), column chromatography, high-speed countercurrent chromatography (HSCCC), and so on. Ethanol extraction and supercritical CO_2 extraction (SC-CO_2) are extraction and preparation techniques used to develop lipid-related products. At present, the development of sea cucumber lipid products focuses on sea cucumber oil soft capsules in the field of health-care products, while also dealing with drinks, lipsticks, etc.

Keywords Lipid · Extraction · Product · Phospholipid · Sphingolipid · Steroid sulfate

1 Introduction

The sea cucumber oil mainly contains phospholipid (PL), sphingolipid (SL), steroid sulfate (SS), glyceride, glycolipid, and other components. The fatty acids (FA) include even and odd carbon chains in sea cucumber oil, in which

J. Xu (✉) · T. Zhang
College of Food Science and Engineering, Ocean University of China, Qingdao, R.P. China
e-mail: xujie9@ouc.edu.cn; zhangtiantian@ouc.edu.cn

T. Sugawara
Division of Applied Biosciences, Graduate School of Agriculture, Kyoto University, Kyoto, Japan
e-mail: sugawara@kais.kyoto-u.ac.jp

T. Koretaro
Hokkaido Univ, Fac Fisheries Sci, Hakodate, Hokkaido, Japan
e-mail: kore@fish.hokudai.ac.jp

C. Xue (ed.), *Advances in Sea Cucumber Processing Technology and Product Development*, Advances in Marine Bioprocesses and Bioproducts,
https://doi.org/10.1007/978-3-031-16512-2_9

branched-chain FAs exist in certain sea cucumber species. The PL content is second only to glycerides, and dominant FAs in PLs are eicosapentaenoic acid (EPA), arachidonic acid (ARA), and docosahexaenoic acid (DHA). The common extraction and separation methods of sea cucumber oil include solvent extraction, supercritical carbon dioxide extraction (SC-CO_2), supercritical fluid extraction (SFE), subcritical fluid extraction (SCFE), column chromatography, and high-speed countercurrent chromatography (HSCCC). Before the extraction of sea cucumber oil, sea cucumber or its processing by-products are usually pretreated. The main steps are selection and cleaning, freeze-drying, crushing, and sieving to obtain sea cucumber powder.

PL takes glycerol as its backbone, the *sn*-3 position is esterified by a polar phosphate group, and the *sn*-1/2 position is replaced by fatty acid. The hydrophilic and lipophilic properties of PL come from its special structure. PL can be divided into three types according to the composition of its acyl groups: (1) The PL obtained by connecting two acyl groups to the PL backbone through an ester bond is called diacylphospholipid; (2) the PL obtained by connecting the acyl group at *sn*-1 position to the PL backbone through an ether bond is called alkoxylated phospholipid; (3) the PL obtained by linking the acyl group at the *sn*-1 position to the backbone through an alkenyl ether bond is called plasmalogen. The hydrolysis product of PL is lysophospholipid (LPL). Notably, the content of PL in sea cucumber is relatively high, accounting for about 35% of total lipids, and PLs are also an important active ingredient in biological tissues. In recent years, commonly used methods for preparing PL include solvent extraction, SFE, adsorption chromatography, etc.

SL, a collective term for sphingosine-containing compounds, has received attention for its potential role in protecting the gut from inflammation and cancer. Sea cucumber SLs mainly include ceramide (Cer), cerebroside and ganglioside (GLS), etc., which are usually extracted by the solvent method, purified, and separated by chromatography. In contrast, sea cucumber long-chain base (LCB) is the product obtained by hydrolysis. Compared with sea cucumber PLs, sea cucumber SLs have a lower content and a complex structure and mainly exist in the form of homologs and isomers, so the separation and purification are more complex. Sea cucumber SLs contain LCB d17:1, which stays at a high ratio, but it is rare in plants and mammals. So, it can be used as the characteristic LCB of sea cucumber SLs. Sea cucumber cerebroside is the highest content of glycosphingolipids (GSL) in sea cucumber, mainly glucocerebroside. Sea cucumber GLSs include monosialoganglioside, disialoganglioside, and trisialoganglioside.

As a main component of biofilms, sterols are natural and practical components and widely exist in nature. With regard to the raw materials, sterols are divided into animal sterols, phytosterols, and fungal sterols. Cholesterol is an important animal sterol in mammals, as well as an integral part of cell membranes. Cell membranes could regulate cholesterol mobility and function. Studies have shown that sea cucumber sterols have rich branched structures and sulfate groups compared with phytosterols and fungal sterols. Steroid sulfates (SS) from sea cucumber have attracted a lot of attention due to the particular physiological activity of the different sulfate groups.

As a nutritious marine product, the sea cucumber and its by-products have a high value with regard to deep processing and application. Related products based on sea cucumber lipids are diverse and involved in many fields. In the field of health products, the products include soft capsules and microcapsules of sea cucumber oil, sea cucumber drinks, lipsticks in the cosmetics industry, sea cucumber oil-based soaps as washing products, and salted duck eggs containing active substances of sea cucumber in food processing.

2 Extraction and Separation Technologies of Sea Cucumber Oil

2.1 Solvent Extraction

Solvent extraction is a traditional method for lipid extraction. Organic solvents can destroy fat cells and protein structures to release lipids, and the principle of "Similar Miscibility" is applied to select appropriate organic solvents according to the polar characteristics of lipids. Commonly used solvents are ethanol, petroleum ether, ethyl acetate, benzene, ether, dichloromethane, and *n*-butanol. The extraction rate can be improved through the adjustment of the solvent concentration, extraction time, extraction frequency, extraction temperature, and solvent dosage, or by the complementarity between polar and non-polar solvents when solvent extraction is used for lipid extraction (Wang et al. 2014a, b). Additionally, the extraction time can be saved effectively, and the extraction process can be simplified if extraction is combined with ultrasonic or microwave assistance (Zhou et al. 2017). Sea cucumbers need to be crushed into powder in advance to increase the contact surface with the extraction solvent when solvent extraction is used for lipid extraction. Then the organic phase is used for the extraction and filtration to obtain sea cucumber oil. The solvent extraction method has the advantages of a short production cycle, large production capacity, good selectivity, and easy automation and is easily achieved with large-scale production. The producer should adjust and optimize the extraction solvent, extraction temperature, extraction time, solid-liquid ratio, and other process parameters according to its own demands when applying the solvent extraction method to industrial production (Wang et al. 2014a, b).

Zhao and Feng used ethanol to extract lipids from sea cucumber intestinal eggs, and the specific steps of the extraction process were as follows (Zhao and Feng 2017): First, distilled water was added to sea cucumber's intestinal eggs for the homogenization process, and the volume ratio of sea cucumber's intestinal eggs to distilled water was 1:(1–3). Subsequently, anhydrous ethanol was added to the stirring, and then the solution was allowed to stand for precipitation of the insoluble matters to be obtained in the extracting solution (the volume ratio of the homogenate to ethanol solution was 1:(1–3)). Then, the extract was filtered and distilled under reduced pressure to obtain a kind of crude yellow or yellowish-brown liquid extract.

Finally, the yellowish-brown paste extract was concentrated by vacuum evaporation to obtain the sea cucumber intestinal's egg oil, which contained PLs, cerebrosides, GLSs, and other unsaturated fatty acids (UFA), with an extraction rate of 15–16%. Additionally, Yi et al. used 95% ethanol, 80% ethanol aqueous solution, and dichloromethane to dissolve, extract, and vacuum-evaporate sea cucumber (*Apostichopus japonicus*) viscera powder to obtain brown lipid extract with a yield of about 7%. The extract was rich in 20C FAs, including EPA and DHA, and the UFA proportion was higher than 70% (Yi et al. 2021).

The hydration of salt ions was used to form an organic solvent/multiphase salting out extraction system in addition to using organic solvents for sea cucumber oil extraction. This method could also separate and extract a variety of active ingredients in the sea cucumbers simultaneously. In this system, sea cucumber oil was located in the upper phase, a small number of soluble proteins and polysaccharides were located in the middle phase, whereas most of the proteins and polysaccharides were distributed in the solid phase between the middle phase and the lower phase, and heavy metals were located in the lower phase. This distribution system realized the rapid extraction and separation of sea cucumber oil (Xiu et al.). Furthermore, the extraction of sea cucumber oil was achieved using a buffer system combined with organic solvents, and the main steps were as follows: First, disodium hydrogen phosphate-citrate buffer (pH = 3) was added to the sea cucumber's intestine homogenate, and the sea cucumber's intestine extract was obtained after shaking and the extraction of chloroform-methanol (2:1, v/v) solution. Then, the lower chloroform layer solution was collected by centrifugation, and sea cucumber intestinal lipid was obtained after nitrogen blowing. The final sea cucumber oil with an EPA content greater than 30% was obtained with an extraction rate of 5.4% (Wu et al. 2017).

2.2 *Supercritical Fluid Extraction*

Supercritical fluid extraction (SFE) is a new type of extraction and separation technology. It uses supercritical fluid as the extracting agent to selectively extract components sequentially with different boiling points, polarities, and molecular weights, thus achieving solute separation. It is worth noting that carbon dioxide, ethane, propane, and dimethyl ether can be used as supercritical fluids (Chen and Yang 2021). CO_2 is generally used as an extracting agent due to its advantages of being in abundance, zero environmental pollution, sound stability, non-toxicity, critical temperature close to room temperature, and moderate critical pressure in the extraction of natural products. In the studies on SFE, a third component, the entrainer, is typically added with altered solubility of the solute. The entrainer maintains good dissolving properties. Commonly used entrainers include methanol, ethanol, acetone, ethyl acetate, etc. Key equipment of SFE include high pressure extractor, separator, heat exchanger, high pressure pump (compressor), storage tank and pipes, valves, and joints connecting the above equipment. The main process of

SFE is to first add raw material to the extraction device, add supercritical fluid to the extraction system, and then apply pressure and adjust the temperature for the supercritical fluid. At this time, soluble components enter the supercritical fluid phase. The supercritical fluid experiences pressure reduction, temperature adjustment, or adsorption after flowing out of the extractor. At this point, oil is selectively separated from the supercritical fluid phase. Then the supercritical fluid is sent back to the extraction device after being tempered and pressurized for the purpose of recycling. This technique has the advantages of being a gentle operation, retention of bioactive substances, high quality of extracted products, zero solvent residue, and zero environmental pollution (Chen and Yang 2021).

Li et al. used supercritical CO_2 (SC-CO_2) extraction to efficiently obtain high-purity UFAs rich in EPA and DHA from sea cucumber raw materials, with the content of 72.8 g/100 g for unsaturated fatty acids in the resultant sea cucumber oil, of which EPA was 18.2 g/100 g and DHA was 0.5 g/100 g. This method completely retained active ingredients such as collagens and polysaccharides in sea cucumber powder with an extraction rate of about 10%; it also realized the comprehensive utilization of sea cucumber powder (Li et al. 2016). Zakharenko et al. also used the SC-CO_2 extraction method to extract sea cucumber lipids under the conditions of 2500 psi pressure, 60 °C temperature, and 20 g/min carbon dioxide flow rate. The extraction rate was 1.8% (dry weight) (Zakharenko et al. 2020).

2.3 *Subcritical Fluid Extraction*

Subcritical fluid extraction (SCFE) is a new technology created after SFE, which mainly overcomes shortcomings of the equipment in SFE such as small volume, high cost, and high energy consumption that is unsuitable for large-scale industrial production. SCFE can extract active components from natural products under low-temperature, oxygen-free, and non-thermal conditions, which has the advantages of guaranteed product quality, large processing volume, zero solvent residue, and zero three-waste emission. SCFE enjoys a wide range of application prospects for obtaining active ingredients from natural animals and plants, active ingredients of traditional Chinese medicine, animal extracts, natural pigments, special oils, etc. and for separating harmful fat-soluble components. Extraction solvents commonly used in SCFE are organic solvents with low boiling points and easy recovery, such as ethane, propane, butane, isobutane, dimethyl ether, etc. The principle of SCFE is to use subcritical fluid as the extraction agent and utilize the molecular diffusion process of extraction material and extracting agent during soaking to transfer fat-soluble components in the solid material to the liquid state in a closed, oxygen-free, low-pressure vessel. Then the extracting agent is separated from the target product by vacuum evaporation, and the final product will then be obtained (Guo et al. 2020). It is worth noting that feedstock particle size, extraction temperature, extraction pressure, and extraction time are the key factors affecting extraction efficiency. Additionally, raising the temperature can increase the volatility and diffusion coefficient of the extracting agent to some degree, resulting in better lipid solubilization and an increased extraction rate. Compared with SC-CO_2 extraction, SCFE allows of lower equipment investment (Xu et al. 2018b). In the actual production process, the parameter ranges can be determined based on production experience, so as to obtain an optimal extraction scheme through experimental validation, response surface, and other mathematical optimization designs (Guo et al. 2020).

Xu et al. obtained a special kind of high-quality sea cucumber oil rich in *n*-3 PUFA such as EPA and DHA efficiently at a low temperature using SCFE technology, and the contents of DHA and EPA in the sea cucumber oil were much higher than what corporate standards stipulate. The specific process was as follows: First, deep-sea *Apostichopus japonicus* was washed, cut into strips, vacuum freeze-dried, crushed, and made into deep-sea *Apostichopus japonicus* particles. Then, the particles were placed in a gauze bag, and the gauze bag was put into an extraction tank to prepare for extraction. Next, the extraction tank was vacuumed, and the extracted solvent was added to the extraction tank for extraction. The extraction solvent used could be recycled, and the sea cucumber particles needed to be extracted three times. Finally, the extract was transferred to an evaporation tank for evaporation to obtain a kind of light red sea cucumber oil, and the sea cucumber oil ought to be isolated from the air and stored in an airtight manner (Xu et al. 2020b).

3 Extraction and Separation Technologies of Sea Cucumber Phospholipids

3.1 Extraction and Separation Technologies of Total Phospholipids

3.1.1 Solvent Extraction

Solvent extraction is a traditional method used to prepare PLs. This method takes advantage of the solubilities of different PLs in solvents to separate PLs from other components. For example, Zhao and Feng extracted phospholipids from dried coarse sea cucumber powder by solvent extraction. First, the coarse powder was stirred and filtered with 60% ethanol solution, and then the obtained clear extract was subjected to vacuum distillation and concentrated to obtain a yellowish-brown paste-like extract with an extraction rate of 8% to 10%. The above extract contained about 30% PLs and about 30% UFAs and was rich in other micronutrients (Zhao and Feng 2018). It is worth noting that solvent extraction is often used to improve the purity of PLs when crude PLs are being purified. Wang et al. extracted total lipids from the body wall of sea cucumber *Cucumaria frondosa* by the Folch method of washing the body wall with cold acetone to remove neutral lipids and glycolipids and finally obtained EPA-phospholipids (EPA-PLs). The EPA-PLs contained 48.7% phosphatidylcholine (PC) and 43.1% phosphatidylethanolamine (PE) (Wang et al. 2020).

3.1.2 Silica Gel Column Chromatography

Silica gel column chromatography is a common method for obtaining high-purity lipids by the principle of separating substances on silica gel with different adsorption forces. For instance, Lou purified and prepared PLs from the body wall of sea cucumber *Cucumaria frondosa* using silica gel column chromatography (Lou 2011). The specific steps were as follows: First, the body wall of *Cucumaria frondosa* was freeze-dried and then directly crushed and sieved to obtain *Cucumaria frondosa* powder. Then, water was added to the powder for swelling, and chloroform-methanol solution (2:1, v/v) was added to obtain an extract. Next, the extract was washed with 0.9% NaCl solution, and the chloroform layer extract was collected, dried with anhydrous sodium sulfate, and concentrated under reduced pressure to obtain *Cucumaria frondosa* total lipids after standing for layers. The extracted total lipids were added with acetone and sonicated, and the acetone was then discarded. This process was repeated three times, and acetone-insoluble substances were collected. Then, the substances were dissolved in chloroform and loaded, and the substances were further eluted with 5 times the column volume of chloroform and acetone in turn. After the elution of acetone, the acetone in the column was replaced with chloroform. Then, the above substances were eluted with the chloroform-methanol solution, and the eluents of different components were obtained with a gradually increased methanol ratio and through the collection in stages. Finally, the collected PL-rich eluate was concentrated under reduced pressure to obtain PLs in *Cucumaria frondosa* via TLC analysis, and the PL content exceeded 90%. Additionally, Wang et al. also used silica gel column chromatography to prepare and purify PLs in *Cucumaria frondosa* dry powder. First, the dry powder of *Cucumaria frondosa* was subjected to extraction with chloroform-methanol (2:1, v/v) and 0.88% potassium chloride solution to obtain the total lipids of the sea cucumber. Then, chloroform-methanol (9:1, v/v), pure acetone, and pure methanol were used as eluents successively, the total lipids were eluted on a silica gel column, and the methanol eluent was collected to obtain a crude PL extract. The crude PL extract was then isocratically eluted with a mixture of chloroform-methanol-water (80:15:1, v/v). Next, the crude PL extract was purified by silica gel column, the eluate was collected, and finally the eluate was concentrated to obtain EPA-PLs. The yield of EPA-PLs was 50% and the purity was 95% (Wang et al. 2016).

3.2 *Extraction and Separation Technologies of Plasmalogen*

3.2.1 Supercritical Fluid Extraction

The SC-CO_2 extraction method allows the solubility of PLs to vary within a wide range with changes in pressure and temperature based on the principle of abnormal phase equilibrium behavior and transfer performance between fluid and oil near the critical point, thereby achieving PL separation. Notably, extraction pressure, temperature, time, and CO_2 flow rate all affect the extraction effect in addition to the entrainer, and pressure has a greater impact on the extraction rate. At a certain temperature, with the increase of pressure, the ability of the CO_2 fluid to dissolve PLs becomes greater. However, the solubility of the CO_2 fluid for PLs increases slowly or even decreases after the pressure reaches a certain level and pressurization continues going. SFE has a strong oil-dissolving ability but weak PL-dissolving ability and can be used to remove most of the neutral oils and cholesterol in the raw material, but the equipment cost is relatively expensive. Additionally, compared with other extraction and separation methods that require use of toxic organic solvents, SFE is an environment-friendly method for purifying PLs with certain advantages in the production of functional health-care products of PLs.

Xue et al. used SC-CO_2 extraction to obtain sea cucumber plasmalogen, rich in EPA and ARA, with a purity of up to 80% (Xue et al. 2020). The specific operations were as follows: (1) Froze and vacuum-dried the sea cucumber, and then pulverized it to obtain sea cucumber powder. (2) Set extraction tank solvent ratio, working pressure, temperature, time, and separation tank temperature, keep substrate concentration at 2–10 g/mL, and add 2–10% amount of enzyme (phospholipase A1, PLA1) for the enzymatic hydrolysis of sea cucumber powder. After full reaction, an appropriate amount of water was added to the extraction tank, the reaction temperature and stirring speed were controlled, and the sea cucumber powder was fully enzymatically hydrolyzed to obtain the corresponding product. (3) Set extraction temperature, extraction pressure, and extraction time for supercritical extraction of the product. Fully remove free fatty acids and α-GP in the product, and obtain EPA/ARA plasmalogen containing or partially containing plasmenyl phosphocholine (pPC), plasmenylethanolamine (pPE), plasmenyl phosphatidylserine (pPS), plasmenyl phosphatidylinositol (pPI), or plasmenyl phosphatidic acid (pPA).

3.2.2 Silica Gel Column Chromatography

Wang et al. obtained PLs from sea cucumber *Cucumaria frondosa* by silica gel column chromatography. The specific operation was to use chloroform-methanol solution as the eluent and separate different kinds of plasmalogens by changing the two-phase ratio of chloroform-methanol. For example, the eluent of chloroform-methanol (2:1, v/v) mainly contained pPE, while the lipid in the pure methanol eluate was mainly PC. The above silica gel column chromatography could not only extract the plasmalogen rich in EPA but also separate each PL component. The purity of the plasmalogen was as high as 90% (Wang et al. 2018). Liu et al. also used silica gel column chromatography with reference to the Folch method to obtain PLs derived from *Stichopus japonicus*. First, total lipids were obtained through leaching of *Stichopus japonicus* with dichloromethane-methanol (2:1, v/v) solution. Then, the total lipids were washed several times with cold acetone, and insoluble matters were obtained as *Stichopus japonicus* PLs. The *Stichopus japonicus* PLs were dissolved in a small amount of dichloromethane, and the PLs were then purified by silica gel column chromatography. The specific operation mentioned above was to use dichloromethane-methanol (9:1, v/v) to remove neutral lipids, use dichloromethane-methanol-water (65:25:4, v/v) to elute PLs, and then collect PE step by step. PE was evaporated in rotation and dissolved in *n*-hexane. The lipids were extracted with *n*-hexane/isopropanol after the addition of PLA1 to enzymatically hydrolyze PE, and the extract of the hexane layer was collected at the same time. Finally, the collected extract was washed with physiological saline, and the *n*-hexane was evaporated until dry to obtain plasmalogen (Liu et al. 2021). Wang et al. used silica gel column chromatography to separate and purify phospholipids from the body wall of sea cucumber *Cucumaria frondosa*. The specific steps were as follows (Wang et al. 2021): Chloroform-methanol (2:1, v/v) was added to the body wall of the sea cucumber. After continuous stirring for ether-PL extraction, distilled water was added for phase separation, and the bottom phase was collected. The upper phase was re-extracted, and the chloroform phase was combined. After vacuum decompression concentration, the total lipids of sea urchin were obtained from the collected chloroform phase. Ether-PLs were then isolated from the total lipids via silica gel column chromatography with the successive multiple elutions of chloroform, chloroform-methanol (2:1, v/v), and methanol. The chloroform-methanol (2:1, v/v) and methanol eluents were collected. Ether-PLs were further purified via silica gel column chromatography with the eluents of methanol, chloroform-methanol (3:1, v/v), chloroform-methanol (1:3, v/v), and methanol. Purified pPE and plasmanyl phosphatidylcholine (aPC) were collected with a split collector and obtained after the removal of organic solvent under vacuum. The purity of pPE and aPC was higher than 90%.

3.3 Extraction and Separation Technologies of Lysophospholipids

Nishikawa et al. prepared sea cucumber lysophospholipids (LPLs) by solvent extraction combined with silica gel column chromatography. The specific operations were as follows: (1) Sea cucumber *Holothuria atra* was put into a chloroform-methanol (1:2, v/v) mixer at room temperature for homogenizing and mixing to obtain an extract. (2) The extract was filtered under reduced pressure, resulting in a suspension. Meanwhile, the filtered residue was treated with a chloroform-methanol mixture (1:4, v/v) and a chloroform-methanol mixture (1:2, v/v) at room temperature, and the suspension was collected by filtration under reduced pressure. (3) The total suspension of *Holothuria atra* was treated with chloroform-methanol-water (8: 4:3, v/v) overnight to allow the suspension to layer, and the lipid layer was collected to obtain the total lipids of *Holothuria atra*. (4) The total lipids were eluted with chloroform, chloroform-methanol (4:1, 2:1, 1:1 and 1:4, v/v), methanol, chloroform-methanol-water (3:6:1, v/v), and chloroform-water (4:1, v/v) on a silica gel column in sequence. Among them, chloroform-water (4:1, v/v) eluted a mixture of LPLs (Nishikawa et al. 2015).

4 Extraction and Separation Technologies of Sea Cucumber Sphingolipids

4.1 Extraction and Separation Technologies of Cerebroside

4.1.1 High-Speed Countercurrent Chromatography

High-speed countercurrent chromatography (HSCCC) is a highly efficient and rapid liquid-liquid partition chromatography technology with sound reproducibility, high separation efficiency, and simple operation. Additionally, samples prepared by HSCCC are characteristic of significant content, high purity, and low solvent consumption due to the sufficient contact between the separated substances and the liquid stationary phase. At the same time, HSCCC is suitable for preparative separation and has been widely used in recent years. Different from traditional chromatography, HSCCC does not use solid support as the stationary phase, and the separation of substances depends on the difference in the distribution coefficient of the separated substances in the two phases, thus avoiding sample loss, inactivation, denaturation, etc. caused by irreversible adsorption. HSCCC also has a strong separation ability, which means that some samples can get one or more monomers after one separation, and the separation time is also short; generally, a few hours can complete a separation compared with normal pressure and low-pressure chromatography. Because of the above advantages of HSCCC, it has attracted more and more attentions from researchers worldwide. At present, HSCCC can achieve the preparative separation of hundreds of milligrams or even tens of grams, and the preparation capacity has reached the preparation scale of pilot-scale HPLC and has gradually evolved into a more mature and industrialized high-efficiency separation technique (Xu 2011).

Xu et al. used the HSCCC method to purify and prepare complete sea cucumber cerebrosides from the total lipids of sea cucumber *Acaudina molpadioides*, *Cucumaria frondosa*, and *Apostichopus japonicus*, respectively, with petroleum ether-methanol-water (5:4:1, v/v) as the solvent system. The yield of the total cerebrosides was about 7.6–12.4% (calculated by total lipid), and the purity of the cerebrosides was over 90%. The specific operations were as follows: The total lipids of sea cucumber were eluted by normal phase silica gel column chromatography, and the eluate fraction of chloroform-methanol (80:20, v/v) was collected. Then, sulfate steroid (SS), which seriously interfered with the purification of sea cucumber cerebroside in the above elution fractions, was removed using HSCCC with ethyl acetate-ethanol-water (5:3.5:3.5, v/v) as the solvent system. Finally, three kinds of sea cucumber cerebrosides were prepared, the total yield was 1.1–6.4%, and the purity was above 90% (Xu 2011).

4.1.2 Supercritical Fluid Extraction

Sun et al. proposed a technique to extract sea cucumber cerebroside using the SC-CO_2 extraction method belonging to SFE by optimizing various conditions in SC-CO_2 extraction and finally determined the best process conditions for SC-CO_2 extraction: the extraction temperature was 32–35 °C, the extraction pressure was 25–30 MPa, the flow rate of CO_2 was 20–25 L/h, the extraction time was 3–3.5 h, and the particle size was 20–100 mesh (Sun et al. 2017). The specific steps were as follows: (1) Boiled fresh sea cucumbers in water until the sea cucumber surface became harder and was not easily deformed; (2) slice the boiled sea cucumbers, put them on a plate after slicing, and put them into a vacuum freeze dryer to dry to constant weight; (3) put the cucumbers into a pulverizer for pulverization and sieving; and (4) put the sieved sea cucumbers into an extraction kettle for dynamic extraction to collect the cerebroside extract. Finally, open the extraction kettle equipment, and then pressurize and heat CO_2, so that the CO_2 could extract the cerebrosides from sea cucumber in a supercritical liquid state. Xu et al. used the SC-CO_2 extraction method to extract cerebroside from sea cucumber. They used a single-factor and orthogonal experiment to study the effect of various supercritical extraction conditions on the extraction rate of sea cucumber cerebrosides. The results showed that suitable supercritical extraction conditions were as follows: extraction time was 3 h, extraction temperature was 38 °C, extraction pressure was 20 MPa, and particle size was 60 mesh. Under the optimized conditions mentioned above, the highest extraction rate of cerebrosides in sea cucumber could exceed 0.1% (Xu et al. 2018a, b).

4.1.3 Column Chromatography

Yamada used column chromatography to separate and prepare two kinds of GSL, sea cucumber cerebroside, and GLS. Chloroform-methanol and *n*-hexane were used to extract sea cucumber cerebrosides. For GLSs, it was necessary to extract with *n*-butanol again and use silica gel column chromatography (Yamada 2002). Significantly, although silica gel column chromatography is relatively simple, it is often impossible to obtain a single cerebroside component. For example, Kisa et al. used silica gel column chromatography to separate and purify three cerebroside series from sea cucumber *Apostichopus japonicus*. Compared with the HPLC method, the yield of some cerebrosides was more than double (Kisa et al. 2005).

4.1.4 Preparative Liquid Chromatography

Xu et al. used high-performance reversed-phase liquid chromatography (RPLC) to obtain four pure sea cucumber cerebroside monomers with more than 90% purity (Fig. 1). The specific operations were as follows: After the purified sea cucumber

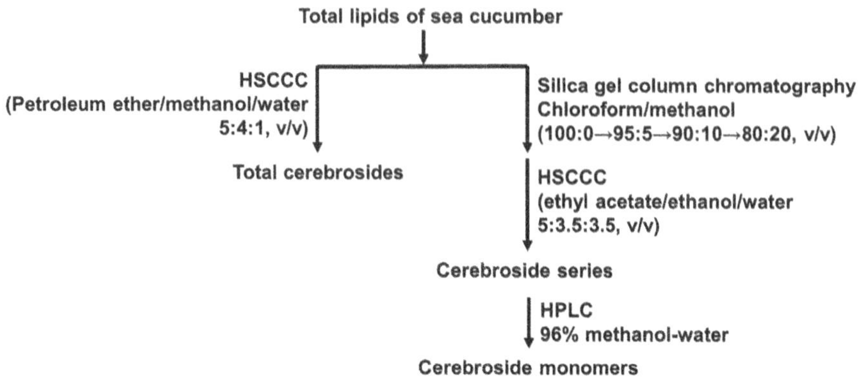

Fig. 1 Flow chart of purification for sea cucumber cerebrosides

total cerebrosides were entirely dissolved with methanol, they were separated by a C18 semi-preparative column. Then, with 96% methanol aqueous solution as the mobile phase, the total sea cucumber cerebrosides were detected by an ultraviolet detector (detection wavelength 220 nm), and the chromatographic peaks of different sea cucumber cerebroside monomers were collected according to retention time. Finally, the collected solution of each chromatographic peak was blown dry with nitrogen to obtain four pure sea cucumber cerebroside monomers (Xu 2011). Remarkably, high-purity cerebroside monomers could be obtained by preparative liquid chromatography but at a high cost.

4.2 Extraction and Separation Technologies of Gangliosides

GLSs refer to GSLs that contain sialic acid in their structure. GLSs are widely involved in normal cells' growth, differentiation, regulation, and information transmission, especially in neuron maturation and repair and regeneration of damaged nerves. The extraction and separation process of GLSs is cumbersome and complicated, and it is challenging to obtain pure GLSs due to the complexity of its sugar chain structure, which often requires a combination of multiple chromatographic methods for the extraction and separation of GLSs.

The extraction of GLSs generally includes leaching with chloroform-methanol solution to obtain GLS extract. A crude GLS sample is used before loading it to a column and is obtained by the principle of water and organic solvent distribution after the extract is concentrated and dried. Common distribution systems for GLS extraction are *n*-hexane-water, *n*-butanol-ethyl acetate-water, chloroform-methanol-water, etc. The hydrophilic sugar chains in the structure of GLSs could easily enter into the aqueous phase, so the aqueous layer of the extract was concentrated and dried and then dissolved in chloroform-methanol solution (1:1, v/v) to obtain the soluble material as the crude sample containing GLSs. Folch established a method for extracting GLSs from tissues based on a chloroform-methanol-water system in 1957 (Folch et al. 1957). The specific operations were as follows: First, use chloroform-methanol (2:1, v/v) to extract sample lipids, and then add water, which accounted for one-fifth of the total volume of chloroform-methanol, to the extract for two-phase distribution, so the GLSs were extracted into the upper aqueous layer after being shaken and let stand. The extraction efficiency of the extraction and distribution method by Folch was not considerable, and it was easy to introduce sulfatide, phosphatidylserine (PS), and some non-lipid impurities. Therefore, Sevennerholm improved the extraction method of GLSs based on the chloroform-methanol-water distribution system proposed by Folch. The specific operations were as follows: Add a certain amount of water to the extract to set the final ratio of chloroform-methanol-water (4:8:5.6, v/v), and a two-phase system was generated. When chloroform-methanol-water (4:8:3, v/v) was used to extract GLSs in the sample, the upper system in the two phases was collected as a crude sample of GLSs (Svennerholm and Fredman 1980). Emulsification often occurred when GLSs were separated by a chloroform-methanol-water extraction system, which was not easy to separate due to the complexity of the matrix in the sample. In practice, a certain concentration of salt solution (such as 0.1 mol/L KCl) could be used to replace water and inhibit the occurrence of emulsification.

Commonly used methods for separating and purifying GLSs include reversed-phase column chromatography and column chromatography with adsorption, ion exchange, and gel mixing. Xu et al. used column chromatography to separate and purify sea cucumber polar glycolipids from the total lipids solution obtained after leaching sea cucumber by chloroform-methanol. The specific operations were as follows: First, water was added to extract the total lipids from sea cucumber, and the upper layer methanol-aqueous phase solution and the lower layer chloroform solution were separated by centrifugation. The methanol-water solution was then extracted with an activated C8 solid-phase extraction cartridge and eluted with methanol, and then the eluate was collected. Next, the eluate was subjected to thin-layer chromatography (TLC) with the chloroform-methanol-water solution as a developing solvent, and the bands with Rf values of 0.8–0.9 were collected. Crude sea cucumber glycolipid powder was dissolved in chloroform-methanol-water solution and subjected to ion-exchange column chromatography after concentration of the bands to obtain crude sea cucumber glycolipid powder. Then, the sample was linearly eluted with chloroform-methanol-water solution containing ammonium acetate, and the eluate was concentrated to obtain the final polar glycolipid component of sea cucumber (Xu et al. 2016). Additionally, Yamada et al. separated and purified GLSs in sea cucumber *Cucumaria echinata* (Yamada et al. 2000), and the specific steps were as follows: Raw sea cucumbers were minced and then extracted with chloroform-methanol (1:2, v/v). The residual aqueous suspension was extracted with ethyl acetate-*n*-butanol (2:1, v/v). After the leaching mixture was concentrated under reduced pressure to remove the organic solvent, the organic phase was removed, and then the aqueous phase was partially washed with *n*-butanol. The solids were dissolved in chloroform-methanol (1:1, v/v) to obtain water-soluble polar lipid components, and then the polar lipids were separated by a

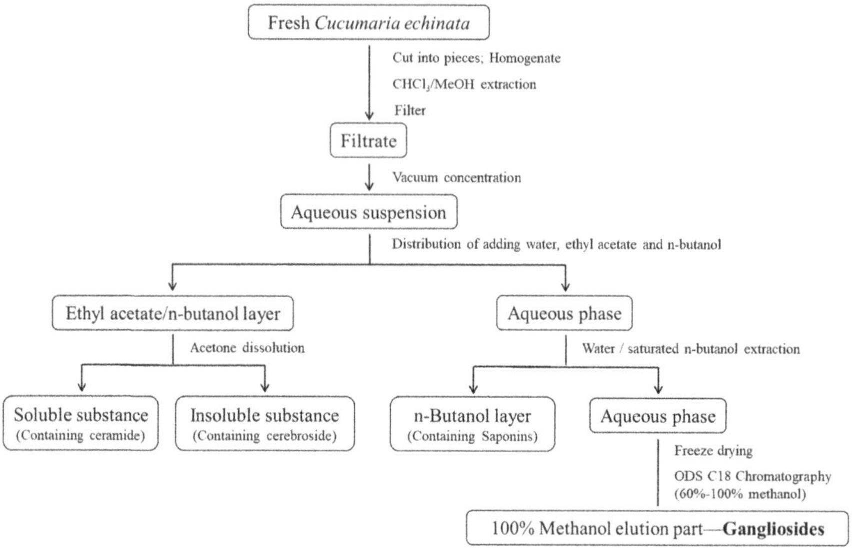

Fig. 2 General separation and purification process for sea cucumber GLSs (Cong et al. 2011)

reversed-phase C18 column (elution solvent: 60%, 80%, and 100% methanol). The fractions containing GLSs (100% methanol elution fraction) were further separated by normal phase silica gel column chromatography, and the GLS fractions were eluted with a chloroform-methanol-water solvent system with a volume ratio of 7: 3:0.5 to 6:4:1 (v/v) to obtain *Cucumaria echinata* GLSs (Yamada et al. 2000). Yamada also reported the separation and preparation of sea cucumber GLSs by using three kinds of mixed column chromatography: normal phase silica gel column, DEAE-TOYOPEARL ion exchange column, and Sephadex LH-20 gel column (Yamada 2002). Cong et al. also studied the reversed-phase column chromatography separation and purification process for sea cucumber GLSs similar to the above (Fig. 2) (Cong et al. 2011).

4.3 Extraction and Separation Technologies of Ceramide

At present, the separation methods of Cer in sea cucumbers often adopt column chromatography. For example, Feng et al. used column chromatography to separate Cer from *Acaudina molpadioides*. The specific operations were as follows: First, dry powder of *Acaudina molpadioides* was extracted with chloroform-methanol-water solution (4:4:1, v/v), the extracts were combined, and the liquid extract was obtained by concentration under reduced pressure. Next, the liquid extract was dispersed in water and extracted with ethyl acetate-*n*-butanol (2:1, v/v), and the organic phase was concentrated to dryness under reduced pressure. The above extract was

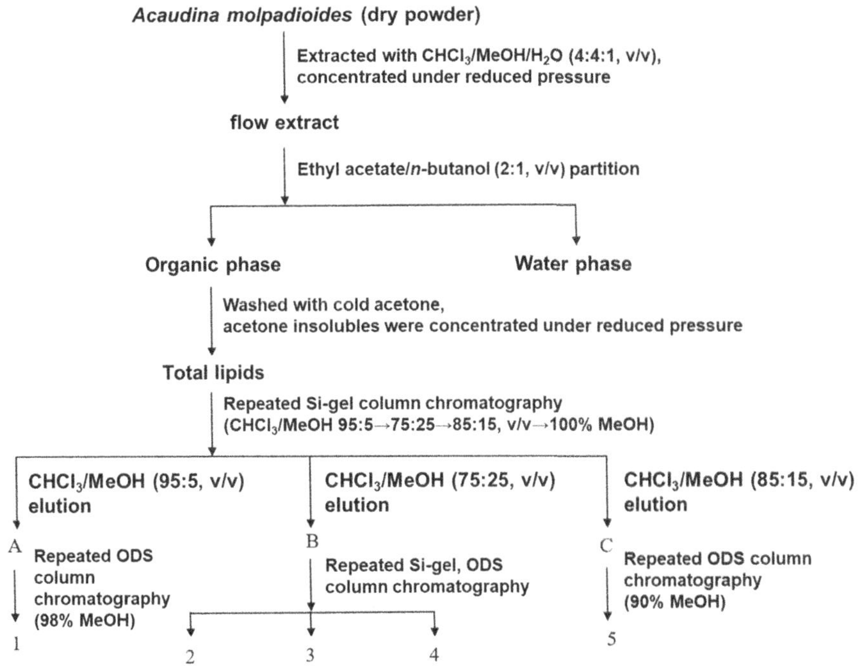

Fig. 3 Purification flow chart of sphingolipids from *Acaudina molpadioides* (Feng 2011)

repeatedly washed with cold acetone, and the total lipids of *Acaudina molpadioides* were obtained after the cold acetone insoluble matter was dried. Then, the total lipids were eluted with different ratios of chloroform-methanol, and different lipid fractions were obtained by repeated forward- and reverse-phase silica gel column chromatography. In reverse-phase C18 silica gel chromatography, the eluted fraction 1 was Cer; fractions 2, 3, and 4 were cerebrosides; and fraction 5 was SS. The specific purification process is shown in Fig. 3 (Feng 2011; Yu et al. 2013).

4.4 Extraction and Separation Technologies of Long-Chain Base

The hydrolysis preparation of LCB is usually carried out according to the steps of total lipids extraction, hydrolysis, and column chromatography purification preparation. Hydrolysis methods include the following:

1. Acid hydrolysis: Total lipids are dissolved in 10% hydrochloric acid/methanol solution and react at 70 °C for 18 h.
2. Alkaline hydrolysis: After a specific treatment, total lipids are dissolved in 10% $Ba(OH)_2$-dioxane (1:1, v/v) solution and react at 110 °C for 24 h.

There are many ways to separate and prepare LCB, including the following:

1. HSCCC: After the total lipids of three kinds of sea cucumbers (*Acaudina molpadioides*, *Cucumaria frondosa* and *Apostichopus japonicus*) were subjected to saponification and acidolysis, the LCBs of the three kinds of sea cucumbers were separated and prepared by HSCCC using the solvent system *n*-hexane-methanol-water (1:2:1, v/v). Their yields ranged from 6.3% to 10.9%, and the purity exceeded 92% (Xu 2011).
2. Silica gel column chromatography: The sample obtained by hydrolysis was loaded into a silica gel column after a certain treatment. According to different polarities of individual components, different solvent systems were used to elute the sample. LCB was generally separated by normal phase silica gel column. For example, Wang et al. used KOH saponification and acid hydrolysis of hydrochloric acid to extract crude LCB with ether. Then, normal-phase silica gel column chromatography was used to obtain LCB, a hydrolyzed product of sea cucumber cerebroside (classical method) (Wang et al. 2020). The above method required lowering experimental conditions, but the selection of conditions was difficult and time-consuming, and it was hard to obtain pure products when LCB and neutral lipids were eluted at the same time.
3. SPE: The total lipid sample was hydrolyzed by weak base and then loaded directly. The lipid mixture was purified by LC-WCX weak anion exchange column, and the recovery rate of LCB was as high as 95%. The advantages of SPE included fast detection and low solvent consumption, but the preparation volume was also limited (Bodennec et al. 2000).
4. Preparative thin-layer chromatography (Sambasivarao and Mccluer 1963).
5. Preparative liquid chromatography (Hu et al. 2017; Hu et al. 2019).

Although all of the approaches enjoyed good separation effects, the preparation amount was always limited.

5 Extraction and Separation Technologies of Sterol Sulfate from Sea Cucumber

5.1 Supercritical Fluid Extraction (SFE)

Xu et al. used the SC-CO_2 method to extract SS from sea cucumbers (Xu 2011). Fine powdered sea cucumber samples were loaded into a closed extraction kettle for static extraction. The extraction conditions were 150 bar extraction pressure, 40 °C extraction temperature, and 5% ethanol entrainer. Dynamic extraction was performed after the flow rate of CO_2 was adjusted to 8 L/h. Neutral lipid extract of sea cucumber, mainly composed of SS, entered the separation kettle (50–70 °C) with CO_2, and the product obtained in the separation kettle was SS from sea cucumber.

5.2 *Column Chromatography*

Goad et al. used column chromatography to separate SS. First, the total lipids of the sea cucumber *Psolus fabricii* were extracted with benzene. SS was separated by normal phase silica gel column chromatography after the total lipids were spin-dried and dissolved in chloroform. Chloroform-methanol solution gradient elution was then used to collect chloroform-methanol (75:25, v/v) and chloroform-methanol (60: 40, v/v) fractions. The above components were separated on an alumina column after they were dissolved with 1,4-dioxane-acetic acid (99:1, v/v). The SS fraction was first eluted with petrol-Et_2O (80:20, v/v) and then extracted with gasoline (Goad et al. 1986). Zeng et al. used a chloroform-methanol system to extract and separate the total lipids of sea cucumber, and the obtained total lipids were separated by silica gel column chromatography using chloroform, chloroform-methanol (9:1, v/v), and acetone in sequence. The above eluent was further separated by silica gel column chromatography and eluted with chloroform-methanol (5:1, v/v) to obtain SS from sea cucumber (Zeng et al. 2020).

Sea cucumber lipid extraction and separation techniques currently reported in the literature are summarized in Table 1, and the development of sea cucumber lipid products is summarized in Table 2.

6 Development of Sea Cucumber Lipid-Related Products

6.1 *Soft Capsule of Sea Cucumber Oil*

The soft capsule of sea cucumber oil is the most developed product among all sea cucumber lipid-related goods. Ke et al. proposed a preparation method for the soft capsule of sea cucumber intestinal egg oil and its compounds (Ke and Qiu 2019). The sea cucumber intestinal egg oil product was composed of sea cucumber intestinal egg oil, algae oil, perilla seed oil, taurine, folic acid, zinc gluconate, zinc lactate, vitamin E, lecithin, and soy lecithin. The capsule material was prepared through a mixing of glycerin, pure water and gelatin, stirring, heating, and vacuum defoaming; then the mixed content was put into the dressing hopper to make soft capsules. The rolling die was selected, and the parameters of the shot were adjusted according to the amount of mixed content. The product could be effectively used to prevent against arteriosclerosis, thrombosis, and other diseases caused by increased body fat. These products could provide opportunities for health-care product/beauty product. Jiang prepared a sea cucumber oil soft capsule by blending sea cucumber oil and vegetable oil from by-products generated in the processing of sea cucumber (sea cucumber boiling substance, sea cucumber intestine, and sea cucumber viscera) (Jiang 2011). It was characterized by the weight ratio of sea cucumber oil to vegetable oil as 1:9–3:7; the vegetable oil was any one among palm oil, olive oil, and soybean salad oil. Specific production steps were as follows: Homogeneous

Table 1 Extraction and separation of sea cucumber lipids

Lipid type	Sea Cucumber species	Pretreatment	Preparation method	Solvent system	Yield/purity	References
Sea cucumber oil	Sea cucumber	Sea cucumber intestinal eggs thawed and homogenized	Solvent extraction	95% ethanol	15–16% yield	Zhao and Feng (2017)
	Apostichopus japonicus Selenka	Washed; freeze dried	Solvent extraction	95% ethanol→80% ethanol→dichloromethane	7% yield, unsaturated fatty acid content over 70%	Yi et al. (2021)
	Sea cucumber	Sea cucumber cooking liquor roughly filtered, centrifuged, desalted, sterilized, enzyme-inactivated, and centrifuged	Salting out extraction	*n*-hexane-anhydrous sodium carbonate-ethanol	63.30% yield	Xiu et al. (2010)
		Sea cucumber intestines washed and homogenized	Solvent extraction	Disodium hydrogen phosphate-citric acid buffer solution-chloroform-methanol-water (1:2:1, v/v)	5.4% yield, 31.13% EPA content	Wu et al. (2017)
		Washed; freeze dried; crushed; sieved	SFE	CO_2	Unsaturated fatty acids 72.8 g/100 g; EPA 18.2 g/100 g	Li et al. (2016)
	Cucumaria frondosa japonica semper	Freeze dried; crushed	SFE	CO_2	1.80% yield	Zakharenko et al. (2020)
	Apostichopus japonicus	Washed; freeze dried; crushed	SCFE	Butane	–	Xu et al. (2020a, b)
Phospholipid	Sea cucumber	Dried; crushed	Solvent extraction	60% ethanol	8%–10% yield	Zhao and Feng (2017)

	Cucumaria frondosa	Dried; crushed	Solvent extraction	Chloroform-methanol (2:1, v/v) → vold acetone	–	Wang et al. (2020)
		Freeze dried; crushed and sieved; extracted in chloroform-methanol (2:1, v/v)	Silica gel column chromatography	Chloroform elution → acetone elution → chloroform-methanol (gradually increasing methanol concentration) elution	Purity>90%	Lou (2011)
		Freeze dried; crushed and sieved; extracted in chloroform-methanol (2:1, v/v) + 0.88% potassium chloride solution	Silica gel column chromatography	Chloroform-methanol (9:1, v/v), acetone, methanol elution → chloroform-methanol-water (65:25:4, v/v)	50% yield, 95% purity	Wang et al. (2016)
Plasmalogen	Sea cucumber	Vacuum dried; crushed	SFE	CO_2	Purity>80%	Xue et al. (2020)
	Cucumaria frondosa	Freeze dried; crushed and sieved; extracted in chloroform-methanol (2:1, v/v) + 0.88% potassium chloride solution	Silica gel column chromatography	Chloroform-methanol (9:1, v/v), acetone, chloroform-methanol (2:1, v/v) elution → add phospholipase D for reaction → *n*-hexane-isopropanol (4:1, v/v) to extract reaction product → chloroform-methanol (2:1 → 0:1) elution	Purity>90%	Wang et al. (2018)
		Dried; crushed; washed by cold acetone; extracted in chloroform-methanol (2:1, v/v)	Silica gel column chromatography	Dichloromethane-methanol (9:1, v/v) to remove neutral lipids → dichloromethane-methanol-water (65:25:4, v/v) for elution	–	Liu et al. (2021)
		Dried; crushed; extracted in chloroform-methanol (2:1, v/v) + 0.15 M NaCl solution	Silica gel column chromatography	Chloroform, chloroform-methanol (9:1, v/v), acetone, chloroform-methanol (2:1, v/v), methanol elution	95% purity	Wang et al. (2021)

(continued)

Table 1 (continued)

Lipid type	Sea Cucumber species	Pretreatment	Preparation method	Solvent system	Yield/purity	References
Lysophospholipid	*Holothuria atra*	Homogenated by chloroform-methanol (2:1, v/v)	Silica gel column chromatography	Chloroform-methanol (4:1 → 2:1 → 1:1 → 1:4, v/v) → methanol→chloroform-methanol-water (3:6:1, v/v) → Chloroform-water (4:1, v/v)	–	Nishikawa et al. (2015)
Cerebroside	*Apostichopus japonicus; Cucumaria frondose; Acaudina molpadioides*	Extracted in chloroform-methanol; extracted in *n*-hexane	HSCCC	Ethyl acetate-ethanol-water (5:3.5:3.5, v/v)	1.1%–6.4% yield	Xu (2011)
	Sea cucumber	Boiled; freeze dried; crushed	SFE	CO_2	2.54% yield	Sun et al. (2017)
					Yield>0.10%	Xu et al. (2018a, b)
	Apostichopus japonicus	Extracted in chloroform-methanol (1:4, v/v)	Silica gel column chromatography	Chloroform-methanol-water (4:6:1, v/v)	–	Yamada (2002), Kisa et al. (2005)
	Apostichopus japonicus; Cucumaria frondose; Acaudina molpadioides	Dissolved in methanol	Preparative liquid chromatography	96% methanol-water	0.42%–5.92% yield	Xu (2011)

Ganglioside	*Cucumaria echinata*	Dried; crushed; leached by chloroform-methanol (1:1–5, v/v)	Column chromatography	Chloroform-methanol-water (1: 1–4:3–10, v/v)	–	Xu et al. (2016), Yamada et al. (2000)
Ceramide	*Acaudina molpadioides*	Dried; crushed; leached by chloroform-methanol-water (4:4:1, v/v)	Column chromatography	98% methanol-water	–	Feng (2011), Yu et al. (2013)
Long-chain base	*Apostichopus japonicus; Cucumaria frondose; Acaudina molpadioides*	Dried Sea cucumber crushed; leached with chloroform-methanol (2:1, v/v), and spin-dried; then saponified with 4 mol/L KOH and acid hydrolyzed with 1 mol/L hydrochloric acid	HSCCC	*n*-hexane-methanol-water (1: 2:1, v/v)	6.3%–10.9% yield	Xu (2011)
	Sea cucumber		Silica gel column chromatography	Ether extraction	–	Wang et al. (2018)
	Acaudina molpadioides		Preparative liquid chromatography	*n*-hexane and ether extraction	1.47% yield, 96.5% purity	Hu et al. (2017)
	Cucumaria frondosa			Ether extraction	1.23% yield, >95% purity	Hu et al. (2019)
Steroid sulfate	Sea cucumber	Dried; crushed	SFE	CO_2 + 5% ethanol		Xu (2011)
	Psolus fabricii	Dried; crushed; extracted with benzene	Column chromatography	Eluted on silica gel column with $CHCl_3$-MeOH (75:25 → 60: 40); solvolysed by reflux followed by 1,4-dioxan-acetic acid (99:1, v/v); chromatographed on a column of alumina eluted with petrol-Et_2O (80:20, v/v)	–	Goad et al. (1986)

(continued)

Table 1 (continued)

Lipid type	Sea Cucumber species	Pretreatment	Preparation method	Solvent system	Yield/purity	References
	Cucumaria frondosa	Dried; crushed; extracted with chloroform-methanol (2:1, v/v)	Silica gel column chromatography	Eluted on silica gel column with chloroform→chloroform-methanol (9:1, v/v) → acetone; acetone eluent fraction obtained; and then eluted on silica gel column with chloroform-methanol (5:1, v/v)	–	Zeng et al. (2020)

Table 2 Development of sea cucumber lipid products

Category	Product	Raw material	Preparation method/formulation	Product efficacy/characteristics	References
Health-care products	Soft capsules of sea cucumber oil	Sea cucumber intestinal eggs	By weight, the product includes 20 parts of sea cucumber intestinal egg oil, 10 parts of algae oil, 5 parts of Perilla seed oil, 0.5 parts of taurine, 0.5 parts of folic acid, 5 parts of zinc gluconate, 0.2 parts of zinc lactate, 7 parts of vitamin E, 3 parts of lecithin, and 15 parts of soybean phospholipid. Gelatin, glycerin, and pure water were used to prepare capsule shell before filling and granulation	The ingredients of the capsules are diversified and can be used to prevent arteriosclerosis, thrombosis, and other diseases caused by fat accumulation	Ke and Qiu (2019)
		Sea cucumber boiling substances and intestinal eggs	Sea cucumber oil and vegetable oil were mixed in a weight ratio of 1:9–3:7, and grinding was performed for a homogeneous liquid. Gelatin, glycerin, pure water, and chocolate brown food coloring were used to prepare capsule shell before filling and granulation	High nutritional value; widely applied in health-care and beauty products	Jiang (2011)
		Body wall of *Cucumaria frondosa*	The sea cucumber oil was extracted by SC-CO_2 technology (35 °C, 5 MPa) and then mixed with salmon oil at a mass ratio of 1:1 to prepare soft capsules	A kind of nourishing oil with significant value in health care, with very stable properties	Jiao and Shao (2012)

(continued)

Table 2 (continued)

Category	Product	Raw material	Preparation method/formulation	Product efficacy/characteristics	References
		Sea cucumber intestinal eggs	Sea cucumber intestinal oil was obtained through thawing, homogenizing, ethanol extraction, filtration, decompression distillation and decompression, concentration, and then mixed with vegetable oil in a ratio of 1:(2–4) to make soft capsules	An extraction process of sea cucumber oil from intestinal eggs	Zhao and Feng (2018)
		Sea cucumber viscera	Crude sea cucumber oil was washed with 18% ethanol, 6.0% sodium carbonate, 5.5% sodium citrate solution, 6.3% sodium citrate solution, and 4.4% citrate solution by weight, stirred, and then dewatered to make soft capsules	Not allergic, pure in taste; widely used as an oral health food to prevent cardiovascular and cerebrovascular diseases	Song and Ning (2021a)
		Sea cucumber viscera	Crude sea cucumber oil was obtained by SC-CO_2 extraction, which was washed with $NaHCO_3$ aqueous solution and HCl and then washed with water to pH 6–7 to make soft capsules	The method can extract sea cucumber oil efficiently with low histamine content	Song and Ning (2021b)
		Fresh Sea cucumber	Sea cucumber powder was dissolved in clean water, and protease enzyme mixture (papain, neutral protease and E-64 cysteine protease inhibitors, 2:1:1) was added to obtain a paste concentrate through decompression concentration (90.0–95.0 kPa, 50–55 °C, 30–50 min). Then, the paste concentrate was stirred continuously until it dissolved in	Higher in nutritional value, the product is not easy to cause allergy to the user versus dried sea cucumber or regular sea cucumber oil capsules	Lin et al. (2019)

			vegetable oil (olive oil/soybean oil) to obtain a mixed oil. Finally, 30–40 mg vitamin E was added into every 1 g mixed oil to make soft capsules		
	Microcapsules of sea cucumber oil	Sea cucumber intestinal eggs	Wall materials (one or more of maltodextrin, corn starch, and whey protein powder), emulsifiers (sodium caseinate, octyl sodium starch succinate, monosaccharide), and purified water were mixed and stirred together and then heated to form an emulsion in a container. Sea cucumber oil and stabilizer (citric acid or sodium citrate) were added into the emulsion to make microcapsules of sea cucumber oil after full stirring, high pressure homogenization, spray drying, and vacuum freeze drying	Prevents arteriosclerosis, thrombosis, and other diseases caused by accumulated fat in the blood	Ke and Qiu (2020)
	Sea cucumber drinks	Sea cucumber body wall	300–500 g sea cucumber cerebroside; 60–70 g cortex periplocae, 30–40 g coptis, and 20–30 g *Siraitia grosvenorii* were added to water (the mass ratio of water to all plant ingredients was 1:1.5) after necessary cleaning to boil for 30–40 min at 100 °C. After cooling to room temperature, a clear liquid was obtained through filtration, mixed with sea cucumber cerebroside, and then packaged after sterilization	Antitumor effect, good taste, and ease for absorption by human body; suitable for commercialized large-scale production	Qin (2020a)
			300–500 g sea cucumber cerebroside; 60–70 g cassia seed, 30–40 g hawthorn, and 20–30 g wolfberry were	Effectively prevents and improves fatty liver	Qin (2020c)

(continued)

Table 2 (continued)

Category	Product	Raw material	Preparation method/formulation	Product efficacy/characteristics	References
			added to water (the mass ratio of water to all plant ingredients was 1:1.5) after necessary cleaning to boil for 30–40 min at 100 °C. After cooling to room temperature, a clear liquid was obtained through filtration, mixed with sea cucumber cerebroside, and then packaged after sterilization		
			300–500 g sea cucumber cerebroside; 20–50 g *Siraitia grosvenorii*, 3–10 g *Scutellaria baicalensis*, 3–10 g ginger, 10–30 g hawthorn and 30–60 g wolfberry were added to water (the mass ratio of water to all plant ingredients was 1:1.5) after necessary cleaning to boil for 30–40 min at 100 °C. after cooling to room temperature, a clear liquid was obtained through filtration, mixed with sea cucumber cerebroside, and then packaged after sterilization	Effectively prevents and improves Alzheimer's disease	Qin (2020b)
Cosmetics	Lipsticks	Sea cucumber body wall	2–4 parts of emulsifier, 10–40 parts of sea cucumber fat-soluble extract, 5–10 parts of marigold oil, 5–12 parts of grape seed oil, 20–65 parts of yellow beeswax, 8–12 parts of water, 1–3 parts of glycerol, and 5–12 parts of argan oil were mixed evenly, the mixed liquor was injected to the mold, and the lipstick containing sea cucumber	Outstanding recovery effect; repairs the phenomenon of dry lips in a short time	Song et al. (2018)

			fat-soluble extract was finally obtained after cooling		
Cleaning products	Soaps	Sea cucumber viscera	Sea cucumber viscera were used as the raw material and petroleum ether as the extraction medium to prepare sea cucumber oil. Add 50–150 parts of lye to 50–150 parts of sea cucumber oil, and stir to obtain soap base; add 0–3 parts of essential oil and 0–20 parts of sugar solution to the melted soap base, remove the foam, and finalize the shape to obtain the cucumber oil soap	Compared with coconut oil or palm oil as soap base, the economic cost of sea cucumber viscera oil is significantly reduced. It maintains good permeability and good effect on skin care	Fang et al. (2020)
Feeds	Fish feeds	Sea cucumber viscera	Fresh sea cucumber intestine was treated with acid to adjust the pH to 4.3–4.7. Acetone was mixed with the raw material in a ratio of 3:1 (v/v) and then stirred for 24 h or until enough pigment was fully released from the intestinal material into the solvent. Pigment lipids were obtained from the solvent by centrifugation and then added to standard pet fish feed	Pigmentation in aquatic species treats or improves diseases caused by lipoxygenase active products or leukotriene reactions	Collin (2002)

capsule contents were mixed with sea cucumber oil and vegetable oil according to their weight ratio after grinding; glycerin, pure water, and chocolate brown pigment were mixed and stirred at a high temperature to prepare the gelatin. The soft capsule was made through rolling die selection and the adjustment of pellet parameters and filling quantity. The soft capsule of sea cucumber oil could be widely used in health-care and beauty products, and the market sales price could reach RMB 1.5 yuan/capsule, with low production cost, rich nutrition, and high market value. Jiao and Shao reported a lipid-related product derived from sea cucumber (Jiao and Shao 2012), which included the following steps: *Cucumaria frondosa* Gunnerus oil was obtained by SC-CO_2 (35 °C, 5 MPa) after washing, eviscerating, drying, and grinding. Mixed health oil was further obtained through the mixing of cucumber oil and salmon oil at a mass ratio of 1:1. This product is a kind of health preservation oil product with significant value in nourishment and health care. Its properties were very stable (after 3 months at room temperature, the residual rates of EPA and DHA in the capsule were 95% and 92%, respectively). Zhao et al. published the extraction process of sea cucumber intestine oil (from guts and gonad, also called sea cucumber viscera) and the method of preparing soft capsule of sea cucumber intestine oil (Zhao and Feng 2017). Specifically, distilled water was added to sea cucumber intestine for homogenation; ethanol was then added and the mixture was stirred to obtain the extract after standing. A yellow or yellowish-brown crude liquid extract was obtained by vacuum distillation. The crude extract was further decompressed and concentrated to obtain a kind of yellowish-brown paste oil from sea cucumbers. Sea cucumber oil and vegetable oil were mixed according to the mass ratio of 1:(2–4). The extracted oil was obtained after stirring and standing, and then the soft capsule of sea cucumber oil was finally obtained. Song et al. published a preparation method for soft capsule of sea cucumber oil (Song and Ning 2021a). The steps were as follows: Add the first washing solution (ethanol, sodium carbonate, sodium citrate, water) to the sea cucumber oil, and then stir and remove the first water lotion. After ultrasonic treatment and stirring under vacuum conditions, the second water washing solution (sodium citrate, citric acid, water) was added. The second water washing solution was removed by liquid separation after stirring. Sea cucumber oil was obtained after washing with distilled water and the addition of a molecular sieve for water removal. The sea cucumber oil was filled into soft capsules in the final step. The obtained soft capsules of sea cucumber oil featured non-allergic and had a pure taste, which could be widely used as an oral health food to soften heart and cerebral vessels. Meanwhile, Song et al. disclosed a method of preparing sea cucumber oil capsules from sea cucumber viscera. The obtained sea cucumber oil products had a very low content of histamine (Song and Ning 2021b). The method included the following steps: Fresh sea cucumber viscera were cleaned, drained, and cut into pieces; water, salt, magnesium chloride, and polypropylene fiber non-woven strips were added for stirring the mixture evenly; the obtained mixture was heated in a high-pressure reactor, and then a solid was obtained by vacuum evaporation; crude sea cucumber oil was obtained by SC-CO_2 extraction, which was washed with a $NaHCO_3$ aqueous solution and HCl, and then washed with water to pH 6–7. The sea cucumber oil was filled into soft capsules to prepare sea cucumber oil capsules. This method was characterized by that vitamin E was added to reach 50–70 mg per gram of sea cucumber oil before filling. Lin et al. invented a preparation method for concentrated sea cucumber oil soft capsule (Lin et al. 2019). Sea cucumber powder was added into clean water, and protease enzyme mixture was also added to obtain a kind of paste concentrate through reduced pressure concentration. Then, the paste concentrate was stirred continuously until it dissolved in vegetable oil (soybean oil, rapeseed oil, or olive oil), yielding a mixed oil. Finally, the mixed oil was separated through washing and added with vitamin E and then filled into soft capsules. Compared with dried sea cucumber or ordinary sea cucumber oil capsule that is directly consumed, the prepared sea cucumber oil capsule product has higher nutritional value and does not cause allergy to the consumer easily.

6.2 Microcapsules of Sea Cucumber Oil

Ke et al. invented a kind of microcapsule containing sea cucumber oil, including sea cucumber oil, wall material, emulsifier, and stabilizer (Ke and Qiu 2020). The preparation method included the following steps: Wall material, emulsifier, and purified water were put into a container to obtain the first solution. The first solution was thoroughly stirred and heated to form an emulsion; sea cucumber oil and stabilizer were added to the emulsion to obtain the second solution after thorough stirring. The second solution was homogenized at least once in high-pressure unit to obtain the third solution. After spray drying and vacuum freeze drying, microcapsules of sea cucumber oil were finally obtained. This kind of sea cucumber oil microcapsule could prevent arteriosclerosis, thrombosis, and other diseases caused by accumulated fat in the blood, as well as exhibited satisfactory functions of human immunity regulation, antibacteria, antivirus, blood sugar reduction, and anti-fatigue, which could be directly applied to ordinary food, health food, cosmetics and other related sectors.

6.3 Sea Cucumber-Related Drinks

Qin developed a sea cucumber drink with antitumor effect (Qin 2020a). This sea cucumber drink contained sea cucumber cerebroside, cortex periplocae, coptis, and *Siraitia grosvenorii*. During preparation, fresh sea cucumbers were cleaned, gutted, dried, and crushed at a low temperature. The hydrolysate of the sea cucumbers was obtained with compound hydrolase. The sea cucumber cerebroside was isolated and purified from the hydrolysate by HSCCC with chloroform and methanol (1:2, v/v) solvent. Cleaned cortex periplocae, coptis, and *Siraitia grosvenorii* were added to water for cooking; after cooling to room temperature, a clear liquid was obtained through filtration and mixing with the sea cucumber cerebroside and then packaged after sterilization. The drink exhibited an antitumor effect, which brought opportunities for the clinical application of novel medical or health-care functions of sea cucumber. Additionally, this sea cucumber drink maintained good taste and was easy to absorb, which was suitable for commercialization and large-scale production. By the same method, Qin developed another sea cucumber drink to prevent and improve fatty liver (Qin 2020c). The ingredients included sea cucumber cerebroside, cassia seed, hawthorn, and wolfberry. In addition, Qin released a sea cucumber drink to prevent and improve Alzheimer's disease (Qin 2020b). The ingredients included sea cucumber cerebroside, *Siraitia grosvenorii*, *Scutellaria baicalensis*, ginger, hawthorn, and wolfberry. The results showed that the combination of sea cucumber cerebroside and other plant bioactive ingredients could significantly improve the learning and memory ability of mice.

6.4 Lipsticks

Song et al. invented a lipid-soluble extract from sea cucumbers and provided the preparation method for a lip balm containing the extract (Song et al. 2018). The preparation method included the following steps:

1. Prepare a sea cucumber homogenate.
2. Enzymatic hydrolysis: The sea cucumber homogenate in step (1) was enzymatically hydrolyzed by pepsin and trypsin successively to obtain the hydrolysate.
3. Adjust the pH of enzymatic hydrolysate to neutral, take the supernatant for concentration, add sesame oil with set proportion, stir and extract several times, centrifuged the extract after extraction, and then combine with the supernatant to obtain a fat-soluble extract of sea cucumber. The mixture obtained in the preceding steps was injected into a mold, cooled, and frozen to obtain lipsticks containing lipid-soluble extract from sea cucumber. This kind of lipstick could repair dry and chapped lips in a short time.

6.5 Sea Cucumber Oil-Based Soaps

Fang et al. reported on a preparation method for sea cucumber oil-based soap and its associated applications (Fang et al. 2020). With sea cucumber oil, alkali, essential oil, and sugar as the raw materials, dried sea cucumber viscera were put into the extraction device and immersed with petroleum ether. Sea cucumber oil was extracted by water bath reflux and decompression concentration. The sea cucumber oil was heated in a water bath, with sodium hydroxide solution slowly added several times; the mixture was stirred until the soap thickened to obtain the sea cucumber oil soap base after cooling. The above sea cucumber oil soap was heated and melted, and then the sucrose solution and lavender essential oil were added and stirred to remove the foam. Lastly, low temperature molding was carried out. In this application, the weight ratio of sea cucumber oil, sodium hydroxide solution, sucrose aqueous solution, and essential oil was 51:34:12:3. The antibacterial test results showed that compared with ordinary palm oil and coconut oil-based soaps, sea cucumber oil soap could effectively inhibit the activity of *Candida albicans* and *Escherichia coli*. The sea cucumber oil-based soap maintained good permeability. The nutrients in sea cucumber combined with the essential oil with sterilization and caring effects could be better absorbed by the skin to ensure superior cleaning as well as skincare effects, exhibiting a good market prospect.

6.6 Feed

Collin's invention involved lipids extracted from *Holothuriidae* or *Holothuroidea*, and the extracted sea cucumber oil could be used for the pigmentation of aquatic species after being added to feed, as well as for the treatment or improvement of diseases caused by lipoxygenase active products or leukotriene reactions (Collin 2002). It was an objective of the present invention to provide a means to recover usable pigmented lipid fractions and pigment-free lipid fractions from the intestinal gut material from sea cucumbers, especially *Cucumaria frondosa*. Fresh sea cucumber viscera were mixed with acetone and agitated at room temperature for 3 h. The acetone was drained from the mixture into a containment vessel, and fresh acetone was washed over the gut material until little pigment color was obtained. The acetone was pumped through a wiped film evaporator (Pope Still), and the water included in the acetone mixture and the acetone were separated from the pigmented lipid fraction and reclaimed. The acetone was salvaged, and the water phase was discarded. The resulting lipid fraction had a heavy grease consistency, and the heavy lipid fraction containing carotenoid material (RED GREASE) was determined to contain 4700 parts per million of canthaxanthin and was incorporated into standard ornamental fish feed. In this patent, the sea cucumber oil containing carotenoids, zeaxanthin, and astaxanthin was suitable for addition to commercial fish feeds to bring color deposits to the muscle or skin of the particular species. All lipid-related products of sea cucumber are shown in Table 2.

7 Conclusion

There is no doubt that the lipid components from sea cucumber have attracted people's attention due to their diversified biological activities, involving plasmalogen, cerebroside, ganglioside, and steroid sulfate. It has been proven that sea cucumber phospholipids can improve metabolic syndrome and lipid metabolism disorder, inhibit hepatic fat accumulation, improve Alzheimer's disease, etc. Sea cucumber sphingolipids have exhibited an effect on improving fatty liver and antitumor activity. Moreover, there are many methods for lipid extraction and separation from sea cucumber, including solvent extraction, supercritical fluid extraction, subcritical fluid extraction, high-speed countercurrent chromatography, etc., with high purity of the target components achieved. Additionally, the extraction process of sea cucumber oil is relatively mature. There is a large number of extraction processes for sea cucumber oil, and the preparation methods of sea cucumber oil soft capsule and the industrial production have both been realized. Drinks, lipsticks, soaps, and feed products containing sea cucumber oil have also achieved excellent results. However, apart from sea cucumber oil, the research and development of other sea cucumber lipids are still in the research stage, and have not yet achieved large-scale application in production.

References

Bodennec J, Famy C, Brichon G et al (2000) Purification of free sphingoid bases by solid-phase extraction on weak cation exchanger cartridges. Anal Biochem 79:245–248

Chen XW, Yang HQ (2021) Discussion on supercritical fluid extraction. Fujian Anal Test 30:43–48

Collin PD (2002) Sea cucumber carotenoid lipid fraction products and methods of use. US Patent Appl 6399105

Cong PX, Li ZJ, Xu J et al (2011) Research progress of gangliosides in marine echinoderms. Chin J Marine Med 30:65–70

Fang X, Zhang PP, Wang KP (2020) Preparation method and application of sea cucumber oil soap base. CN Patent. Appl. 111471540A

Feng TY (2011) Isolation and identification of active lipids and preliminary study on their activity in *Acaudina molpadioides*. Dissertation, Ocean University of China

Folch J, Lees M, Stanley GH (1957) A simple method for the isolation and purification of total lipids from animal tissues. Biol Chem 226:297–302

Goad LJ, Garneau FX, Simard JL et al (1986) Composition of the free, esterified and sulphated sterols of the sea cucumber *Psolus fabricii*. Comparative Biochemistry and Physiology Part B: Comparative Biochemistry 84:189–196

Guo TT, Wan CY, Huang FH et al (2020) Research progress in oil extraction and trace components by subcritical fluid. Chin J Oil Crops 42:154–160

Hu SW, Wang JH, Wang JF et al (2017) Long-chain bases from sea cucumber mitigate endoplasmic reticulum stress and inflammation in obesity mice. J Food Drug Anal 25:628–636

Hu SW, Xu YL, Gao X et al (2019) Long-chain bases from sea cucumber alleviate obesity by modulating gut microbiota. Mar Drugs 17:455

Jiang YB (2011) A soft capsule of sea cucumber oil and its preparation method. CN Patent. Appl. 101991113A

Jiao J, Shao JJ (2012) Preparation of lipid preparation from sea cucumber and its preparation method. CN Patent. Appl. CN102823666A

Ke YF, Qiu DY (2019) Composition of sea cucumber intestine oil, soft capsule of sea cucumber intestine oil and preparation method. CN Patent. Appl. 110236197A

Ke YF, Qiu DY (2020) A preparation method of sea cucumber oil microcapsule. CN Patent. Appl. 111096457A

Kisa F, Yamada K, Kaneko M et al (2005) Constituents of *Holothuroidea*, 14. Isolation and structure of new glucocerebroside molecular species from the sea cucumber *Stichopus japonicus*. Chem Pharm Bull 53:382–386

Li HR, Xu QH, Li WK (2016) Extraction process of deep-sea cucumber oil. CN Patent. Appl. 106479663A

Lin Y, Chen XL, Sun HM (2019) Preparation method of soft capsule of concentrated sea cucumber oil. CN Patent. Appl. 110192625A

Liu YJ, Cong PX, Zhang T et al (2021) Plasmalogen attenuates the development of hepatic steatosis and cognitive deficit through mechanism involving p75NTR inhibition. Redox Biol 43:102002
Lou QM (2011) Lipid analysis of several marine animals and biological activity of EPA-phospholipid in sea cucumber. Dissertation, Ocean University of China
Nishikawa Y, Furukawa A, Shiga I et al (2015) Cytoprotective effects of lysophospholipids from sea cucumber *Holothuria atra*. PLoS One 10:e0135701
Qin DW (2020a) A sea cucumber drink with anti-tumor effect and its preparation method. CN Patent. Appl. 110754597A
Qin DW (2020b) A sea cucumber drink for preventing and improving fatty liver and its preparation method. CN Patent Appl 110876437A3
Qin DW (2020c) A sea cucumber drink for preventing and improving Alzheimer's disease and its preparation method. CN Patent. Appl. 110973420A
Sambasivarao K, Mccluer R (1963) Thin-layer chromatographic separation of sphingosine and related bases. J Lipid Res 4:106–108
Song FF, Ning JC (2021a) A preparation method of sea cucumber oil soft capsules. CN Patent. Appl. 107853704
Song FF, Ning JC (2021b) A method for preparing sea cucumber oil capsules by using sea cucumber viscera. CN Patent. Appl. 107874165B
Song SL, Zhang E, Ji AG (2018) At-soluble extract of sea cucumber, lip balm containing the extract and preparation method. CN Patent. Appl. 105832646B
Sun JF, Xu ZL, Dai XP et al (2017) A method for supercritical extraction of cerebrosides in sea cucumber. CN Patent. Appl. 106977562B
Svennerholm L, Fredman P (1980) A procedure of the quantitative isolation of brain gangliosides. Biochim Biophys Acta 617:97–109
Wang Q, Shi SJ, Han HX et al (2014a) Separation and purification of soybean phosphatidylethanolamine. China Oil Fats 39:72–77
Wang WG, Zhu TY, Yin YJ (2014b) Study on ethanol low temperature extraction and purification of phosphatidylcholine. Food Oil 27:29–31
Wang YM, Wang D, Xu J et al (2016) A sea cucumber phospholipid and its application in products for improving body movement coordination ability. CN Patent. Appl. 106377533A
Wang YM, Ding L, Xue CH et al (2018) A preparation method and application of EPA plasmalogen. CN Patent. Appl. 108276438B
Wang YM, Wang CC, Zhang TT et al (2020) Application of sea cucumber cerebroside and its derivatives in the preparation of medicines for improving blood-brain barrier damage. CN Patent. Appl. 111544438B
Wang XC, Chen QS, Wang XX et al (2021) Lipidomics approach in high-fat-diet-induced atherosclerosis dyslipidemia hamsters: alleviation using ether-phospholipids in sea urchin. J Agric Food Chem 69:9167–9177
Wu HT, Song YK, Song L et al (2017) A preparation method of EPA-rich sea cucumber intestinal lipid. CN Patent. Appl. 105105118B
Xiu ZL, Yang XC, Hui HQ et al (2010) Method for extracting active ingredients from sea cucumber by salting out. CN Patent. Appl. 201010236666A
Xu J (2011) Separation, purification, structural analysis and biological activity of sea cucumber cerebroside. Dissertation, Ocean University of China
Xu J, Wang XX, Cong PX et al (2016) A polar glycolipid extract and its application. CN Patent. Appl. 105920046B
Xu XB, Fan NN, Zong J (2018a) Research progress on extraction technology of Antarctic krill oil. Food Ind Sci Technol 39:306–310
Xu ZL, Zhao XW, Sun JF et al (2018b) Supercritical extraction of sea cucumber cerebroside and its detection and analysis by HPLC. Food Res Dev 39:46–50
Xu J, Li H, Xue CH et al (2020a) A preparation method and application of sterol sulfate from sea cucumber. CN Patent. Appl. 111303233A

Xu QH, Wang GX, Liu FJ (2020b) A subcritical fluid extraction process of sea cucumber oil. CN Patent. Appl. 111808673A

Xue CH, Tian YY, Liu YJ et al (2020) A preparation method and application of EPA/ARA plasmalogen. CN Patent. Appl. 111560405B

Yamada K (2002) Chemo-pharmaceutical studies on the glycosphingolipid constituents. J Pharmaceut Soc Jpn 122:1133–1143

Yamada K, Harada Y, Miyamoto T et al (2000) Constituents of *Holothuroidea*. 9. Isolation and structure of a new ganglioside molecular species from the sea cucumber *Holothuria pervicax*. Chem Pharm Bull 48:157–159

Yi YH, Li XP, Li TP et al (2021) Imitation Sea cucumber viscera lipid extract and its preparation method and application. CN Patent. Appl. 112451546A

Yu KJ, Xu J, Feng TY et al (2013) Isolation, identification and activity study of ceramides from *Haiti melon*. Chin J Food Sci 13:181–186

Zakharenko A, Romanchenko D, Thinh PD et al (2020) Features and advantages of supercritical CO_2 extraction of sea cucumber *Cucumaria frondosa* Japonica Semper, 1868. Molecules 25: 4088

Zeng BB, Zhang LY, Chen C et al (2020) Sea cucumber sterol alleviates the lipid accumulation in a high-fat-fructose diet fed mice. J Agric Food Chem 68:9707–9717

Zhao B, Feng WS (2017) An extraction process of sea cucumber intestinal egg oil and a method for preparing sea cucumber intestinal egg oil soft capsules by using sea cucumber intestinal egg oil. CN Patent. Appl. 107699359A

Zhao B, Feng WS (2018) A joint extraction method of various nutrients in sea cucumber. CN Patent. Appl. 107955081A

Zhou X, Wu H, Zhang C (2017) Research progress on extraction, separation and purification of lecithin. Beverage Ind 20:61–65

The Extraction, Separation Technology, and New Product Development of Collagen Peptides from Sea Cucumber

Yanchao Wang and Jingfeng Wang

Abstract The body wall is the main edible part of sea cucumber, which is mainly composed of epithelial tissue and dermal connective tissue. The extracellular matrix between the dermal connective tissues of the sea cucumber mainly consists of collagens, proteoglycans, and other glycoproteins, about which the collagen content accounts for more than 70% of the total protein on a dry weight basis. In addition to terrestrial biological resources, marine biological resources such as sea cucumber serve as an important raw material for the extraction and preparation of collagens and collagen peptides. Presently, there has been considerable development in the research and application of sea cucumber collagens and collagen peptides worldwide. This chapter mainly introduces the progress of sea cucumber collagen processing technology focusing on the extraction and separation of sea cucumber collagen fibrils, solubilized collagens, and collagen peptides and discusses the application progress of sea cucumber collagens and collagen peptides in the development of food, cosmetic, and biomedical material products.

Keywords Sea cucumber · Collagen · Peptide · Technology · Product

1 Introduction

Collagen is a biopolymer synthesized by animal cells, which is widely found in animal bone, tendon, cartilage, skin, and other connective tissues. Collagen is the basic functional unit for the formation of connective tissue that consists of collagen fibrils, which further aggregate to form collagen fibers (Ricard-Blum 2011). There is a repeat sequence Gly-X-Y in collagen molecules where Gly represents glycine, X is often proline, and Y is often hydroxyproline or hydroxylysine. Collagens derived

Y. Wang · J. Wang (✉)
College of Food Science and Engineering, Ocean University of China, Qingdao, R.P. China
e-mail: wangyanchao@ouc.edu.cn; jfwang@ouc.edu.cn

C. Xue (ed.), *Advances in Sea Cucumber Processing Technology and Product Development*, Advances in Marine Bioprocesses and Bioproducts,
https://doi.org/10.1007/978-3-031-16512-2_10

from aquatic invertebrates such as sea cucumbers are generally less soluble compared to those from vertebrates. Collagens derived from vertebrates are mainly comprised of type I, II, and III, while collagens in the sea cucumber mainly consist of type I, V, IX, etc. (Sorushanova et al. 2019; Wang et al. 2020b). Collagen types I and V, belonging to fibrillar collagens, could provide tissues with tensile strength (Theocharis et al. 2016). Collagen type IX, which belongs to fibril-associated collagens with interrupted triple helices (FACITs), could interact with fibrillar collagens and link collagen fibers together and with other ECM components (Theocharis et al. 2016). Meanwhile, due to the high content of sulfated polysaccharides in the sea cucumber body wall, the yield of solubilized collagens from sea cucumbers is generally very low using acidic and enzymatic treatment methods. This may be attributed to the presence of electrostatic binding interactions between positively charged collagens and negatively charged glycosaminoglycans under acidic pH condition, preventing the effective dissolution of sea cucumber collagens. Therefore, the extraction methods for sea cucumber collagen fibrils and solubilized collagens have been optimized and modified from conventional collagen extraction, improving the yield of sea cucumber collagens.

Sea cucumber collagen peptides generally refer to peptides derived from collagens through the process of protein hydrolysis, refinement, or purification. Due to the low solubility and protease digestibility sensitivity induced by the covalent cross-linking and the triple-helix structure of collagen, it is often required to degrade the telopeptides and disrupt the triple-helix structure of collagen by the pretreatment of the substrate before enzymatic hydrolysis. Meanwhile, the fishy smell and turbidity of sea cucumber collagen peptides can affect product quality, significantly reduce the acceptance of consumers, and thereby prevent the development of new products. For these reasons, the refinement of sea cucumber collagen peptides is generally required to improve the purity and quality of collagen peptides by removing the non-collagenous substances, pigments, and fishy smell. This chapter provides a review of the extraction and separation technologies for sea cucumber collagens and collagen peptides and their applications in the development of novel products.

2 Extraction and Separation of Sea Cucumber Collagens

Sea cucumber collagen fibril, comprising of triple-helix structural regions, is one type of insoluble fibrillar structural protein formed by multi-level aggregation of collagen molecules, which is connected with proteoglycans and glycoproteins in the extracellular matrix of the body wall (Trotter et al. 1995; Wang et al. 2018). The basic principle of collagen extraction is to separate collagens from other non-collagen components according to the characteristics of collagens by changing the environmental conditions where collagens are located. The purpose of sea cucumber collagen extraction is to obtain collagen extracts with the triple-helix structure still intact. The extraction of sea cucumber collagen can be divided into two processes: the separation of collagen fibrils and the extraction of solubilized collagen. To avoid the denaturation and degradation of collagens during the extraction process, the entire extraction process should be performed at a low temperature, generally below 10 °C.

2.1 Extraction and Separation of Sea Cucumber Collagen Fibrils

The extraction and separation of sea cucumber collagen fibrils refer to the process of extracting and separating collagen fibrils from the complex matrix using sea cucumber body wall as the raw material. The fucosylated chondroitin sulfate is covalently associated with collagen fibrils via *O*-glycosidic linkage, and the covalent interactions remain stable when subjected to either denaturants or disulfide reducing agents (Trotter et al. 1995; Wang et al. 2018). Therefore, sea cucumber crude collagen fibrils are generally treated with a low-concentration alkali solution to hydrolyze the glycosidic linkage, resulting in the separation of collagen fibrils and sulfated polysaccharide and eventually obtaining high purity collagen fibrils.

Trotter et al. (1995) identified macromolecules composed of collagen fibrils and sulfated polysaccharides from the sea cucumber (*Cucumaria frondosa*) body wall by repeated water-EDTA-water-guanidine-HCl extraction. The protocol is as follows (Fig. 1): Fresh sea cucumber is gutted, and the inner muscle layer is removed, resulting in the sea cucumber body wall. The body wall is cut into small pieces and stirred in distilled water for 30 min, and then the water is discarded and replaced with fresh distilled water. The mixture is stirred for 60 min, and the distilled water is replaced with fresh water. The above extraction process is repeated once. The precipitate is collected through filtration with filter gauze. The precipitate is resuspended in 0.1 M Tris-HCl buffer (4 mM EDTA, pH 8.0) and stirred overnight, and then the liquid is discarded and replaced with distilled water. The mixture is

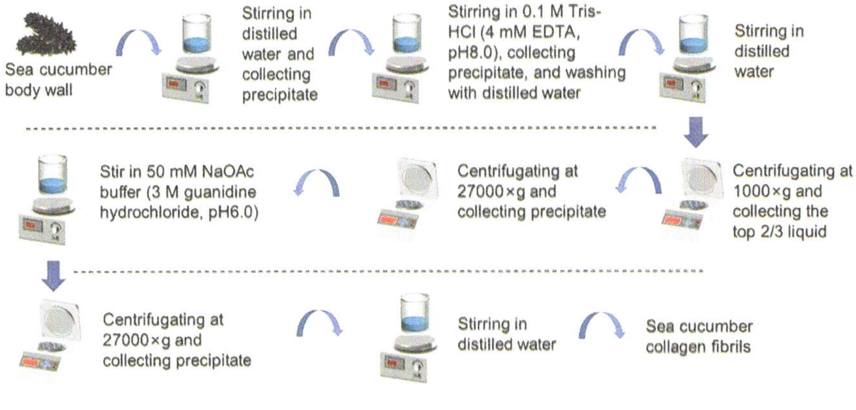

Fig. 1 The flow chart of sea cucumber collagen fibril extraction

slowly stirred for 15 min and the liquid is discarded. The precipitate is washed with distilled water twice or more, resuspended in distilled water, and then slowly stirred overnight. The fluid consisting of cloudy collagen fibrils is separated and centrifuged at 1000 × g for 30 min. The top 2/3 of the liquid containing collagen fibrils is collected, the rest is mixed with distilled water and stirred overnight, and then the collagen fibrils are collected with the repeated extraction steps as described above. The collected liquid containing collagen fibrils is centrifuged at 27,000 × g for 30 min, and the precipitate is collected. The precipitate is mixed with 50 mM sodium acetate buffer (3 M guanidine-HCl, pH 6.0) and stirred slowly, and the mixture is centrifuged at 27,000 × g for 30 min to collect the precipitate. The precipitate is resuspended in distilled water and stirred slowly for 60 min. The abovementioned process is repeated twice to obtain sea cucumber collagen fibrils.

Saito et al. (2002) isolated high purity collagen fibrils from sea cucumber (*Stichopus japonicus*) body wall via disaggregation with the NaCl-EDTA-mercaptoethanol-Tris-HCl method, which yields coarse collagen fibrils, and then via purification with NaOH treatment to break down the covalent bonds between sulfated polysaccharide and collagen fibrils. The protocol is as follows: Fresh sea cucumber is gutted and the inner muscle layer is removed, and then the body wall is cut into small pieces, mixed with 0.1 M Tris-HCl buffer (0.5 M NaCl, 50 mM EDTA, 0.2 M β-mercaptoethanol, pH 8.0) and stirred for 72 h. The fluid containing cloudy collagen fibrils is centrifuged at 7500 × g for 30 min, and the precipitate is washed with distilled water to obtain crude collagen fibrils. The crude collagen fibrils are mixed with 0.1 M NaOH solution and stirred for 72 h, and the precipitate is collected after centrifugation. The precipitate is washed with distilled water to obtain purified collagen fibrils.

Based on the two methods mentioned above, Cui et al. (2007) employed repeated extraction with water-EDTA-Tris-HCl-water to extract collagen fibrils from sea cucumber (*Stichopus japonicus*) body wall. The protocol is as follows: Fresh sea cucumber is gutted and the inner muscle layer is removed, and then the body wall is cut into small pieces, mixed with distilled water, and stirred for 30 min. The precipitate is collected through filtration with gauze, washed with distilled water, and then mixed with 0.1 M Tris-HCl buffer (4 mM EDTA, pH 8.0) and stirred overnight. The precipitate is collected, washed with distilled water twice, and then stirred for 48 h in distilled water. The fluid containing cloudy collagen fibrils is centrifuged at 9000 × g for 5 min, and the supernatant is collected. The extraction process described above is repeated. The solution containing collagen fibrils is centrifuged at 10,000 × g for 30 min, and the precipitate is collected to obtain crude collagen fibrils. The crude collagen fibrils are treated with 0.1 M NaOH solution through stirring for 72 h, and the precipitate is collected and then washed with distilled water, leading to collagen fibrils.

2.2 *Extraction and Separation of Solubilized Sea Cucumber Collagens*

The extraction and separation of solubilized sea cucumber collagens is a process using sea cucumber collagen fibrils as the raw material to further improve collagen solubility without breaking down the collagen triple-helix structure. The collagen molecule is composed of triple-helix regions and telopeptides including the N-telopeptide and C-telopeptide. Collagen fibrils are formed by collagens through covalent cross-linking between histidine and lysine, which further aggregate to form collagen fibers. Covalent cross-linking generally occurs between the C- or N-telopeptides of collagens (Wess 2005). The collagen fibrils can be partially disaggregated with increased solubility by acid or enzymatic treatment and be precipitated by the salting-out method or organic solvent precipitation method, leading to solubilized collagens.

Acid extraction is a process to extract solubilized collagens by using acid solutions under certain conditions, mainly using acidic conditions with low ionic strength to break down intermolecular ionic bonds and Schiff bases, causing fibril swelling and dissolution; the resulting product is generally known as acid-solubilized collagen (Coppola et al. 2020). Acetic acid is the most commonly used acid in the extraction process of sea cucumber acid-solubilized collagen. Acid extraction can dissolve collagen molecules that are not cross-linked, as well as collagen fibrils containing aldehyde-amine cross-linked bonds. Acid extraction can maintain the triple-helix structure of collagens, and acid-solubilized collagen can be used for biomedical materials and raw materials, but the extraction yield is lower than that of enzymatic extraction. Gao et al. (2008) extracted acid-solubilized collagens from collagen fibrils by optimizing the acid extraction method, resulting in a yield of 7.35% of collagen fibrils. The protocol is as follows: Sea cucumber collagen fibrils are resuspended in 0.5 M acetic acid solution and extracted for 72 h to obtain acid-solubilized collagens.

Enzymatic extraction is a process used to extract solubilized collagens by treating the raw materials with proteases, which is performed under mild reaction conditions focusing on the telopeptides other than the triple-helix structure, leading to increased solubility of collagens (Matinong et al. 2022). Pepsin is the most commonly used protease in sea cucumber collagen extraction, and the product is generally known as pepsin-solubilized collagen. Enzymatic extraction is characterized by a fast hydrolysis reaction, environmental-friendliness, and a high purity of extracted collagens with consistent physicochemical properties. Pepsin can be used in combination with acetic acid or alone after the extraction of acid-solubilized collagens. Pepsin-solubilized collagen possesses an intact triple-helix structure and low antigenicity, making it suitable for use as biomedical material and raw material for other collagen-derived products such as collagen peptides.

To separate the non-collagenous substances, powdery salt (commonly sodium chloride) or concentrated sodium chloride solution is added to the crude collagen extract for the precipitation of collagens, which is generally known as salting-out.

The salt concentrations required for collagen precipitation under neutral or acidic conditions differ. Generally, a concentration of 4 M or 20% NaCl is required for collagen precipitation under neutral conditions, while a concentration of 2 M or 10% NaCl is sufficient under acidic conditions (Matinong et al. 2022). It is worth noting that when the collagen concentration is low, the solution needs to be incubated for a period to allow for collagen precipitation after salting out. Generally, the solution should be stirred thoroughly and stand for 12–24 h after the salt is added to allow for collagen precipitation. The precipitated collagen is collected after centrifugation, dissolved with 0.5 M acetic acid, and dialyzed against 0.5 M acetic acid in order to remove the salt from the collagens, and the operation is repeated 2–3 times to reduce the content of non-collagenous substances to below 5% in the final product.

Trotter et al. (1995) treated sea cucumber (*Cucumaria frondosa*) collagen fibrils with pepsin in acetic acid solution at pH 2.5, and approximately 1% of the collagens were separated from the extracted fibril matrix. This might be due to the electrostatic binding interaction between the positively charged collagens and the negatively charged glycosaminoglycans under low pH conditions, which resulted in a low extraction yield. Therefore, the extraction method has been further improved as follows: Collagen fibrils are extracted in 0.5 M acetic acid solution (pH 2.5) for 24 h. The mixture is centrifuged at 100,000 × g for 60 min, and the precipitate is collected. The precipitate is resuspended in 0.5 M acetic acid solution (pH 2.5), and the mixture is extracted for 24 h with the addition of pepsin. The mixture is centrifuged at 100,000 × g for 60 min, the precipitate is resuspended in 50 mM Tris-HCl solution (1 M NaCl, pH 7.8) and extracted for 24 h, and the mixture is centrifuged at 100,000 × g for 60 min. The supernatant and the precipitate are collected separately, and the precipitate is repeatedly extracted according to the abovementioned procedure. The combined supernatant is dialyzed until the NaCl concentration is below 1 mM, and the mixture is centrifuged at 27,000 × g for 60 min to collect the precipitate. The precipitate is dissolved in 50 mM Tris-HCl solution (1 M NaCl, pH 7.8) and dialyzed. The NaCl concentration of this mixture is adjusted to 0.2 M, and the solution is centrifuged at 27,000 × g for 60 min to obtain pepsin-solubilized collagen.

Saito et al. (2002) reported a method for extracting pepsin-solubilized collagen from sea cucumber (*Stichopus japonicus*). Pepsin participates in the extraction process, but the collagen fibrils need to be in a disaggregated state, resulting in the full release and increased extraction yield of solubilized collagens. The protocol is as follows: Sea cucumber collagen fibrils are resuspended in 0.5 M acetic acid solution, and the mixture is extracted for 48 h with the addition of pepsin. The solution is centrifuged at 70,000 × g for 60 min. The supernatant is collected and adjusted to a NaCl concentration of 0.8 M with 4 M NaCl solution, and the collagen precipitate is collected after low-speed centrifugation. The precipitate is resuspended in 0.5 M acetic acid solution and dialyzed against 0.02 M Na_2HPO_4 solution (pH 8.0), and the collagen precipitate is collected after centrifugation. The pepsin-solubilized collagen is obtained through resuspension of the precipitate in 0.5 M acetic acid solution and dialysis against 0.1 M acetic acid solution.

Saallah et al. (2021) compared the differences between dialysis and ultrafiltration separation methods for the extraction of sea cucumber pepsin-solubilized collagens. The ultrafiltration separation method resulted in a higher yield in comparison with the dialysis separation method, and no significant difference was observed in the physicochemical properties of sea cucumber collagens. It should be mentioned that the solution should be stirred continuously throughout the collagen extraction process, which is beneficial to the disaggregation and release of collagens. Meanwhile, the extraction temperature should be kept moderate to avoid the denaturation of collagens, which may result in a reduced extraction yield. In addition, there are multiple centrifugation steps in the collagen extraction process, and the centrifugation rate is one of the major factors affecting the purity of collagen. Therefore, an appropriate centrifugation rate and sufficient centrifugation time should be selected.

2.3 *Purification of Solubilized Sea Cucumber Collagens*

Purification of sea cucumber collagen is a process to obtain collagens with a higher level of purity based on the abovementioned collagen extraction protocols. Due to the existence of glycosaminoglycans in the sea cucumber body wall, the solubilized collagens obtained by acid extraction or enzymatic extraction generally contain a certain amount of glycosaminoglycans. Further purification by ion exchange chromatography is required to obtain high purity sea cucumber collagens.

Ion exchange chromatography separates sea cucumber collagens and glycosaminoglycans based on the discrepancy of surface charges in collagens and sulfated glycosaminoglycans, using a suitable solvent as the eluent to exchange the exchangeable ions on the surface of ion exchanger with sulfated glycosaminoglycans. The commonly used ion exchanger for the separation of collagens and glycosaminoglycans is diethylaminoethyl (DEAE)-cellulose. For example, Trotter et al. (1995) purified pepsin-solubilized collagens from glycosaminoglycans using a DEAE-cellulose ion exchange column and obtained high purity sea cucumber collagens with NaCl solution at different concentrations as the eluent.

3 Extraction and Separation of Sea Cucumber Collagen Peptides

Sea cucumber collagen peptides generally refer to peptides derived from collagens through the process of protein hydrolysis, refinement, or purification using body wall or collagen as the raw material, which has the advantages of low molecular weight and good solubility (Ahmed et al. 2020). Due to the low solubility and protease digestibility sensitivity induced by covalent cross-linking of collagen, collagen peptides are generally prepared using pepsin-solubilized collagen as the substrate, which is produced through the degradation of telopeptides by pepsin. The preparation of collagen peptides mainly includes three processes: pretreatment of the substrate, degradation of collagens, and refinement or purification of collagen peptides.

3.1 Pretreatment of Sea Cucumber Collagens

Due to the high resistance of many proteases toward the triple-helix structure of collagens, a substrate pretreatment process is generally required to disrupt the triple-helix structure. Heat treatment and enzymatic treatment are often employed for the pretreatment of sea cucumber collagens. Ren et al. (2010) demonstrated the preparation of sea cucumber collagen peptides with the high-temperature treatment of collagen fibrils at 121 °C for 30 min, which was followed by enzymatic hydrolysis by proteases. Liu et al. (2011) pretreated sea cucumber (*Parastichopus californicus*) pepsin-solubilized collagens with collagenase, and the experimental conditions were as follows: enzyme to substrate ratio of 1:99, pH 7.0, 37 °C, and 1 h.

In addition, the pretreatment of the sea cucumber body wall can remove non-collagenous substances such as saponins, lipids, and inorganic salts, which is beneficial to improving the purity of sea cucumber collagen peptides using the body wall as the substrate. Cui (2007) developed a pretreatment method for the preparation of sea cucumber collagen peptides from the sea cucumber body wall. Denaturation of sea cucumber collagens was induced through soaking of the body wall in water at 60 °C for 48 h. Saponin is removed through continuous extraction with ethanol at 40 °C, which is conducive to reducing the saponin content in the final collagen peptide product.

3.2 Degradation of Sea Cucumber Collagens

Sea cucumber collagens are degraded into collagen peptides with low molecular weight and good hydrophilicity by physical, chemical, or biological treatment. Compared with the acid or alkali treatment method, enzymatic hydrolysis has the advantages of mild reaction conditions, easy control, and high specificity; therefore, the degradation of sea cucumber collagens is mainly performed using the enzymatic hydrolysis method.

3.2.1 Enzymatic Hydrolysis Method

The proteases used for the preparation of sea cucumber collagen peptides can be divided into three categories, including animal-origin proteases (trypsin and pepsin), plant-origin proteases (papain and bromelain), and microbial-origin proteases

Table 1 Preparation of sea cucumber collagen peptides by enzymatic hydrolysis

Sea cucumber species	Substrate	Enzymatic hydrolysis method	References
Paracaudina chilensis	Collagen	Substrate concentration 18 mg/100 mL (as hydroxyproline); bromelain, 240 U/mL; 40 °C; pH 5.2; 4 h	Xiao and Zeng (2006)
Acaudina molpadioides	Gelatin	(1) Substrate concentration 1%; bromelain, enzyme to substrate 1% (w/w); pH 7.0; 45 °C; 3 h (2) Alcalase, enzyme to substrate 2% (w/w); pH 7.5; 55 °C; 3 h	Zhao et al. (2007)
Stichopus japonicus	Collagen fiber	Substrate concentration 5.5%; papain, 2500 U/g; pH 6.7; 67.8 °C; 4 h	Ren et al. (2010)
Parastichopus californicus	Collagen	Alcalase, enzyme to substrate 0.064; 54.9 °C; 1.76 h	Liu et al. (2011)
Parastichopus californicus	Fresh body wall	Solid-liquid ratio 1:5 (g/mL); neutrase and flavourzyme, enzyme to substrate 1.05%; pH 7.5; 55 °C; 3 h	Yang et al. (2016)
Stichopus horrens	Fresh body wall	Solid-liquid ratio 1:2 (g/mL); alcalase, 2103 U/mL; pH 7.5; 55 °C; 5 h	Forghani et al. (2016)
Stichopus japonicus	Lyophilized body wall	Solid-liquid ratio 1:2 (g/mL); flavourzyme, enzyme t substrate 5% (w/w); pH 7.0; 50 °C; 8 h	Song et al. (2016)
Stichopus japonicus	Fresh body wall	Solid-liquid ratio 1:6 (g/mL); protamex, enzyme to substrate 0.8% (w/w); pH 7.0; 40 °C; 6 h	Hua et al. (2018)
Acaudina molpadioides	Acid-solubilized collagen (ASC)	Solid-liquid ratio 1:100 (g/mL); alcalase, 200 U/mg; pH 10.0; 45 °C; 1 h	Li et al. (2019)
Stichopus japonicus	Lyophilized body wall	Alcalase, enzyme to substrate 0.06 (w/w); pH 8.0; 50 °C; 5 h	Liu et al. (2019)
Apostichopus japonicus	Lyophilized body wall	(1) Solid-liquid ratio 1:20 (g/mL); trypsin, 3000 U/g; pH 8.0; 45 °C; 4 h (2) Papain, 3000 U/g; pH 6.0; 60 °C; 4 h	Guo et al. (2020)
Stichopus variegates	Dried body wall	Solid-liquid ratio 1:4 (g/mL); protamex, enzyme to substrate 0.4% (w/w); pH 7.0; 55 °C; 4 h	Lin et al. (2020)

(flavourzyme, neutrase, and alcalase). Sea cucumber collagen peptides could be prepared with single enzymatic hydrolysis or combined enzymatic hydrolysis methods. Compared with single enzymatic hydrolysis, combined enzymatic hydrolysis is beneficial to obtain a higher proportion of collagen peptides with low molecular weight, but the cost and hydrolysis conditions are more difficult to control. Collagen peptides with different molecular weights and activities can be prepared through the adjustment of substrate concentration, the type of protease, and the reaction time during enzymatic hydrolysis. Some examples of sea cucumber collagen peptides by enzymatic hydrolysis are shown in Table 1.

3.2.2 Other Methods

In order to improve the efficiency for the preparation of sea cucumber collagen peptides by enzymatic hydrolysis, researchers have made various attempts to shorten the hydrolysis time and improve hydrolysis efficiency by using physical auxiliary treatments such as ultrasound-assisted enzymatic hydrolysis and microwave-assisted enzymatic hydrolysis technologies.

The principle of ultrasound-assisted enzymatic hydrolysis is primarily attributed to the mechanical and cavitation effects generated when ultrasound propagates in the medium. It is a very promising processing technology characterized by high pertinence. The improvement effect of ultrasound on enzymatic hydrolysis is determined by the characteristics of the enzymes used and substrates as well as the parameters used for ultrasound. Ultrasound treatment and enzymatic hydrolysis can be performed simultaneously or sequentially, and appropriately, the ultrasound conditions can improve hydrolysis efficiency to a certain extent, thus increasing the yield of collagen peptides produced. Ultrasound treatment might accelerate the enzymatic reaction by changing the spatial structure of protein substrates, generating micro-resonance on the surface to render protein molecules disordered, and then improving the binding between the enzymes and the substrates. Xu et al. (2018) prepared collagen peptides from sea cucumber (*Acaudina molpadioides*) acid-solubilized collagens by using ultrasound-assisted enzymatic hydrolysis. The preparation of sea cucumber collagen peptides with ultrasound-assisted enzymatic hydrolysis by pepsin, trypsin, neutrase, and papain was significantly improved. For protein hydrolysates by neutrase, the content and yield of peptides with molecular weight below 1 kDa increased to 75.4% and 52.9% under the following condition, including reaction temperature of 60 °C, reaction time of 20 min, enzyme to substrate of 1: 20, and ultrasound powder of 100 W. The DPPH and ABTS radical scavenging abilities for collagen peptides were 19.0% and 87.2%, respectively, at the concentration of 8 mg/mL.

A microwave is a high-frequency electromagnetic wave that acts through the energy transfer by the polarization of molecules and the rapid absorption of microwave energy. This process can considerably shorten the time for complete hydrolysis. Microwave-assisted enzymatic hydrolysis can transfer energy to all reaction active centers of the medium at the same time, so there are more active centers compared to other heating methods, thus increasing the enzymatic hydrolysis efficiency. Li et al. (2019) optimized the process parameters of microwave-assisted enzymatic hydrolysis for the preparation of sea cucumber (*Acaudina molpadioides*) collagen peptides using acid-solubilized collagen as the raw material. The content of collagen peptides with molecular weight below 1 kDa was increased from 58.0 to 60.3%, and the OH· scavenging activity of collagen peptides was significantly increased from 70.3% in non-microwave assisted to 96.2% with the microwave power of 250 W for 30 min.

3.3 Refinement of Sea Cucumber Collagen Peptides

Refinement of sea cucumber collagen peptides is a process to improve the purity and quality indicators of peptides. The fishy smell, turbidity, and color of sea cucumber collagen peptides can affect product quality, significantly reduce the consumer acceptance, and thereby hinder the development of related products. Refinement of sea cucumber collagen peptides mainly includes two aspects: one is to improve the purity of collagen peptides by removing non-collagenous substances, and the other is to improve the quality of collagen peptides by removing the pigments and fishy smell and others.

The decolorization and deodorization of sea cucumber collagen peptides is mainly fulfilled by the adsorption methods and the biological methods. The adsorption method is a process to remove undesirable flavor substances using adsorbents. Commonly used compounds include activated carbon, diatomite, ion exchange resin, and macroporous adsorption resin. Zheng et al. (2013) investigated the effect of various adsorbents on the decolorization and deodorization of sea cucumber (*Apostichopus japonicus*) enzymatic hydrolysates. The result showed that the strong alkaline anionic resin D280 exhibited the best decolorization and deodorization effects with the volume ratio of enzymatic hydrolysates to resin as 3:1 at a flow rate of 2 mL/min. As a result, the decolorization and deodorization ratios were 69.78% and 67%, respectively, with the recovery rate of 82.49%.

The biological deodorization method, mainly referring to the biological fermentation method, is a process to remove the fishy and off-flavored substances from the product using fermentation technology and to generate a special flavor for the purpose of deodorization. Common biological deodorizers mainly include yeast, *Lactobacillus*, and *Acetobacter*. Wang et al. (2020a) developed a biological fermentation method with a bacterial mixture at a ratio of 3:1 (*Lactobacillus rhamnosus*/*Lactobacillus acidophilus*, v/v) for the deodorization of sea cucumber enzymatic hydrolysates. After 7 h of fermentation, the percentage of compounds with an unpleasant odor in the enzymatic hydrolysates decreased from 12.65 to 6.36%, while the percentage of compounds with pleasant odor increased from 31.46 to 44.03%.

In addition, due to the presence of other components such as sulfated polysaccharides and lipids in the sea cucumber body wall, sea cucumber enzymatic hydrolysates generally contain a certain amount of lipids, polysaccharides, and ash, which result in the reduced purity of collagen peptides, and can induce turbidity during preparation and storage. Thus, it is beneficial to reduce these compounds in the final product.

Cui (2007) prepared sea cucumber (*Stichopus japonicus*) enzymatic hydrolysates containing collagen peptides with Flavourzyme using the body wall as the raw material under the following conditions: substrate concentration of 4%, enzyme to substrate ratio of 1:99, pH 6.5, 45 °C, and hydrolysis time of 4 h. Using this as a starting material, polysaccharide substances were removed from the enzymatic hydrolysates through organic solvent precipitation, leading to an improved purity of sea cucumber collagen peptides. Briefly, cold ethanol was added to the enzymatic hydrolysates at a volume ratio of 3:1 to precipitate the proteins with high molecular weight and sulfated polysaccharide; as a result, sea cucumber enzymatic hydrolysates containing a large amount of collagen peptides were obtained, and the collagen peptides showed predominant molecular weight distribution below 2 kDa.

Su et al. (2013) investigated the cause of the turbidity of sea cucumber (*Acaudina molpadioides*) enzymatic hydrolysates using body wall as the raw material. The turbid particles in the sea cucumber enzymatic hydrolysates are mainly composed of protein, lipid, and ash with a surface potential of −46.3 mV and an average particle size of 597 nm. The turbidity of sea cucumber enzymatic hydrolysates is mainly caused by the interaction between the protein and lipid components via hydrophobic interactions in the turbid substances. Filtration with diatomite AG-800# could effectively remove turbid substances from enzymatic hydrolysates.

3.4 Purification of Sea Cucumber Collagen Peptides

Purification of sea cucumber collagen peptides is a process of enriching collagen peptides with a specific structure and biological activities from collagen peptide mixtures. Purification of sea cucumber collagen peptides is mainly based on the discrepancy of physical and chemical properties, such as molecular weight, charge, polarity, and interactions, through membrane separation, gel filtration chromatography, ion exchange chromatography, and reverse phase chromatography, in which the first three types of purification methods can be scaled for industrial production. The above methods have been widely applied to the purification of collagen peptides, and multiple purification methods are usually combined to achieve better purity in practice.

3.4.1 Membrane Separation Technology

According to the difference in molecular weight for collagen peptides, the membrane separation technology is a process which allows peptides with low molecular weight to pass through the membrane with a certain molecular weight cut off, while fractions with a high molecular weight are retained by the membrane, for the purpose of separation and purification (Vandanjon et al. 2007). Common membrane separation methods include microfiltration, ultrafiltration, and nanofiltration. The molecular weight cut offs (MWCOs) commonly used for collagen peptide separation and purification are 10 kDa, 5 kDa, 3 kDa, 1 kDa, and 500 Da. Generally, membrane separation is considered as the initial step for the purification of sea cucumber collagen peptides and through a combination of membrane separation and other purification methods can be performed according to the purification requirements.

Zhao et al. (2007) purified sea cucumber (*Acaudina molpadioides*) collagen peptides with molecular weights below 10, 5, and 1 kDa through the use of ultrafiltration separation and demonstrated that collagen peptides with molecular weights below 1 kDa showed the strongest ACE inhibitory effect. Quaisie et al. (2022) investigated the technical parameters for the production of sea cucumber (*Apostichopus japonicus*) collagen peptides using an enzyme membrane reactor coupled to a nanofiltration desalination unit consisting of a temperature-controlled enzymatic hydrolysis bioreactor, an ultrafiltration unit and a 5 kDa ultrafiltration membrane. Peptide products with a molecular weight lower than 5 kDa were found in the permeate, while the substrate and water were continuously replenished into the reactor during the enzymatic hydrolysis process. Meanwhile, the online desalination of sea cucumber peptides could be realized through coupling of a nanofiltration unit with the enzymatic membrane reactor, which means that the desalination and enzymatic hydrolysis processes are conducted simultaneously. Compared with the conventional offline membrane separation method, the enzyme membrane reactor method increased the protein conversion rate by 60.39%.

3.4.2 Gel Filtration Chromatography

Gel filtration chromatography is also known as size exclusion chromatography or molecular sieve chromatography. The basic principle of gel filtration chromatography is to separate and purify peptides according to their molecular weights. Peptides are separated by the gel packing in the chromatographic column, which results in different retention times depending on the molecular weight of the compound. The gel chromatography packing materials used for the separation and purification of sea cucumber collagen peptides are mainly dextran gels with different cross-linking degrees designed according to the required molecular weight ranges.

Xiao and Zeng (2006) purified sea cucumber collagen peptides with molecular weights below 5 kDa using a Sephadex G-25 column, and four fractions were isolated, among which the free radical scavenging ability of fraction 2 was significantly improved after purification. Guo et al. (2020) prepared sea cucumber peptides with significant antioxidant and anti-aging effects after desalination using a Sephadex G-10 column.

3.4.3 Ion Exchange Chromatography

Ion exchange chromatography is a purification method based on the difference between the magnitude of the binding force between charged groups within the compounds in the mobile phase and oppositely charged groups on the column phase. The binding of collagen peptides to ion exchangers depends on several parameters, including the pH, ionic strength, and the type, quantity and distribution of electric charge. Ion exchangers used in collagen peptide purification include cation exchanger and anion exchanger, and NaCl solutions at different concentrations are generally used as the eluents.

Zhao et al. (2007) isolated antihypertensive peptides from sea cucumber (*Acaudina molpadioides*) collagen peptides with a molecular weight below 1 kDa using a SP Sephadex C-25 cation exchange column with 0–0.15 M NaCl solution at an elution rate of 0.4 mL/min. Liu et al. (2019) purified bioactive peptides with a molecular weight below 5 kDa from sea cucumber peptides using a DEAE Fast Flow anion exchange column, with 20 mM Tris-HCl solution (1 M NaCl, pH 8.0) as the eluent at a flow rate of 2 mL/min.

3.4.4 Reverse Phase Chromatography

The basic principle of reverse phase chromatography is to separate and purify collagen peptides with different polarities at different retention times according to the difference in adsorption capacity between peptide fractions at the hydrophobic stationary phase, with high polarity fractions eluted first and low polarity components eluted later. Reverse phase chromatography is mainly used for the separation and purification of collagen peptides with molecular weight below 5 kDa, especially those with a molecular weight below 1 kDa. Reverse phase chromatography is usually performed as the last step in a series of combined purification approaches during the separation and purification of sea cucumber collagen peptides.

Zhao et al. (2012) purified and separated a novel ACE inhibition peptide MEGAQEAQGD with the IC_{50} value of 15.9 μmol/L from sea cucumber (*Acaudina molpadioides*) collagen peptides using the combination of gel filtration chromatography, cation exchange chromatography, gel filtration chromatography, and Zorbax C_{18} reverse phase chromatography.

Forghani et al. (2016) developed a sea cucumber (*Stichopus horrens*) peptide purification workflow using reverse phase chromatography separation as the first step on a ZORBAX 300SB C_{18} reverse phase column and collected peptide fractions with the strongest ACE inhibition activity. The collected fractions were further purified with the isoelectric focusing technology, and finally three bioactive peptides, i.e., EVSQGRP, CRQNTLGHNTQTSIAQ, and VSRHFASYAN, with strong ACE inhibition activity were purified from the sea cucumber peptides with the IC_{50} values of 0.05, 0.08, and 0.21 mM, respectively.

4 Quality Control of Sea Cucumber Collagens and Collagen Peptides

At present, there is still a lack of international standards specifically issued for sea cucumber collagen peptides, and the production of sea cucumber collagen peptides usually still follows to the relevant standards for general collagen peptides. The Chinese standard DB61/T 1202-2018 (2018) defines the general technical

Table 2 General technical requirements of type I collagen

Item	Index
Purity	≥95% (m/m)
Hydroxyproline	≥5% of total protein content
Fat	≤1%
Total heavy metals (as Pb)	≤10 mg/kg
Residue on ignition	≤10 g/kg
Total aerobic count	≤100 cfu/g, no pathogenic bacteria
Cytotoxic response	≤Grade 2
Mean difference of intradermal responses	≤1.0
Acute systemic toxicity	None
Hemolysis rate	≤5%
Allergic reaction	None
Tissue response after 2 and 4 weeks of muscle implantation	No significant difference with negative control
Genotoxicity	None
Immunogenicity	Refer to GB/T 16886.20

requirements of type I collagen (Table 2), which can be employed for the preparation of tissue repair products.

Chinese standard GB 31645-2018 (2018) defines that collagen peptides for food processing are fractions with a molecular weight below 10 kDa, which are produced by extraction, hydrolysis, and refinement using fresh animal tissues such as skin, bone, sinew, tendon, and scale, rich in collagen, as the raw materials. The standard defines that collagen peptides are powdery or granular products with white or light-yellow color and are free from unpleasant odors, lumps, and foreign matter visible by regular visual activity. The proportion of collagen peptides with a molecular weight below 10 kDa must be higher than 90%, and the hydroxyproline content (on dry basis) has to be higher than 3%. The total nitrogen content (on dry basis) is no less than 15%, while the ash and moisture contents are no more than 7% and 7%, respectively. In addition, the heavy metal and microbial content should meet the relevant requirements.

5 New Product Development of Sea Cucumber Collagens and Collagen Peptides

5.1 Food and Functional Food Product Development

In recent years, the global demand for collagens and collagen peptides has been increasing; collagens and collagen peptides are suitable for applications in food and functional food product development due to their nutritional and functional properties. Food regulations specify no limitation for the addition of collagens and collagen peptides in most food; therefore, the demand for collagens can be satisfied with the administration of food and functional food which are rich in collagens and collagen peptides. In addition, collagens and collagen peptides have shown various beneficial physiological activities such as antihypertension, immunomodulation, antiaging, and melanin synthesis inhibition. However, the industrial application of sea cucumber collagens and collagen peptides remains relatively limited compared to those collagens and collagen peptides derived from mammalian sources and other marine sources.

Liu et al. (2009, 2019) developed a workflow for the degradation of sea cucumber proteins into solubilized collagen peptides with a low molecular weight by gutting, blanching, pulping, inactivation, and enzymatic hydrolysis using fresh sea cucumber as the raw material. According to the theory held within traditional Chinese medicine, traditional medicine extracts from herbs like American ginseng, jujube, cinnamon, and wolfberry were mixed with sea cucumber enzymatic hydrolysates to develop a sea cucumber compound oral liquid with anti-fatigue and immunity enhancement functions. In addition, there have been many other foods and functional foods containing sea cucumber collagens or collagen peptides, mainly including jelly, drinks, soft capsules, pressed candy, and alcoholic beverages. Table 3 lists some of the product development patents relating to sea cucumber collagens or collagen peptides. Although most of these functional effects have been determined by using in vitro or in vivo models, much more studies are needed to further validate and prove the described biological effects in these literatures.

5.2 *Cosmetic Product Development*

The water binding capacity of collagen is the basis for its successful performance as a cosmetic ingredient. Studies have shown that sea cucumber (*Holothuria cinerascens*) collagen provides excellent moisture absorption and water binding capacity compared to glycerol and has the potential as a moisturizer for cosmetic development (Li et al. 2020). In addition, collagen peptides have been applied in a variety of cosmetic formulations due to their low molecular weight, high solubility, and ability to penetrate into deep layers of the skin. Collagen peptides can penetrate into the skin through three potential pathways, including sweat glands, hair follicles, and sebaceous glands, as well as stratum corneum. Both external local application and oral intake of collagen peptides can effectively improve skin elasticity and delay skin aging (Aguirre-Cruz et al. 2020), and it is expected that the synergistic use of externally applied collagen peptides nutritional supplement will become one of the promising development trends in the future. Nowadays, the cosmetic products containing sea cucumber collagens or collagen peptides in the market are focused on mask products only.

Table 3 Product development patents on sea cucumber collagen and collagen peptide

Product Name	Formula	Function	References
Sea cucumber liquor	Uses dried sea cucumber as the raw material; applies soaking, cleaning, enzymatic hydrolysis, and fermentation. Compounds sea cucumber peptides with herbal extracts	Antitumor, immunity enhancement, anti-coagulation, antithrombosis, anti-radiation, antivirus, and liver protection	Guo et al. (2012)
Sea cucumber collagen peptide tablet candy	Compounds sea cucumber powder, collagen powder, oyster powder, black fungus powder, maltodextrin, xylitol, corn starch, and oligomeric isomaltose	Skin nourishing and detoxification, anti-osteoporosis, and removal of heavy metals	Chen (2017)
Sea cucumber peptide chewable tablet	Compounds casein phosphopeptide, sea cucumber peptide, colostrum powder, soy peptide, whey protein concentrate, lutein ester, vitamin C, fruit-vegetable powder, resistant dextrin, lactose, erythritol, sorbitol, DL-malic acid, oligosaccharide, magnesium stearate, and targeted slow-release biocompatible microcapsules	Crosses the blood-brain barrier, acting on brain cells to repair cardiovascular and post-oxidative stress damage, and assists in the digestion of colostrum powder and whey protein concentrate in the stomach to enhance the immune system	Cui and Wan (2020)
Sea cucumber collagen peptide solid beverage	Compounds sea cucumber collagen peptide and non-spore probiotics.	Immunity enhancement, detoxification, and intestinal function improvement	Wang (2020)
Sea cucumber peptide stabilization preparation	Compounds sea cucumber peptide, low molecular hyaluronic acid, seaweed sugar, and water	Anti-wrinkling, moisturization, and skin elasticity enhancement	Song and Zhang (2014)
Sea cucumber collagen for skin repair	Uses sea cucumber as the raw material; applies ultra-high pressure treatment, enzymatic hydrolysis, deodorization, membrane separation, concentration, and drying	Anti-oxidation, anti-inflammation, moisturization, anti-wrinkling, and whitening	Zhang (2018)
Acne repair mask	Compounds hydroxyethylcellulose, pentylene glycol, butylene glycol, allantoin, nicotinamide, deep-sea fish skin collagen peptide, sea cucumber collagen peptide, fish roe peptide, dipotassium glycyrrhizinate, aloe vera lyophilized powder, and water	Acne treatment	Jia and Zhou (2018)

5.3 Biomedical Material Product Development

Collagen is considered as a successful biomedical material due to its biodegradability, weak antigenicity, good biocompatibility, bioresorbability, durability, and non-toxicity. The diversity of collagen resources and phenotypes also contributes to the application of collagen as a biomedical material. Previous studies have shown that sea cucumber (*Stichopus japonicus*) pepsin-solubilized collagen has the ability to promote wound healing and shows a positive effect on cell migration and proliferation of human keratinocyte cell lines compared to conventional collagens, demonstrating the potential application of sea cucumber collagens in biomedical materials (Park et al. 2012). The mutable collagenous tissue of echinoderms can undergo extreme changes in passive mechanical properties under the control of the nervous system, and the mutable collagenous tissue can be used to prepare fibrous collagen membranes, which have been proposed as an intelligent and dynamic biomaterial for tissue engineering and regenerative medicine application (Ferrario et al. 2017).

Zhang et al. (2015) investigated into the method of producing a sea cucumber medical collagen sponge. The protocol is as follows: Pepsin-solubilized collagen solution is transferred into the stainless-steel storage tank of a tunnel-type automatic belt instant freezer (−40 °C). The sea cucumber collagen solution is added into the stainless-steel batching tank, and then food-grade glutamine transaminase is added to the solution as the cross-linking agent with constant stirring. A high-pressure spray pump is turned on, and misted sea cucumber collagen solution is sprayed into the stainless-steel mold basin on the conveyor belt of the instant freezer. The instantaneously quick-frozen collagen sponge is transferred into a −40 °C freezer, and the material is freeze-dried until the water content is in the range of 10–15%. The collagen sponge with 10–15% water content is transferred to a vacuum dryer and dried at a rate of 10 °C/30 min with temperature risen from 60 to 100 °C for 2 h and then dried at 118–120 °C until the water content declines to 1–3%. After drying, the collagen sponge is taken out, transferred to the cooling box, cooled, sorted, packaged, irradiated, and inspected, resulting in the medical collagen sponge.

Ferrario et al. (2017) produced a method for the preparation of sea cucumber (*Holothuria tubulosa*) collagen membranes and the obtained membranes had the potential to act as collagen membranes for guided tissue regeneration. The protocol is as follows: The collagen suspension is dried overnight in a mold at 37 °C, and the remaining suspension is centrifuged at 50 × g for 10 min to remove the precipitated debris and then centrifuged at 4000 × g for 20 min. The precipitate is resuspended in sterile water, and the suspension is placed in a rubber silicone mold and dried overnight at 37 °C. The resultant collagen membranes are immersed in 1-ethyl-3-(3-dimethylaminopropyl)-carbodiimide/*N*-hydroxysuccinimide (EDC/NHS) cross-linking solution for 4 h at room temperature and then washed with phosphate buffer, deionized water, and a 70% ethanol solution for the final product.

6 Conclusion

Research on sea cucumber collagens and collagen peptides is an emerging and rapidly developing field. Inspired by the continuous understanding of the structural characteristics and functional properties of sea cucumber collagen molecules, more and more novel products containing sea cucumber collagen or collagen peptide are expected to emerge in the near future. However, many factors such as more advanced and intelligent processing technologies, product stability, product sensory acceptance, market competitiveness, and production cost, etc. that affect the promotion of products containing sea cucumber collagen or collagen peptide should be taken into consideration, thereby driving sea cucumber collagen and collagen peptide products to meet the market demands for food, cosmetics, and biomedical materials.

References

Aguirre-Cruz G, Leon-Lopez A, Cruz-Gomez V, Jimenez-Alvarado R, Aguirre-Alvarez G (2020) Collagen hydrolysates for skin protection: oral administration and topical formulation. Antioxidants 9:181

Ahmed M, Verma AK, Patel R (2020) Collagen extraction and recent biological activities of collagen peptides derived from sea-food waste: a review. Sustain Chem Pharm 18:100315

Chen C (2017) Sea cucumber collagen peptide tablet candy. CN Patent Application 201711225379.6

Coppola D, Oliviero M, Vitale GA, Lauritano C, D'Ambra I, Iannace S, de Pascale D (2020) Marine collagen from alternative and sustainable sources: extraction, processing and applications. Mar Drugs 18:214

Cui F (2007) Study on the biochemical characterization of collagen and activity of collagen peptide extracted from the body wall of sea cucumber *Stichopus Japonicus*. Dissertation, Ocean university of China

Cui L, Wan P (2020) An intestinal slow-release colostrum sea cucumber peptide chewable tablet and its preparation method CN Patent Application 202010729596.4

Cui FX, Xue CH, Li ZJ, Zhang YQ, Dong P, Fu XY, Gao X (2007) Characterization and subunit composition of collagen from the body wall of sea cucumber *Stichopus japonicus*. Food Chem 100:1120–1125

Ferrario C, Leggio L, Leone R, Di Benedetto C, Guidetti L, Cocce V, Ascagni M, Bonasoro F, Caterina AM, Carnevali M, Sugni M (2017) Marine-derived collagen biomaterials from echinoderm connective tissues. Mar Environ Res 128:46–57

Forghani B, Zarei M, Ebrahimpour A, Philip R, Bakar J, Abdul Hamid A, Saari N (2016) Purification and characterization of angiotensin converting enzyme-inhibitory peptides derived from *Stichopus horrens*: stability study against the ACE and inhibition kinetics. J Funct Food 20:276–290

Gao Y, Dong XP, Xiao GH, Jiang HM, Zhu BW (2008) Study on the extraction and amino acid composition analysis of acid soluble collagen from the body wall of sea cucumber *Stichopus japonicus*. Food Ferment Indust 34:171–174

Guo JJ, Xu JC, Liu XL (2012) A preparation method of full nutrition sea cucumber bio wine. CN Patent Application 201110246904.9

Guo K, Su L, Wang Y, Liu H, Lin J, Cheng P, Yin X, Liang M, Wang Q, Huang Z (2020) Antioxidant and anti-aging effects of a sea cucumber protein hydrolyzate and bioinformatic characterization of its composing peptides. Food Funct 11:5004–5016

Hua X, Sun LC, Wan CJ, Tang Y, Weng L, Liu GM, Cao MJ (2018) Preparation and antihypertensive activity of angiotensin I-converting enzyme (ACE) inhibitory peptides derived from sea cucumber (*Stichopus japonicas*). Food Sci 39:125–130

Jia K, Zhou SS (2018) A marine biological peptide acne repair mask and its preparation method. CN Patent Application 201811015005.6

Li Y, Li J, Lin SJ, Yang ZS, Jin HX (2019) Preparation of antioxidant peptide by microwave-assisted hydrolysis of collagen and its protective effect against H2O2-induced damage of RAW264.7 cells. Mar Drugs 17:642

Li PH, Lu WC, Chan YJ, Ko WC, Jung CC, Huynh LDT, Ji YX (2020) Extraction and characterization of collagen from sea cucumber (*Holothuria cinerascens*) and its potential application in moisturizing cosmetics. Aquaculture 515:734590

Lin L, Zhu Q, Zheng L, Zhao M, Fan J, Liu S (2020) Preparation of sea cucumber (*Stichopus variegates*) peptide fraction with desired organoleptic property and its anti-aging activity in fruit flies and D-galactose-induced aging mice. J Funct Food 69:103954

Liu CH, Wang XJ, Yuan WP, Meng XM, Xia XK, Zhang MS, Tang JD, Hu W, Sun YJ, Liu JM (2009) Anti-fatigue and immune functions of sea cucumber oral liquid. Modern Food Sci Technol 25:1115–1119

Liu ZY, Chen D, Su YC, Zeng MY (2011) Optimization of hydrolysis conditions for the production of the angiotensin I converting enzyme inhibitory peptides from sea cucumber collagen hydrolysates. J Aquat Food Prod Technol 20:222–232

Liu X, Wang Z, Zhang J, Song L, Li D, Wu Z, Zhu B, Nakamura Y, Shahidi F, Yu C, Zhou D (2019) Isolation and identification of zinc-chelating peptides from sea cucumber (*Stichopus japonicus*) protein hydrolysate. J Sci Food Agric 99:6400–6407

Matinong AME, Chisti Y, Pickering KL, Haverkamp RG (2022) Collagen extraction from animal skin. Biology 11:905

National Health Commission of the People's Republic of China, State Administration for Market Regulation (2018) Food safety national standard collagen peptide. GB Patent Application 31645-2018

Park SY, Lim HK, Lee S, Hwang HC, Cho SK, Cho M (2012) Pepsin-solubilised collagen (PSC) from red sea cucumber (*Stichopus japonicus*) regulates cell cycle and the fibronectin synthesis in HaCaT cell migration. Food Chem 132:487–492

Quaisie J, Ma H, Guo Y, Tuly JA, Igbokwe CJ, Ekumah JN, AkpabliTsigbe N, Ding Y, Liu D (2022) Highly stable, antihypertensive, and antioxidative peptide production from *Apostichopus japonicus* by integrated enzymatic membrane reactor and nanofilter-purification mechanism. Food Funct 13:2306–2322

Ren TT, Dong XP, Zhu BW, Zhou DY, Qin L, Zheng J (2010) Optimization on collagen peptides preparation from sea cucumber by response surface analysis. Food Machin 26:120–123

Ricard-Blum S (2011) The collagen family. Cold Spring Harb Perspect Biol 3:a004978

Saallah S, Roslan J, Julius FS, Saallah S, Razali UHM, Pindi W, Sulaiman M, Khairul F, Kamal SMM (2021) Comparative study of the yield and physicochemical properties of collagen from sea cucumber (*Holothuria scabra*), obtained through dialysis and the ultrafiltration membrane. Molecules 26:2564

Saito M, Kunisaki N, Urano N, Kimura S (2002) Collagen as the major edible component of sea cucumber (*Stichopus japonicus*). J Food Sci 67:1319–1322

Shaanxi Provincial Bureau of Quality and Technical Supervision (2018) General technical requirements for type I collagen. CN Patent Application DB61/T 1202-2018

Song SL, Zhang E (2014) Sea cucumber peptide stabilizing preparation and its preparation method and Application. CN Patent Application 201410135479.X

Song J, Li T, Cheng X, Ji X, Gao D, Du M, Jiang N, Liu X, Mao X (2016) Sea cucumber peptides exert anti-inflammatory activity through suppressing NF-kappa B and MAPK and inducing HO-1 in RAW264.7 macrophages. Food Funct 7:2773–2779

Sorushanova A, Delgado LM, Wu ZN, Shologu N, Kshirsagar A, Raghunath R, Mullen AM, Bayon Y, Pandit A, Raghunath M, Zeugolis DI (2019) The collagen suprafamily: from biosynthesis to advanced biomaterial development. Adv Mater 31:1801651

Su W, Cui HB, Xue CH, Zhang CY, Yu L, Chang YG (2013) Character study and elimination of turbidity of *Acaudina Molpadioides* enzymatic hydrolysate. J Chin Inst Food Sci Technol 13: 130–134

Theocharis AD, Skandalis SS, Gialeli C, Karamanos NK (2016) Extracellular matrix structure. Adv Drug Deliv Rev 97:4–27

Trotter JA, Lyons-Levy G, Thurmond FA, Koobt TJ (1995) Covalent composition of collagen fibrils from the dermis of the sea cucumber, *Cucumaria frondosa*, a tissue with mutable mechanical properties. Comp Biochem Physiol A Mol Integr Physiol 112A:463–478

Vandanjon L, Johannsson R, Derouiniot M, Bourseau P, Jaouen P (2007) Concentration and purification of blue whiting peptide hydrolysates by membrane processes. J Food Eng 83: 581–589

Wang BS (2020) Probiotic sea cucumber collagen complex peptide solid drink. CN Patent Application 202010386306.0

Wang J, Chang YG, Wu FX, Xu XQ, Xue CH (2018) Fucosylated chondroitin sulfate is covalently associated with collagen fibrils in sea cucumber *Apostichopus japonicus* body wall. Carbohydr Polym 186:439–444

Wang T, Wen YQ, Yu J, Xue Y, Xue CH (2020a) Preparation of deodorized sea cucumber peptide powder rich in lactic acidbacteria. Food Ferment Indust 46:187–191

Wang YC, Tian M, Chang YG, Xue CH, Li ZJ (2020b) Investigation of structural proteins in sea cucumber (*Apostichopus japonicus*) body wall. Sci Rep 10:18744

Wess TJ (2005) Collagen fibril form and function. Adv Protein Chem 70:341–374

Xiao F, Zeng MY (2006) Study on preparation of free radical scavenging active peptides from collagen hydrolysates of *Paracaudina chinens var.* Food Sci 27:336–339

Xu HP, Xie H, Liang JH, Yi X, Jin HX (2018) The preparation and activity of oligopeptides by ultrasonic-assisted enzymatic hydrolysis of *Acaudina molpadioides* collagen. J Zhejiang Ocean Univ (Nat Sci) 37:388–393

Yang Y, Wang QK, Gu Y, He YH, Song YF, Cong HH, Liu S, Qi YX (2016) Study on the hydrolyzing technologies of basic flavoring materials from *Parastichopus californicus*. Sci Technol Food Ind 37:190–200

Zhang LY (2018) Collagen for skin repair and its preparation method. CN Patent Application 201811019575.2

Zhang YB, Ye ZF, Ye ZZ, Ye C (2015) A method of manufacturing medical collagen sponges from East China sea black cucumber. CN Patent Application 201310323629.5

Zhao Y, Li B, Liu Z, Dong S, Zhao X, Zeng M (2007) Antihypertensive effect and purification of an ACE inhibitory peptide from sea cucumber gelatin hydrolysate. Process Biochem 42:1586–1591

Zhao YH, Li BF, Ma JJ, Dong SY, Liu ZY, Zeng MY (2012) Purification and synthesis of ACE inhibitory peptide from *Acaudina molpaioidea* protein hydrolysate. Chem J Chin Univ 33:308–312

Zheng JW, Wang QK, He YH, Liu S, Ren DD (2013) Technique optimization of decolorization and deodorizatin for sea cucumber hydrolysates. J Dalian Ocean Univ 28:303–306

The Quality Management Systems and Standards of Sea Cucumber Products

Qingjuan Tang and Yueru Li

Abstract In China, the sea cucumber has a long history as a food. The nutrients of sea cucumbers abound in polysaccharides, lipids, proteins, and a variety of vitamins and minerals. In recent years, the sea cucumber industry has been growing rapidly, and sea cucumber products are becoming increasingly diversified, entailing the establishment of high-quality standards and detection systems. These standards and systems are crucial for increasing production, market supervision, combating counterfeit and shoddy products, and thus protecting consumers' rights and interests. Rapid and precise identification of sea cucumber species, traceability of production techniques (covering origin, breeding procedure, etc.), and detection and quantification of practical components in sea cucumber products serve as the foundation for constructing standard quality-control systems. This can help to standardize the whole value chain of the sea cucumber business, from breeding to processing, deep processing, and sales. This chapter will discuss how to identify sea cucumber species, track their origins, detect the practical components, and establish the relevant product quality-control systems and standards.

Keywords Sea cucumber · Nutritional value · Species identification · Quality management system · Standard

1 Introduction

There are around 1400 species of sea cucumbers known globally, found in all the world's oceans. There are approximately 140 types of sea cucumbers in China, of which more than 20 are edible. The structure and composition of nutritional and functional components such as collagen, sulfate polysaccharide, lipid, and saponin

Q. Tang (✉) · Y. Li
College of Food Science and Engineering, Ocean University of China, Qingdao, Shandong, P. R. China
e-mail: tangqingjuan@ouc.edu.cn; liyueru@ouc.edu.cn

C. Xue (ed.), *Advances in Sea Cucumber Processing Technology and Product Development*, Advances in Marine Bioprocesses and Bioproducts,
https://doi.org/10.1007/978-3-031-16512-2_11

vary considerably across sea cucumber species. Traditional sea cucumber processing relies heavily on dried sea cucumbers, ready-to-eat sea cucumbers, and other sea cucumber-like items. Sea cucumbers are often dried and sliced, making it difficult for consumers, distributors, and regulatory agencies to determine the species and origin of the sea cucumbers, or the production techniques used, based on their appearance and shape. Various species, sources, and production techniques (e.g., wild and farmed) might result in a price differential of tens of times between sea cucumber products.

With the development of deep processing technology for sea cucumbers, new product forms that do not retain the sea cucumber's shape have emerged, such as sea cucumber capsules and orally consumed liquid. For deep-processed sea cucumber products, the species, composition, nutritional levels, and efficacy components are critical factors impacting product quality. Therefore, the technology for identifying raw materials and determining the content of sea cucumber products serves as the foundation for developing a standard quality-control system for sea cucumber. Clarifying commercial sea cucumber species, their origin, and production techniques are vital to maintaining customer confidence in sea cucumber products. As a kind of technical assistance for the purposes of regulation and control, traceability offers technical verification for relevant authorities. Utilizing traceability technologies and developing quality-control systems and standards for sea cucumber products mark a critical growth path for the sea cucumber sector for the time being and for the future.

2 Identification of Sea Cucumber Species

The conventional approaches to sea cucumber morphological identification are based primarily on the exterior appearance and anatomical properties of the different species of sea cucumbers (Arndt et al. 1996). External morphological features include the shape of the tentacles, the presence or absence of ambulacral feet and parapodia, their shape and distribution characteristics, etc. The anatomical characteristics of sea cucumber mainly refer to the morphology and number of sea cucumber ossicles. Liao (2001) categorized 134 species of sea cucumbers in China based on their morphology. They have also studied the compositional features for the ossicles of more than 50 species of sea cucumber in China and regarded sea cucumber ossicles as a crucial foundation of sea cucumber identification. Li et al. (2008) observed and described the morphology of the ossicles of 15 sea cucumber species from China and other countries. The types of ossicles for each of the sea cucumber species and the relative proportions for each type of the ossicles were compared to provide a basis for the identification of these 15 species of sea cucumbers. Wen et al. (2009), Fan et al. (2010), and Yao et al. (2017) used scanning electron microscopy to examine the composition and structure of the dorsal ossicles of more than ten different species of sea cucumber (*Holothuria leucospilota*, *Stichopus variegatus*, *Thelenota ananas*, *Actinopyga lecanora*, *Actinopyga lecanora*, *Actinopyga echinites*, *Stichopus horrens*, etc.). This work has complemented the shortage of morphological research on sea cucumber ossicles and has established a foundation for further categorization and identification of sea cucumbers.

However, conventional morphological identification methods have the following defects: (1) The external morphological characteristics (phenotypes) of species are influenced not only by genotype but also by the growth environment, which leads to inaccurate identification results. (2) The morphological methods are not capable of identifying the pervasive concealed taxon. (3) The identification results are limited by biological sex and developmental stage. (4) The identification and statistics for the type and proportion of ossicles in each piece of sea cucumber pose high technical requirements and require professional identification. Moreover, the process is cumbersome and time-consuming and cannot be used as a fast and accurate method to identify sea cucumber species.

Physical and chemical testing procedures are also frequently employed to identify sea cucumber species, mostly based on their chemical composition. By analyzing the distribution and structural features of triterpenoid saponin and using morphological approaches, Levin et al. (1986) established an evolutionary link between *Apostichopus japonicus* and *Holothuriidae* from the North Pacific. Kalinin et al. (2008) established a taxonomy of sea cucumbers by analyzing the structure and distribution of triterpenoid saponins from several sea cucumber species. Yu et al. (2011) analyzed the saponin components of *Apostichopus japonicus* using an HPLC fingerprint. The similarity of fingerprints between different batches of samples was greater than 97%. This demonstrates that both the saponin content of *Apostichopus japonicus* and its structure are stable. It might be used to determine the authenticity of *Apostichopus japonicus* as a quality evaluation index. However, the preceding procedure required first separating and purifying the chemical components unique to sea cucumber species, followed by physical and chemical analysis for the purpose of species identification. It was still difficult to perform specific analysis and identification of sea cucumber species that had undergone extensive processing. Furthermore, current physical and chemical tests demand large-size instruments, sophisticated operation, and a high workload in most cases; in addition, this process carries significant risks or variances in the outcome.

With advances in modern molecular biotechnology, the primary approach for identifying species is to use DNA sequence fragments. As a strategy for identifying genetic markers, molecular marker technology is critical for genetic diversity study and species identification. Its benefits include excellent specificity, high sensitivity, superior heat resistance to protein, and simplicity and speed. In recent years, molecular marker technology has increasingly been used to identify marine creatures.

2.1 Mitochondrial Sequence for Identifying Sea Cucumber Species

Mitochondria are the cytoplasmic organelles found in all eukaryotic cells. Mitochondria are semi-autonomous organelles that are both mitochondrial and nuclear genomes controlled. Since NassM and NassS first detected delicate filamentous DNA within mitochondria in the 1960s, the existence of genetic material within mitochondria has progressively gained acceptance. Since the 1980s, mitochondrial DNA has grown in importance as a tool for examining the genetic differentiation of closely related and intraspecific populations. Mitochondrial DNA analysis techniques have swiftly spread into conventional study domains such as categorization, phylogeography, population genetics, and anthropology, gradually displacing prior isoenzyme and experimental immunology approaches based on analysis of protein characteristics.

The phenomenon of variation in mitochondrial DNA within or across populations is referred to as mitochondrial DNA polymorphism, and it could be examined using sequence analysis and restriction enzymes. This polymorphism is classified into two types: site polymorphisms and length polymorphisms. The analysis of mitochondrial DNA data from currently accessible animals demonstrated that the primary characteristic of mitochondrial DNA evolution was the increase in base substitutions combined with a slight decrease in insertions and deletions. At varying rates, base substitutions occurred mostly in the intergenic and regulatory regions. In general, direct sequencing analysis of the full mitochondrial DNA sequence could detect variation in DNA nucleotide sites directly, resulting in the most complete and trustworthy dataset available for comparing species or individuals. However, early sequencing technologies were prohibitively expensive and were deemed inadequate for studying genetic evolution in large populations and datasets. Since the advent of PCR and the development of "universal primers," direct sequencing technologies have become increasingly widespread in the study of mitochondrial DNA.

During molecular phylogenetic analysis, the choice of genes or DNA fragments is crucial. At present, the most frequently utilized genetic markers for identifying sea cucumber species are the mitochondrial DNA CO I and CO II genes, the ND 1 and ND 5 genes, the 16S rRNA gene, and the 12S rRNA gene. In general, 16S rRNA and 12S rRNA are better suited for studying the systematic connections between genera, between species, and between species at earlier divergence dates. In addition, CO I, CO II, ND 1, and ND 5 are frequently employed to examine the systematic connections of closely related species and subspecies, as well as regional populations.

Shan et al. (2005) amplified mitochondrial DNA from six species of dried sea cucumbers, acquired six CO I sequence, and constructed NJ and MP molecular system trees using the neighbor-joining and maximum parsimony methods. The categorization of sea cucumbers was achieved through the computation of the Kimura-2 parameters genetic distance, base composition, and length range among 16 haplotypes. Liang (2008) collected mitochondrial DNA from seven different species of sea cucumbers from 11 different origins and aligned the 16S rRNA sequences. They constructed NJ and ME molecular system trees using the neighbor-joining approach and the minimum evolution method. The categorization findings were compared to those previously published by Shan et al. (2005). The research findings for CO I were comparable to those previously reported in echinoderms.

To demonstrate the practicality of using mitochondrial sequence technology for identifying sea cucumber species, the sea cucumber nutrient solution from the deep processed product was also recognized (Liang et al. 2008). By comparing the CO I sequence of the mitochondrial DNA in the sea cucumber nutrition solution to that of *Apostichopus japonicus*, it was determined that the DNA was genuinely from a sea cucumber and that the DNA sequence was 100% similar to that of *Apostichopus japonicus*. PCR-RFLP and FINS were used to identify sea cucumber products on the market, and 11 species' 16S rRNA genes were amplified using species-specific PCR (Wen et al. 2011). These findings provided simple, practical, and academically molecular markers for proper product labeling, traceability, and other issues relating to the regulation of sea cucumber fishery.

In general, the mitochondrial DNA CO I and 16S rRNA genes are frequently utilized to investigate the taxonomy and species identification in sea cucumbers. Since the 16S rRNA phase is less conservative and specific than the CO I genes, researchers prefer the CO I genes for their specificity.

2.2 *Identification of Sea Cucumber Species Based on DNA Barcoding*

DNA barcoding is a novel technology for rapidly identifying species by extracting DNA sequences from microscopic tissue samples of live creatures. The process of DNA barcoding is divided into two essential steps: (1) Create a DNA barcoding library for recognized species and (2) identify new species by matching DNA barcoding sequences to the DNA barcoding library (Kress and Erickson 2012). Following the introduction of the idea of DNA barcoding, molecular biology techniques based on DNA barcoding have advanced rapidly and have assumed increasing importance in species identification. Kress et al. (2005) and Taberlet et al. (2007) proposed the following ideal criteria for DNA barcoding: (1) the target gene sequence exhibits sufficient variation across species and sufficient intra-species variation; (2) the existence of relatively conservative regions that can be used to design universal primers; and (3) the target gene fragment being short enough to avoid the influence of gene degradation during processing on DNA extraction and PCR amplification.

Mitochondrial DNA (mtDNA) is frequently employed in molecular evolutionary research because it is devoid of repetitive sequences, unequal exchanges, recombination, inversions, translocations, and other abnormalities and lacks spacer regions and introns. By examining the differences between identical genes in various species, their evolutionary connection could be deduced. On mitochondrial DNA,

Table 1 Studies on identification of sea cucumber species by DNA barcoding technology

Sea cucumber species	Amplified target genes	Fragment length/bp	References
Acaudina molpadioides	*COI*	676	Amin et al. (2016)
16 species of sea cucumber	16SrRNA and *COI*	/	Wen et al. (2011)
Apostichopus japonicus, etc.	16SrRNA	570	Wen et al. (2010)
Stichopus hermanni, etc.	16SrRNA	570	Wen et al. (2012)
11 species of sea cucumber in the genus *Holothuria*	16SrRNA	451	Kamarudin (2015)
Stichopus horrens, etc.	16SrRNA	570	Kamarudin et al. (2016)
Actinopyga lecanora, etc.	16SrRNA	570	Ling et al. (2018)
Parastichopus regalis	16SrRNA and *COI*	471 and 561	Maggi and Gonzalez-Wanguemert (2015)
Holothuria leucospilota	*COI*	550	Idris et al. (2011)
Apostichopus japonicus	16SrRNA and *COI*	550 and 652	Jo and Park (2016)
Holothuria leucospilota etc.	16SrRNA and *COI*	560 and 690	Hu et al. (2019)
Psolus phantapus	*COI*	658	Lee et al. (2017)
Holothuria etc.	*COI*	257	Rrx et al. (2021)
16 species of sea cucumber	18SrRNA	1716	Lacey et al. (2005)
Stichopus horrens	12SrRNA	360	Kamarudin et al. (2017)

cytochrome oxidase subunit I (CO I) is a 692 bp protein-coding gene. Due to its high diversity and ease of amplification, it is currently commonly employed as a DNA barcode for identifying animal species. Hebert et al. (2003) found the mitochondrial cytochrome c oxidase component for the first time in 13,320 species from 11 phyla in 2003. The results of this work established the feasibility and validity of employing the mitochondrial CO I gene as a DNA barcode to identify species.

At present, the mitochondrial CO I gene (Hu et al. 2019), 16S rRNA (Wen et al. 2010), 18S rRNA (Lacey et al. 2005), and 12S rRNA are the primary DNA barcoding molecular markers used to identify sea cucumber species (Kamarudin et al. 2017). Because the data for the 12S rRNA and 18S rRNA genes of some sea cucumbers are incomplete, mitochondrial DNA CO I and 16S rRNA gene sequences are frequently utilized to identify sea cucumber species. Table 1 highlights some relevant research on sea cucumber species identification using DNA barcoding.

Lyv et al. (2011) assessed the genetic distance between different sea cucumber species and between individuals of the same sea cucumber species by collecting sea cucumbers from various origins. Researchers found that the mitochondrial CO I gene in sea cucumbers was both variable and stable enough to be utilized as a DNA barcode for species identification. 16S rRNA gene fragments and mitochondrial COI gene fragments were chosen as the DNA barcodes for sea cucumber species identification by Hu et al. (2019). This study revealed that the approach might be used to identify sea cucumber species by comparing the capacity of the above two target genes to identify different species of sea cucumber. In addition, they suggested that identification findings from 16S rRNA and CO I be merged for closely related sea cucumber species to compensate for the constraints of a single target gene and thereby improve the accuracy of identification results. Sea cucumber import and export inspection and market supervision can be supported by DNA barcoding technology, which is used to identify distinct species of sea cucumber.

Sea cucumbers are identified using the conventional biological categorization approach. One of the most important uses of this technique is to discover new relationships between species by examining the physical features and morphology of diverse organisms and then comparing them to known physiological and genetic traits. The biological categorization approach, however, is cumbersome and time-consuming. As a result, the approach is unable to fulfill market demand for food identification due to its sensitivity to environmental factors including temperature and stage of development. The fast development of molecular biology technology has enabled a more direct, accurate, and straightforward approach of species identification. To ensure the safety of sea cucumber products and to defend customers' rights, researchers have devised a molecular biology approach for identifying sea cucumber species quickly and accurately.

Dotblot, multiplex-PCR, and PCR-RFLP techniques have been developed for the rapid identification of sea cucumber species using DNA barcoding technology (Lyv 2012; Zuo et al. 2012; Zeng et al. 2018).

2.2.1 Dotblot

Dotblot is a method for identifying nucleic acids by hybridizing with tagged probes and measuring the subsequent color shift. In this method of detection, enzyme-catalyzed reactions were paired with the prominent selectivity enabled by nucleic acid hybridization. Dotblot has three advantages as follows. First, it achieves high-throughput identification because it can simultaneously identify multiple unknown targets in a sample through one single reaction. Second, it is easy to judge, and the results are more intuitive. Third, no special equipment is needed, and the operation is relatively simple. Dotblot has found widespread applications in a variety of sectors, including pathogen detection (Yang et al. 2019) and antibody screening (Svobodova et al. 2021).

To determine the species of sea cucumber, Lyv et al., developed a dotblot for this purpose. This study could realize the sequence alignment analysis of sea cucumber and provide the recognition of specific regions. The approach has a sensitivity of up to 100 pg. Dotblot has created the theoretical groundwork for the development of gene chips that will allow sea cucumber species identification. However, due to its lengthy operating time, the dotblot is not fit for fast identification purposes.

2.2.2 Multiplex-PCR

In the technique known as multiplex-PCR, several particular pairs of primers are used in one PCR reaction, meaning that many nucleic acid fragments of various lengths can be amplified at the same time (Bang et al. 2021). Multiplex-PCR is a technology that uses a single PCR reaction and electrophoresis to detect several target fragments at the same time, saving chemicals, PCR reaction time, and electrophoresis time. The creation of primers and the optimization of reaction parameters such as primer concentration and annealing temperature are crucial phases in this entire process. When it comes to accuracy, efficiency, and cost, multiplex-PCR is superior to single PCR.

Multiplex-PCR was employed by Lyv and Zuo (Lyv 2012; Zuo et al. 2012) to identify sea cucumber species. The mitochondrial CO I gene sequences of sea cucumber were first examined. On the basis of the findings from the investigation, primers were created, and their sensitivity and specificity were assessed. Finally, 40 individual samples of four species of sea cucumber were successfully identified with high accuracy, and no abnormal conditions such as cross-reaction, primer dimer, and non-specific amplification were found in the identification results.

Sea cucumber species could be accurately identified using the multiplex-PCR approach, which is highly specific, sensitive, and repeatable. Multiple sea cucumber species can be identified at the same time. However, because of a large number of samples involved, as the number of primers rises, the possibility of dimer formation between primers increases as well, adding to the difficulty of detection. Furthermore, multiplex-PCR experiences difficulties in identifying closely related species because of the constraints of primer design and the specificity of PCR reactions.

2.2.3 PCR-RFLP

PCR-RFLP (polymerase chain reaction-restriction fragment length polymorphism) is a method that combines PCR and RFLP. There are several steps involved in this process. These include amplification of a target region, digestion using restriction endonucleases, electrophoretic separation of the resultant fragments, and final sequencing and analysis of the sequence variations across samples. This approach is faster, simpler, and more accurate since it allows for the study of base differences among individual samples (Amin et al. 2016; Wilwet et al. 2018). As of now, this approach has been commonly employed in the identification of species.

According to Lyv et al., sea cucumbers might be identified using PCR-RFLP technique. Restrictive endonuclease profiles were established for BamHI, KpnI, PstI, XbaI, and Eco31I, which were eventually chosen. All the five sea cucumbers were readily distinguished from each other using PCR-RFLP, and there were no non-specific enzyme bands. Both speckle hybridization and multiplex-PCR approaches were less specific than PCR-RFLP, especially for species that were closely related. This technique provides a new form of technological support for the quality control of sea cucumber products, compensating for the shortcomings of dotblot and multiplex-PCR.

Further to the research conducted by Lyv et al., Zeng also employed PCR-RFLP to identify 16 commercially available species of sea cucumber (Zeng et al. 2018). Dde I, Hae III, and Sty I endonucleases were employed for enzymatic cleavage of sea cucumbers, and the combination of enzymatic haplotypes including 16 commercially available seafood products was synthesized to accomplish an appropriate categorization of the species.

2.3 Identification of Sea Cucumber Species Based on Fatty Acid Composition

Different species of sea cucumbers differ in lipid contents and fatty acid compositions. The NPLC-Triple-TOF-MS/MS approach was used by Wang et al. (2020) to identify the phospholipids (PLs) in six edible sea cucumber species. *Cucumaria frondosa* had the greatest phospholipid content of 8.05 mol/g, according to the data. At least 295 molecular species from seven PL classes were reliably found in all sea cucumbers, according to researchers' study. Sea cucumbers contain large volumes of 1-alkyl-2-acetylcholine and 1-enyl-2-acetylethanolamine. C20:5 and C20:4 were the two most common polyunsaturated fatty acids esterified to PLs in this study. A significant percentage of odd-chain fatty acids (OCFAs) was found in all sea cucumbers, ranging from 35 to 90% of the total fatty acid content. Phospholipid molecular species composition varied sharply across the six sea cucumber species, and this allowed for a precise categorization of all the six species.

2.4 Identification of Sea Cucumber Species Based on Polysaccharide Structure

There are two types of active polysaccharides in the body wall of sea cucumber, one of which is sea cucumber chondroitin sulfate (SC-CHS), which is composed of D-N-acetylgalactose, D-glucuronide, and L-fucose. The other is sea cucumber fucoidan sulfate (SC-FUC), which consists of L-fucose. The structural complexity of sea cucumber polysaccharides is influenced by changes in the degree of sulfation, relative molecular mass, microstructure, and the number of residues.

Chen et al. (2010) analyzed the SC-CHS of eight sea cucumbers, namely, *Parastichopus californicus*, *Acaudina molpadioides*, *Holothuria mexicana*, *Cucumaria frondosa*, *Apostichopus japonicus*, *Holothuria scabra*, *Thelenota anax*, and *Parastichopus parvimensis*, using 1H-nuclear magnetic resonance (1H-NMR) technology. Through comparison of the heterocapsid hydrogen signals on the sulfated rockulose branched chains in the SC-CHS and determination of the type of substitution of the sulfate group, different species of sea cucumbers could be distinguished.

From the standpoint of biological small molecules, glycomics analysis enables a more thorough identification of molecular markers for food traceability analysis. Glycomics analysis, on the other hand, is more difficult than proteomic or lipidomic analysis. Firstly, because of the variety and complexity of their structures, oligosaccharides and polysaccharides are more difficult to separate than monosaccharides; secondly, the greater polarity of sugars means that it is more difficult to separate samples.

3 Geographical Origin Traceability Technology of Sea Cucumber

The environment in which sea cucumbers grow has an impact on their nutritional level and composition. Moreover, a substantial price discrepancy is the result of the varied concentrations of active components in each kind of product. Difficulties in distinguishing between fake and genuine products can occur due to their similar visual characteristics. To ensure that the sea cucumber industry continues to thrive, a reliable procedure for identification must therefore be developed. A sea cucumber's origin and nutritional worth may be determined by where the sea cucumber comes from.

Modern biotechnology, including microbiome technology, lipidomics technology, stable isotope analysis, inorganic element analysis, and other technologies, are being used to track the origin of sea cucumbers (Table 2).

3.1 Microbiomics (Gut Microbiota) Analysis

In recent years, the link between gut microbiota, food, and the development of illness has gained increasing attention in life sciences. People in various geographical locations have considerably diverse gut microbiotas because of different dietary structures, which is one of the key variables impacting gut microbiota composition. The intestinal flora structure of *Apostichopus japonicus* may also be variable in different culture regions due to the varying water composition.

Zhao et al. (2022) constructed the biogeographic distribution for the gut microbiota of Chinese *Apostichopus japonicus* by high-throughput sequencing of 287 intestinal samples from different habitats. Then, a random forest model was established using 20 key intestinal bacterial genera, and attempts were made to trace the source of cultured *Apostichopus japonicus* by gut microbiota. The results showed that the accuracy of predicting the origin of *Apostichopus japonicus* reached 97.6%. Zhao et al. analyzed the contribution of random and deterministic factors to

Table 2 Geographical origin traceability technology of sea cucumber

Sea cucumber species	Identification method	Place of origin	Experimental results	References
Apostichopus japonicus	Microbiome method	East coast of Dalian, west coast of Dalian, coast of Jinzhou, coast of Qinhuangdao, coast of Dongying, coast of Yantai, and coast of Xiapu, China	There were significant differences in intestinal bacterial community composition among different places of origin. The abundance of microorganisms in the same region was significantly correlated	Zhao et al. (2022)
Holothuria edulis, Holothuria scabra	Lipidomics analysis	Malaysian waters	The main saturated fatty acids (SFA) of *H. edulis* and *H. scabra* accounted for 83.95% and 98.60%, respectively, and palmitic acid was the main fatty acid. For polyunsaturated fatty acid (PUFA), 16.05% of arachidonic acid was found only in *H. edulis*	Al Azad et al. (2017)
Apostichopus japonicus	Fatty acid composition analyzed by meta-analysis	Changhai in Liaoning Province, Zhangzidao in Liaoning Province, Muping in Shandong Province, Laizhou in Shandong Province, Danzidao in Shandong Province, and Xiapu in Fujian Province, China; Peter the Great Bay, Sea of Japan; Mediterranean Sea; Gulf of Asia; Portugal; Sabah, Malaysia	The percentages of EPA and DHA decreased with the decrease of latitude for the same species from different latitudes. The results of the study on *Stichopus japonicus* from different latitudes also showed the same variation trend: the percentages of EPA and DHA in *Stichopus japonicus* decreased with the decrease of latitude	Li et al. (2022)
Apostichopus japonicus	Fatty acid analysis combined with	Laizhou, Danzidao,	The fatty acid composition of	

(continued)

Table 2 (continued)

Sea cucumber species	Identification method	Place of origin	Experimental results	References
	stable isotope analysis	Muping, Rushan, Wafangdian, Pikou, Changhai, Zhangzidao, and Xiapu, China	organisms in different sea areas was directly related to salinity	Zhang et al. (2017)
Apostichopus japonicus	Proteomics based on SWATH-MS technology	Xiapu, Jiaonan, Weihai, Yantai, and Dalian, China	Samples from five places were classified by OPLS-DA model	Zhang et al. (2019)
Pearsonothuria graeffei/ Holothuria vagabunda/ Stichopus tremulus/ Isostichopus badionotus	Fourier transform infrared spectroscopy analyzer and high temperature 1H-NMR technology	Indo-Pacific, coast of Norway, Western Indian Ocean, Atlantic Ocean	The fingerprints of sea cucumbers in different sea areas could be established according to SC-FUC anomeric hydrogen signal	Wu et al. (2014)
Apostichopus japonicus	Stable isotope analysis	Laizhou, Danzidao, Rushan, Wafangdian, Pikou, Changhaidao, and Zhangzidao, China	According to $\delta^{13}C$ and $\delta^{15}N$ values, the origin of *Apostichopus japonicus* could be distinguished	Zhang et al. (2017)
Apostichopus japonicus	Elemental analysis	Jinzhou, Dalian, Lyushun and Zhangzidao in Liaoning Province, Qingdao, and Qingjiang and Haili of Wenzhou in Zhejiang Province, China	The differences of Al, Fe, Co, Mn, Mo, As, Cd, Co, Ni, Se, Cd, Cu, and Cr in the body wall of *Apostichopus japonicus* from different places were significant	Liu et al. (2011)
Apostichopus japonicus	Determination of element content by atomic absorption method and ICP-MS	Dalian, Qingdao, Yantai, and Fujian, China	For *Apostichopus japonicus*, the content of Zn and Mn was the highest in Dalian, that of B and Cu was the highest in Yantai, and that of Al and Cr was the highest in Fujian. The contents of B, Cr, and Cu in the body wall of deep sea *Apostichopus japonicus* were higher than those	Gao et al. (2016)

(continued)

Table 2 (continued)

Sea cucumber species	Identification method	Place of origin	Experimental results	References
			in cultured *Apostichopus japonicus*	
Stichopus japonicus	Stable isotope analysis	Hongyanhe Town, Xietun Town, Yongning Town, Liguan Town and Santaixiang Town under the jurisdiction of Wafangdian in Dalian, China	There were significant differences in $\delta^{13}C$ and $\delta^{15}N$ values among different places for *Stichopus japonicus*, and the accuracy of discriminant analysis reached 71.5%	Zhou (2016)
Holothuriida	Stable isotope analysis	Qinhuangdao, Dalian, Dongying, Qingdao, and Fujian, China	There were significant differences in the stable isotopes of C, N, O, and H in sea cucumbers, and LDA had a high recognition rate of 93.4%	Kang et al. (2020)
Stichopus japonicus	Amino acid relative contents and $\delta^{13}C$ values	The Yellow Sea and the Bohai Sea waters in Liaoning and Shandong Provinces	The $\delta^{13}C$ values of non-essential amino acids were significantly higher than those of essential amino acids	Liu et al. (2018)
Apostichopus japonicus	Monomer stable isotope analysis of amino acids	Fujian, Liaoning, and Shandong, China	The total accuracy and cross-validation rate of the eight samples from the $\delta^{13}C$ values of amino acids were 100%, and the $\delta^{13}C$ Ser showed a significant ability to distinguish the samples	Zhao et al. (2018)

the intestinal microbiota community construction of *Apostichopus japonicus* from different habitats, confirming that geographical constraints were the main driving force of the intestinal microbiota change.

3.2 Lipidomics Analysis

The fatty acid content of the same species of sea cucumber in different places may vary owing to factors such as aquaculture, climate, and temperature. For this reason, experts believe that lipidomic analysis could be utilized to differentiate between sea cucumbers from various places. Lipidomic analysis relies heavily on chromatographic and spectroscopic techniques to examine the fatty acid content and composition of the samples and employs chemometrics to identify the source of sea cucumber. Food safety inspection and food nutrition analysis are increasingly relying on lipidomic analyses due to advances in experimental approaches.

A novel compound-specific isotope analysis (CSIA) approach for fatty acid compounds was developed by Li et al. (2017). Combined with principal component analysis and discriminant analysis, Li et al. used CSIA to determine the origin of *Apostichopus japonicus* in coastal China's Laizhou and Danzidao regions as well as Muping, Rushan, and Wafangdian. First of all, *Apostichopus japonicus* total lipids were extracted and methyl-esterified, and the fatty acid types were identified by GC-MS (gas chromatography-mass spectrometer). A mass spectrometer equipped with a stable isotope ratio was used to pinpoint the stable carbon isotopes of the fatty acids. Sea cucumbers from various origins were then separated using PCA for fatty acid $\delta^{13}C$. With the combination of CSIA and PCA, it was shown that *Apostichopus japonicus* from China's coastal regions might be traced to its geographical origin.

Further research is needed into the link between seasonal change and fatty acid content in sea cucumber since this link might lead to the overlapping of origin and production method. Additionally, a database of the fatty acid distribution of different sea cucumber species is needed to establish the origin of unknown samples. Lipidomic analysis can play a significant role in sea cucumber traceability analyses once these issues are resolved.

3.3 Proteomic Analysis

The primary constituent of sea cucumber is protein, which makes up 80–91.20% of total sea cucumber weight. The protein content of sea cucumber varies depending on the type. To find and screen biomarkers, protein technology could be employed to develop traceability analysis technology for sea cucumbers and to identify the unique characteristics of sea cucumbers found in various places.

Traditional protein technology covers enzyme-linked immunoassay, electrophoresis, and chromatography. However, these traditional techniques have their own limitations (as shown in Table 3); thus new protein analysis techniques need to be developed for sea cucumber traceability.

When the Genome Project began, mass spectrometry became an increasingly popular method for identifying and studying proteins. Food authenticity traceability analysis may now be conducted using mass spectrometry-based proteomics, which

Table 3 Limitations of traditional protein technology

No.	Protein technology	Limitations
1	Enzyme-linked immunosorbent assay	Closely related species tend to produce false-positive results. If the monoclonal antibody is used, the operation will be complicated, and the monitoring cost will increase
2	Protein electrophoresis	Not applicable to the analysis of deeply processed samples
3	Chromatographic analysis	Complex in operation, weak in qualitative ability, unsuitable for separation of hydrophobic proteins

offers a new technical approach for the examination of aquatic goods, meat, dairy products, and other foods.

The SWATH-MS (Sequential Window Acquisition of All Theoretical Mass Spectra) approach was employed by Zhang et al. (2019). Xiapu, Jiaonan, Weihai, Yantai, and Dalian were the five Chinese cities where dried *Apostichopus japonicus* were collected. Data-dependent acquisition (DDA) of all the combined samples produced 101,456 recognized spectra and 7997 distinct peptides for protein identification, resulting in the creation of a spectral library for SWATH-MS quantitative analysis. The developed spectral library was used to process multiplex SWATH-MS data, and protein quantitative and stoichiometric analyses were carried out. Protein composition from distinct *Apostichopus japonicus* strains might be easily discriminated using dimension reduction analysis. Cluster analysis indicated that the proteome levels of samples from various locations differed in a distinct way. In validation testing and real sample analysis, 17 proteins were identified as location-traceable biomarkers with 100% prediction accuracy.

Although proteins are more durable than DNA molecules in various processes, there are still limitations on tracing proteins back to their dietary source. Specifically, the extraction of low abundance marker proteins and complex matrix proteins has become more challenging due to the standardization of protein extraction and enrichment procedures. A high-resolution mass spectrometer must be employed to conduct data mining in advance because of the autolysis of sea cucumbers and the reduction in protein solubility caused by processing. Essentially, progress may be made in bioinformatics and omics databases, both of which are critical to successful proteomic analysis. As more genomes are sequenced, protein data will continue to improve. In addition, advances in data mining technologies will aid in the interpretation of proteomic data in a larger and deeper perspective (Jiang et al. 2021).

3.4 Stable Isotope Analysis

Stable isotope analysis (SIA) is a process of distinguishing between distinct samples by the number of protons in each isotope. Useful isotopes for traceability include $\delta^{13}C/\delta^{12}C$, $\delta^{2}H/\delta^{1}H$, $\delta^{15}N/\delta^{14}N$, and $\delta^{18}O/\delta^{16}O$. Stable isotopes are incorporated into animal tissues through the shunt effect as they move up in the trophic chain.

As a result of characteristics such as origin, production mode, and species differences, isotopic fractionation may trace sea cucumbers back to their sources. Using carbon and nitrogen SIA, Zhang et al. (2017) were able to identify *Apostichopus japonicus* from seven locations in north China. *Apostichopus japonicus* $\delta^{13}C$ and $\delta^{15}N$ values were obtained from 133 samples. The researchers were able to determine the origins of *Apostichopus japonicus* using the $\delta^{13}C$ and $\delta^{15}N$ readings. *Apostichopus japonicus* could thus be identified by its stable carbon and nitrogen isotopic composition.

3.5 Analysis of Element Composition

Inductively coupled plasma emission spectrometry (ICP-ES) and inductively coupled plasma mass spectrometry (ICP-MS) are the most advanced and accurate methods for testing element composition (Liu et al. 2011). Other methods include atomic emission spectrometry, elemental fingerprint analysis, graphite furnace atomic absorption spectrometry, flame atomic absorption spectrometry, and stoichiometry.

Inorganic element content in marine food, such as sea cucumbers (Liu et al. 2015) and oysters, is now the highlight of domestic and foreign publications on the use of inorganic elements in variety distinction or origin categorization (Zhao et al. 2015). *Salvia miltiorrhiza*, starwort *Stellaria*, wolfberry, wine (Cao 2019), beer (Zhou et al. 2012), and honey, as well as animal items including meat and dairy products (Liu et al. 2021), have all been used in traditional Chinese medicine (Peng et al. 2021).

When it comes to food ingestion, *Apostichopus japonicus* relies heavily on its tentacles. Depending on where the sea cucumber is located, it either sweeps or picks up the silt on the bottom of the rocks. Inorganic components in the body have a deeper association with the developing environment because of the distinct living and eating behaviors of *Apostichopus japonicus*, reflecting geological aspects and revealing distribution patterns in the sea. A total of 22 samples were taken from seven habitats by Liu et al. (2011) (3 from Jinzhou, 3 from Lyushun, 3 from Zhangzidao, 5 from Qingdao, 2 from Qingjiang, Wenzhou, Zhejiang, and 6 from Haili, Wenzhou, Zhejiang). With ICP-MS, 15 inorganic elements were identified in the *Apostichopus japonicus* samples and a database of inorganic elements was established, which proved useful in the origin tracking of the species.

4 Determination of Functional Components of Sea Cucumber

4.1 Determination of Sulfated Polysaccharide Content of Sea Cucumber

One of the main functional components of sea cucumber is sea cucumber sulfated polysaccharide, which plays an important role in the bioactivity of sea cucumber. Therefore, the determination of sulfated polysaccharide content in sea cucumber and sea cucumber-related products is of great significance for the evaluation of quality.

Techniques for determining sea cucumber polysaccharide have been partially developed. For example, *Method for Determining Polysaccharide Content in Sea Cucumber* (Patent No. CN200410036250.7) describes how to determine sea cucumber polysaccharide. In recent year, HPLC has been increasingly popular as a tool for determining the concentration of various substances. A technique based on high performance liquid phase technology was developed for measuring the polysaccharide content in sea cucumber by Xue et al. (Patent No. CN200810016622.8). Polysaccharide content in sea cucumber products could be determined using this approach, which is simple, sensitive, fast, and accurate. Particularly, sea cucumber polysaccharides may also be identified using glucuronic acid as a key identification index. The overall amount of sea cucumber polysaccharides can be determined using high performance liquid chromatography; however the sea cucumber fucoidan (SC-FUC) and chondroitin sulfate (SC-CHS) contents are still difficult to quantify correctly. Sea cucumber polysaccharide was hydrolyzed in acid, and the chondroitin disaccharide and fucose generated were detected using HPLC-MS/MS by Zhu et al. (2018). This allowed for the quantitative assessment of the SC-FUC and SC-CHS contents in sea cucumber polysaccharide.

4.2 Determination of Sea Cucumber Cerebroside and Ceramide

Sphingolipids, such as cerebroside, ceramide, and sphingomyelin, are commonly consumed by humans on a regular basis. A diet rich in cerebrosides and ceramides is essential for human health (Wang et al. 2021). Thin layer chromatography, liquid chromatography tandem mass spectrometry, and high-performance liquid chromatography is now the most widely used techniques in the scientific literature for determining sphingolipids. Detecting ceramide concentration in the mouse epidermis could be achieved in a straightforward way using thin-layer chromatography, but it was more difficult to distinguish compounds with complex components and comparable structures. For example, examining lipids with low mass-to-charge ratios using liquid chromatography tandem mass spectrometry, which had a low resolution (Ma et al. 2020), presents a significant challenge. Ceramide and cerebroside content in aquatic goods must thus be determined in a simple manner that is rapid, precise, accurate, and efficient.

Evaporative light scattering detectors (ELSD) could be used to detect lipids quantitatively with low UV absorption in HPLC, and as such, this is one of the most frequently used instruments for testing lipids. The Groener (Groener et al. 2007) reversed-phase HPLC fluorescence detection technique could concurrently assess plasma levels of ceramide, glucocerebroside, and trihexosylceramide, although derivatization is required and the procedure is difficult. For the detection of cerebroside content in wheat and other plant foods, the normal-phase HPLC-ELSD approach developed by Kashima et al. (2002) required burdensome sample preparation procedures. Cerebrosides in aquatic products were detected using HPLC-ELSD, a high-performance liquid chromatography-evaporative light scattering detection technique. Jia et al. (2015) employed HPLC-ELSD to concurrently measure the amounts of cerebroside and ceramide in several sea cucumbers based on the previous research. The separation of sea cucumber cerebroside and ceramide was accomplished in 18 min under the optimal circumstances, and the process was good reproducibility. A quicker, simpler, and more precise means of distinguishing between and quantifying the two glucocerebrosides in biological tissues is provided by the enhanced pretreatment procedure and chromatographic condition (Song et al. 2016). Ceramide deacylase (SCDase) was used to hydrolyze ceramides and cerebrosides, which were subsequently derivatized with OPA to detect OPA-GlcSph and OPA-GalSph using a normal-phase HPLC technique. Glucocerebroside and galactocerebroside were separated in 10 min at the optimal conditions, and the process was stable.

4.3 Determination of Sea Cucumber Gangliosides

Thin-layer chromatography (TLC) and high-performance thin-layer chromatography are the most frequently used methods for ganglioside measurement. Chromogenic reagents are used to develop the color after the plate is exhibited. Many chromogenic reagents, such as thiobarbituric acid, resorcinol, etc., contain a sialic acid-specific reagent. Gangliosides with varying glycocluster structures might be separated and quantified using TLC technology (Tanabe et al. 2009). Ganglioside separation, identification, and quantification are frequently performed using high-performance thin-layer chromatography (HPTLC) (Hayakawa and Hirai 2003). Because of its limited sensitivity and easy interference with contaminants, TLC can only offer information on the mixture's primary components and is difficult to assess when the biomarker value in secondary components is being looked for (Sisu et al. 2011).

The liquid chromatography-mass spectrometry (LC-MS) technology has been widely employed for chemical separation and analysis in recent years, thanks to its excellent sensitivity and precision. LC/MS is superior to TLC in terms of sensitivity and selectivity, and it is less susceptible to interference by contaminants when employed for quantitative analysis. In 2006, Svennerholm performed quantitative determination for two gangliosides, GD3 and GM3, in milk and infant formula for the first time, using reverse phase liquid mass spectrometry (HPLC-MS) after the separation of samples with normal phase HPLC-MS (Sørensen 2006).

The amount of carbon atoms in the amide chain is used to sort gangliosides on a reversed-phase column. Homologue gangliosides may be split into more than one peak during the analysis, and this might alter the results. A technique for quantifying gangliosides in mouse brain and dairy products was devised by Fong et al. considering the aforesaid disadvantages, and this approach has been utilized effectively ever since (Fong et al. 2009). Cong optimized the HPLC separation conditions of sea cucumber gangliosides and devised an MRM approach of triple quadrupole mass spectrum for quantitative analysis of sea cucumber and other marine echinoderms (Cong 2012; Cong et al. 2013). The gangliosides in samples might be quantified down to the nanogram level using this approach.

4.4 Determination of Sea Cucumber Saponins

To determine saponin concentration, many techniques such as colorimetry, thin-layer scanning (TLC), high-performance liquid chromatography, and others have been employed in the research. Vanillin-perchloric acid is a frequently used color developer for triterpenoid saponins in the colorimetric technique. Ginseng's total saponin content was measured by Zhao et al. (2014). Soybean saponins from six distinct cultivars were studied by Li et al. (2020). High-performance liquid chromatography was employed by Xu et al. to analyze 12 ginseng products for the presence of 15 ginsenosides (Xu et al. 2021).

Sea cucumber saponins, the primary secondary metabolites, are structurally varied and complicated. Many saponins have been found to date. Sea cucumber saponin quantification has been hindered by the absence of a standardized method. Methods for the detection of total saponins in the body wall of sea cucumber were developed, and UV-Vis spectrophotometry was used to quantify the total saponin content. Using quinoa content unique in the sea cucumber saponin oligosaccharide chain as the quantitative standard, Dong et al. (2008) devised a PMP pre-column derivatization high-performance liquid chromatography technique that was easier, faster, more accurate, more sensitive, and more stable than the conventional detection using spectrophotometric methods.

5 Quality Standard System of Sea Cucumber Products

A scientific organic whole is generated by the interrelationships between standards in each range. The current standard system for sea cucumber breeding and processing includes germplasm, breeding specifications, processing specifications, product quality standards, testing methods, and processing specifications (Zhu et al. 2011). This involves the quality system for sea cucumber breeding and the quality system for sea cucumber products. In the sea cucumber breeding process, the most important quality requirements cover germplasm standards, inputs, breeding technology, disease diagnosis, and control standards. Product standards, inspection method standards, and processing technical specifications all fall within the purview of the sea cucumber processing quality system.

5.1 Quality Standard System of Sea Cucumber in China

5.1.1 National and Industrial Standards of Sea Cucumber

The sea cucumber has long been prized in China as a delicacy, and as a valuable medical and marine resource (Zhu et al. 2011). Traditional dried sea cucumber has dominated the sea cucumber processing sector in China for a long time. *Dried Sea cucumber* (*Stichopus japonicus*) (GB 8583-1988) was the first national standard for sea cucumber in China, published in 1988. The first national standard for sea cucumbers, "Dried Sea Cucumber (Spiny Cucumber) (GB 8583-1988)," was established by China in 1988. "Dried Sea Cucumber (Spiny Ginseng) (GB 8583-1988)" was amended and published in 2000 as the People's Republic of China's aquatic industry standard "Dried Sea Cucumber (Spiny Ginseng) (SC/T 3206-2000)." In 2009, China established the People's Republic of China aquatic industry standard "Dried Sea Cucumber (Spiny Ginseng) (SC/T 3206-2009)." According to the standard of *Dried Sea Cucumber* (SC/T 3206-2009), food additives other than salt are no longer authorized for use in the production and business of dried sea cucumber. SC/T 3206-2009 *Dried Sea Cucumber*, an industry standard, became an obligatory national standard in 2009.

With time, the quality of sea cucumbers has gradually improved. To ensure the safety of dried sea cucumber products, the National Health and Family Planning Commission of China (NHFPC) established and implemented the *National Standard for Food Safety—Dried Sea Cucumber* (GB31602-2015) in 2015. With this standard, dried sea cucumber products' total water-soluble sugar and dry weight rate will be defined, and the production and manufacturing process for dried sea cucumber products will be standardized (Pan et al. 2021). In 2017, China released the *Dried Sea Cucumber Grade Specification* (GB/T 34747-2017), which defines the grade classification of dried sea cucumbers and specifies the requirements and test procedures for dried sea cucumber grade specification. The *Technical Specification for Dried Sea Cucumber Processing* (SC/T 3050-2017) was released by the Ministry of Agriculture in China in 2017.

5.1.2 Product Standards of Sea Cucumber

Since the 1980s, new processed sea cucumber products have been emerging in the Chinese market, such as salted sea cucumber, semi-dried sea cucumber, salt-pretreated dried sea cucumber, ready-to-eat sea cucumber, and freeze-dried sea cucumber, among which salted sea cucumber is an important sea cucumber product. The forms of sea cucumber products are becoming more and more diversified, and numerous deep-processed products such as canned sea cucumber and sea cucumber liquor have appeared. In addition, high-end products such as sea cucumber milk, sea cucumber oral liquid, sea cucumber nutrition capsules, and sea cucumber jelly have also gradually entered the market. To regulate this market, relevant Chinese authorities have successively issued standards for sea cucumber products. In 2006, the Ministry of Agriculture of the People's Republic of China issued the agricultural industry standard *Pollution Free Food—Sea Cucumber* (NY5328-2006). In 2007, the Ministry of Agriculture issued the aquatic industry standard *Salted Sea Cucumber* (SC/T 3215-2007), which was then revised and issued as the aquatic industry standard *Salted Sea Cucumber* (SC/T 3215-2014) in 2014. In 2007, the Ministry of Agriculture issued the agricultural industry standard *Green Food—Sea Cucumber and Products* (NY/T 1514-2007), and the agricultural industry standard *Green Food—Sea Cucumber and Products* (NY/T 1514-2020) was issued in 2020 upon revision. In 2014, the Ministry of Agriculture issued aquatic industry standards *Ready-to-Eat Sea Cucumber* (SC/T 3308-2014) and *Freeze-dried Sea Cucumber* (SC/T 3307-2014). In 2018, the Ministry of Agriculture issued the aquatic industry standard *Sea Cucumber Powder* (SC/T 3310-2018). The above standards all fall into the category of product standards.

5.1.3 Processing and Cooking Standards of Sea Cucumber

Local administrative authorities in China have also published and amended the necessary criteria for the processing of sea cucumber products. In 2008, market supervision and administration authorities of Shandong Province issued a series of local technical specifications for the processing of sea cucumber products, including general technical conditions for sea cucumber capsules, non-salted dried sea cucumber, and ready-to-eat sea cucumber. As of 2021, local standard operating procedures for tracing important products such as *Operating Procedures for Traceability of Important Products—Dried Sea Cucumber* (DB37/T 4352-2021) were published by the Shandong Administration for Market Regulation. *Ready-to-eat Sea Cucumber* (DB21/2392-2014) was published by the Health and Family Planning Commission of Liaoning Province in 2014. *Technical Specifications on Circulation Management of Sea Cucumber and Products* (DB21/T2674-2016) were published by the Bureau of Quality and Technical Supervision of Liaoning Province in 2016. In addition, several Chinese provinces have set several guidelines for sea cucumber processing and cooking techniques. A variety of preparation methods for specific sea cucumber

Table 4 Existing sea cucumber cooking standards in China

No.	Standard	Published in
DB50/T 449-2012	*Chongqing cuisine—Technical specifications of home-cooked sea cucumber*	Chongqing, China
DB41/T 924-2014	*Changyuan cooking skills—Braised sea cucumber with scallion*	Henan, China
DB43/T 1302.4-2017	*Classic Hunan cuisine—Part 4: Hydrangea sea cucumber*	Hunan, China
DB37/T 1120-2008	*Shandong cuisine—Braised sea cucumber with scallion*	Shandong, China
DB37/T 1866-2011	*Shandong cuisine—Shandong-style sea cucumber*	Shandong, China
DB37/T 1969-2011	*Shandong cuisine—Braised sea cucumber with hoof tendon*	Shandong, China
DB37/T 2658.41-2015	*Shandong cuisine—Sea cucumber meat ball in clear soup*	Shandong, China
DB37/T 2658.65-2015	*Shandong cuisine—Sea cucumber in casserole*	Shandong, China
DB37/T 2658.95-2015	*Shandong cuisine—Braised sea cucumber with hoof tendon*	Shandong, China
DB37/T 2658.96-2015	*Shandong cuisine—Braised sea cucumber with shrimp roes*	Shandong, China
DB37/T 2658.105-2015	*Shandong cuisine—Braised sea cucumber with minced pork*	Shandong, China
DB37/T 2903.79-2017	*Shandong cuisine—Fresh sea cucumber in clear soup*	Shandong, China
DB37/T 2903.86-2017	*Shandong cuisine—Warm mixed sea cucumber*	Shandong, China
DB37/T 2903.19-2017	*Shandong cuisine—Braised sea cucumber with deer tendon*	Shandong, China
DB37/T 2903.70-2017	*Shandong cuisine—Sea cucumber with kelp*	Shandong, China
DB37/T 2903.118-2017	*Shandong cuisine—Braised sea cucumber in soy sauce*	Shandong, China
DB37/T 2903.136-2017	*Shandong cuisine—Braised sea cucumber with scallion*	Shandong, China

dishes, such as braised sea cucumber in soy sauce, warm mixed sea cucumber, and braised sea cucumber with scallion, can be found in these guidelines. They provide consumers with information on how these dishes are made and how the ingredients and methods are used in their preparation (Table 4).

5.1.4 Standards for Testing Methods of Sea Cucumber and Its Products

Determination of sea cucumber polysaccharides in Stichopus japonicus and its derivatives by HPLC (SC/T 3049-2015) was issued by the Ministry of Agriculture of the PRC in 2015. Determination of saponins in sea cucumber and its products by HPLC (GB/T 33108-2016) was published by the Ministry of Agriculture in 2016. GB 2733-2015 Fresh and frozen animal fishery products was issued by the National Health and Family Planning Commission of the People's Republic of China in 2015, which sets requirements for the sensory, physical, and chemical indexes, pollutant limits, shellfish toxin limits, agricultural and veterinary drug residues, and food additives of aquatic products. This standard further standardizes and ensures the safety for the consumption of sea cucumbers. As above mentioned, Chinese authorities have developed a set of sea cucumber quality standards to standardize the production and processing to satisfy the industry's development demands for healthy development and increase the quality of sea cucumber goods (Table 5). The market is regulated and monitored to a certain extent by these criteria.

5.1.5 Development Direction of Sea Cucumber Product Standards

The quality and safety of sea cucumber products are intimately linked to the health and safety of the customer. As the first of the "eight luxuries of seafood," the sea cucumber has grown in popularity in recent years with the improvement of people's living standards. There has been a rapid increase in demand for sea cucumber and its processed goods, as well as a significant expansion of sea cucumber breeding in China due to this growing demand. As a result, China's sea cucumber aquaculture and the booming processing business have seen a tremendous rise in the production, scientific research, processing, and sales of sea cucumbers during the past few decades.

Sea cucumber production in China increased by 24,800 tons (or 14.5%) in 2020, according to *China Fisheries Statistical Yearbook* published in 2021. With a total of 3.64 million *mu* of sea cucumber breeding area, China produced RMB 29.17 billion worth of sea cucumber raw materials in 2020, while the entire industrial chain produced nearly RMB 60 billion in production value. As a result, foreign sea cucumbers are flooding the Chinese market, thanks to their low price and high quality. In 2019, China imported 396.8 tons of sea cucumbers for further processing or preservation. The quality of dried sea cucumbers imported and exported has steadily improved in recent years in China, and the import and export of high-grade dried sea cucumbers have become a general trend in the development of the sea cucumber industry. In China, sea cucumbers produced are dominated by locally farmed ones, with imports of *Isostichopus fuscus*, *Illicium japonicus*, *White Apostichopus japonicas*, and *Talinum paniculata* rounding out the market. A large number of sea cucumber production and processing businesses have been established in response to this enormous demand. In addition, large numbers of new methods for preparing sea cucumber are being developed in the market. Salt-pretreated dried sea cucumber, salted sea cucumber, high-pressure treated sea cucumber, mild-salted sea cucumber, ready-to-eat sea cucumber, freeze-dried sea cucumber, sea cucumber capsule, sea cucumber oral liquid, and other sea cucumber products are now available in the Chinese market (Zhu et al. 2015). The sea

Table 5 Existing quality standards of sea cucumber in China

No.	Standard	Contents	Scope of application
SC/T 3206-2009	Dried sea cucumber (*Stichopus japonicus*)	The requirements, test methods, inspection rules, labeling, packaging, storage, and transportation of dried sea cucumber are specified	It is suitable for dried sea cucumbers made from fresh and live *Stichopus japonicus* through viscerating, cooking, drying, and other processes. It can also be implemented by reference for dried sea cucumber products made from other varieties of sea cucumber
SC/T 3215-2014	Salted sea cucumber	The requirements, test methods, inspection rules, labeling, packaging, transportation, and storage of salted sea cucumber are specified	It is suitable for dried sea cucumbers made from fresh and live *Stichopus japonicus* through viscerating, cooking, drying, and other processes. It can also be implemented by reference for dried sea cucumber products made from other varieties of sea cucumber
SC/T 3308-2014	Ready-to-eat sea cucumber	The products, requirements, test methods, inspection rules, labeling, packaging, transportation, and storage of ready-to-eat sea cucumber are specified	It is suitable for ready-to-eat products made of fresh, frozen, salted, or dried *Stichopus japonicus*. It can also be implemented by reference for ready-to-eat sea cucumbers made from other varieties of sea cucumber
SC/T 3307-2014	Freeze-dried sea cucumber	The requirements, test methods, inspection rules, labeling, packaging, transportation, and storage of freeze-dried sea cucumber are specified	It is suitable for products made from fresh and live, frozen, or salted *Stichopus japonicus* through vacuum freeze-drying and other processes. It can also be implemented by reference for products processed with other varieties of sea cucumber
GB 31602-2015	National food safety standard—dried sea cucumber	–	Suitable for dried sea cucumber
SC/T 3049-2015	Determination of sea cucumber polysaccharides in *Stichopus japonicas* and its products by HPLC	A method for determining sea cucumber polysaccharide in *Stichopus japonicas* and its deep processed products, such as dried sea	It is applicable to the determination of sea cucumber polysaccharide content in dried and ready-to-eat *Stichopus*

(continued)

Table 5 (continued)

No.	Standard	Contents	Scope of application
		cucumber, ready-to-eat sea cucumber, capsule, and oral liquid, by HPLC is established	*japonicus*, as well as sea cucumber capsule, oral liquid, and other deep-processed products made from *Stichopus japonicus*
GB 2733-2015	Fresh and frozen animal fishery products	–	It is applicable to fresh and frozen animal fishery products, including sea-water products and fresh-water products
GB/T 33108-2016	Determination of saponins in sea cucumber and its products by HPLC	A method for determining sea cucumber saponins in sea cucumber and its products by HPLC is established	It is applicable to the determination of sea cucumber saponins (calculated by quinolones) in sea cucumber and its products. Quinolones are characteristic monosaccharides of sea cucumber saponins
DB21/T 2674-2016	Technical specifications for circulation management of sea cucumber and its products	The terms and definitions and the basic requirements for the procurement, transportation, storage, and sales in the circulation of sea cucumber and its products are specified	It is applicable to the circulation of sea cucumber and its products
GB/T 34747-2017	Dried sea cucumber grade specification	The requirements, test methods, inspection rules, labeling, packaging, transportation, and storage for dried sea cucumber grade specifications are specified	It is suitable for dried sea cucumbers made from fresh and live *Stichopus japonicus* through viscerating, cooking, drying, and other processes. It can also be implemented by reference for dried sea cucumber products made from other varieties of sea cucumber
SC/T 3050-2017	Technical specifications for processing dried sea cucumber	The basic conditions, raw and auxiliary material requirements, processes, labeling, identification, storage, production records, and product properties for dried sea cucumber processing are specified	It is applicable to the production process of dried sea cucumbers made from live *Stichopus japonicus* and *Thelenota ananas*. It can also be implemented by reference for the production and processing of dried sea cucumbers from other varieties of sea cucumber

(continued)

Table 5 (continued)

No.	Standard	Contents	Scope of application
SC/T 3310-2018	Sea cucumber powder	The requirements, test methods, inspection rules, labeling, marking, packaging, transportation, and storage of sea cucumber powder are specified	It is suitable for sea cucumber powder made with *Stichopus japonicus* as the raw material through pretreatment, enzymatic hydrolysis, or non-enzymatic hydrolysis, drying, crushing, packaging, and other processes. It can also be implemented by reference for sea cucumber powder processed with other varieties of sea cucumbers
NY/T 1514-2020	Green food—sea cucumber and products	The terms, definitions, requirements, inspection rules, labeling, packaging, transportation, and storage of green food sea cucumber and its products are specified	It is applicable to green food sea cucumber and its products, including live sea cucumber, salted sea cucumber, dried sea cucumber, freeze-dried sea cucumber, and ready-to-eat sea cucumber
DB37/T 4352-2021	Operating procedures for traceability of important products—dried sea cucumber	The basic contents and operation instructions of dried sea cucumber traceability information collection, information recording, information sharing, and information query are specified	It is applicable to the requirements for the operating specifications on establishing a traceability system by businesses involved in the sea cucumber supply chain

cucumber products are also characteristic of flaws such as unpredictable quality, unknown active components, and immature active ingredient detection techniques resulting from the fast development and change in the types of sea cucumber products, especially in recent years. As a result, product standards lag behind product development speed, causing a lack of sea cucumber and product categorization standards: for example, detection technology for crucial active ingredients must be enhanced (Sun et al. 2018).

The quality and technical indicators of many kinds of sea cucumber products vary significantly. The quality of sea cucumber products is affected by the following factors:

1. Sea cucumber texture. Examples include the number of proteins, amino acids, fats, and vitamins, as well as various micro indicators that indirectly represent the quality of aquatic goods such as the number of sulfate groups, sugars, and polysaccharides and their content. Based on risk analysis, in-depth fundamental

research should be conducted on the standards of sea cucumber products as soon as feasible, and necessary standards and technical specifications for sea cucumber products should be established and improved.

2. Sea cucumber quality and safety. Regarding the quality and safety of sea cucumber, high heavy metal content in sea cucumber and pesticide residue can be found when the breeding environment is contaminated or when the use of fishery medications is not standardized.
3. Adulterated or fake sea cucumber and products. Sea cucumber processing, for example, can alter the content and quality of the nutritional and functional components and may involve poisonous and hazardous substances that could be brought in or illegally utilized in production. Additionally, product quality might be affected by adulteration (Sun et al. 2018). Therefore, it is difficult to regulate the production of various sea cucumber products, and the number of standards that must be set is quite large.

In addition to identifying sea cucumber products, it is required to identify the raw materials utilized in the refined processing products and novel forms of sea cucumber products. This is essential for a healthy and peaceful sea cucumber market, particularly for identifying things disguised as pricey sea cucumbers that are actually inexpensive sea cucumbers. This will enable authorities to prevent and punish criminal behaviors that involve fraudulent or substandard practice and to protect the legitimate interests of consumers. Therefore, to assure the food safety of processed sea cucumber products, production process control and the creation of quality and safety standards for processed sea cucumber products should be strengthened. Product standards, inspection criteria, and processing technical parameters that are in urgent need in China are included in Table 6.

The quality, safety, and grade specifications of sea cucumber have been stipulated in relevant national standards, but few inspection method standards and only a few industrial or local standards in processing technical specifications are available at this time. There is still an urgent need to develop and improve industry standards that cover the entire sector in a timely manner. Speeding up the creation of national and industrial standards relating to sea cucumber processing is critical to standardize the operations of sea cucumber processes nationwide, continually improve the quality of processed sea cucumber products, enhance the market competitiveness of the businesses, and protect the legitimate rights and interests of those producers and processors.

5.2 Quality Standard System of Sea Cucumber in Other Countries

Regulations for dried sea cucumber were promulgated by the Indonesian government in 2009: Technical specifications for dried sea cucumber commodities, including environmental hygiene and personal hygiene, food quality, and safety

Table 6 Plan of quality standards for processed sea cucumber products in China

Category	Target	Contents to be specified
Product standard	Sea cucumber capsule	Specify the requirements, test methods, and inspection rules for sea cucumber protein, polysaccharide, saponin, and additives in products
	Sea cucumber oral liquid	Specify the requirements, test methods, and inspection rules for sea cucumber protein, polysaccharide, saponin, and additives in products
	Flavored sea cucumber	Specify microbial indicators, additive dosage, and inspection rules in products
	Canned sea cucumber	Specify product packaging process, microbial indicators, additive dosage, and inspection rules in products
	Rehydrated sea cucumber	Specify microbial indicators, test methods, and inspection rules in products
Inspection method standard	Determination of sphingolipids in sea cucumber	Specify the determination method of sea cucumber sphingolipid in sea cucumber and its products
	Identification of sea cucumber species	Specify the species identification technique and method of available content for sea cucumber in sea cucumber products
Processing technical specification	Technical specification for processing of rehydrated sea cucumber	Production safety control of microbial, chemical, and physical hazards based on HACCP system
	Technical specification for processing of flavored sea cucumber	Production safety control of microbial, chemical, and physical hazards based on HACCP system
	Technical specification for processing of sea cucumber capsule	Production safety control of microbial, chemical, and physical hazards based on HACCP system
	Technical specification for processing of sea cucumber oral liquid	Production safety control of microbial, chemical, and physical hazards based on HACCP system
	Technical specification for processing of canned sea cucumber	Production safety control of microbial, chemical, and physical hazards based on HACCP system
	Technical specification for processing of sea cucumber liquor	Production safety control of microbial, chemical, and physical hazards based on HACCP system

requirements, are specified in Part 1 of the standard *Teripang Kering-Bagian 1: Spesifikasi* (SNI 2732.1-2009). Part 2 of the standard *Teripang Kering-Bagian 2: Persyaratan Bahan baku* (SNI 2732.2-2009) specifies the requirements for raw materials—dried sea cucumber. Part 3 of the standard *Teripang Kering-Bagian 3: Persyaratan Bahan baku* specifies the requirements for dried sea cucumber treatment and processing.

PEPINO DE MAR—ESPECIFICACIONES (NC 1167-2016) was issued by Cuba in 2016, which gives specifications for sea cucumbers. *Beche-de-mer (Processed Sea Cucumber)* (SLS 1574-2017), which specifies the sampling, test requirements, and techniques for processed sea cucumber, was issued by Sri Lanka in 2017.

Except for Japan, Korea, and Southeast Asia, European and American nations do not consume sea cucumbers due to their lifestyle or dietary culture. The sea cucumber, however, is very popular in many Asian countries due to China's influence on their cuisines. Due to the longstanding tradition of eating sea cucumber, China's standards in the field of sea cucumber are almost ideal, and there are few international or foreign standards for comparison. This way, it is imperative that the sectors of sea cucumber seedling rearing, breeding, and processing be standardized globally and that the key national and industrial standards in China be strengthened for their implementation and monitoring. It is also necessary to develop international standards in relevant fields; fill in the technical gaps in the breeding mode, supporting facilities, processing indicators, and product packaging; and provide reference, operable technical guidance, and production specifications for all businesses on the value chain, thus contributing to the improvement of product quality and promotion of the international circulation of sea cucumbers.

6 Summary and Outlook

The culinary and medical benefits of sea cucumber are becoming increasingly well-known as people's living standards rise. Because of the prevalence of fake and inferior sea cucumber products on the market, it becomes an essential matter of research to develop methods of categorizing and identifying sea cucumbers.

In comparison to traditional identification methods, molecular biology identification of sea cucumber species has the benefits of high specificity and sensitivity, and with the advance in biotechnology, molecular biology identification has become increasingly prevalent. As a sequence of mitochondrial DNA, the CO I gene exhibits high sequence diversity, which is essential for the identification of sea cucumber species. In contrast, ribosomal RNA genes are highly conserved, change slowly, and can be employed for species- or higher-level detection. In addition, the identification of sea cucumber species is also possible based on changes in their fatty acid content. Sea cucumber species identification and analysis, as well as varied origins and processing methods, have an essential role in product quality. Genomic, metabolomic, and proteomic technologies have opened new avenues for the development of sea cucumber traceability analysis technology in the era of big data. Rather than relying just on a single analytical methodology, it is more accurate to use a variety of methodologies. It is also required to further expand the associated analytical database, study the best chemometric statistical approaches, and create discriminative models for the traceability analysis of sea cucumbers from various species, different origins, and diverse production processes.

It is essential to create relevant processing standards as well as rules and regulations to standardize all stages of sea cucumber aquaculture and production, including breeding, processing, sale, and consumption. The current identification criteria for sea cucumbers are incapable of keeping pace with the rapid growth of the industry, and the system of standards should therefore be updated. To improve the accuracy of sea cucumber species identification and improve the technical standards for distinguishing different sea cucumber species or origins, new methodologies and technologies should be combined. Moreover, higher level national standards should be created for widespread use, which may then promote the promulgation of international standards in related fields.

References

Al Azad S, Shaleh SRM, Siddiquee S (2017) Comparison of fatty acid and proximate composition between *Holothuria edulis* and *Holothuria scabra* collected from coastal water of Sabah, Malaysia. Adv Biosci Biotechnol 8:91–103

Amin MHFA, Pidada I, Sugiharto, Widyatmoko JN, Irawan B (2016) Sea cucumber species identification of family Caudinidae from Surabaya based on morphological and mitochondrial DNA evidence. International Conference & Workshop on Basic & Applied Sciences 1718: 030001.1–030001.6

Arndt A, Marquez C, Lambert P, Smith MJ (1996) Molecular phylogeny of eastern Pacific sea cucumbers (*Echinodermata*: *Holothuroidea*) based on mitochondrial DNA sequence. Mol Phylogenet Evol 6:425–437

Bang WJ, Kim HC, Ryu J, Lee HS, Lee SY, Kim MS, Chong ST, Klein TA, Choi KS (2021) Multiplex PCR assay for the identification of eight Anopheles species belonging to the Hyrcanus, Barbirostris and Lindesayi groups. Malar J 20:1–7

Cao G (2019) Study on the separation analysis and origin identification of wines. Dissertation, Shenzhen University

Chen S, Li G, Yin L, Huang W, Dong P, Xu J, Chang Y, Xue C (2010) Identification of 8 sea cucumber chondroitin sulfate species by high temperature~1H NMR. J Anal Test 29:782–786

Cong P (2012) Isolation and identification of monosialic acid gangliosides from sea urchin and sea cucumber. Dissertation, Ocean University of China

Cong P, Li Z, Xu J, Yu J, Chang Y, Xue C (2013) Determination of monosialic acid gangliosides in sea cucumbers and sea urchins by liquid chromatography-tandem mass spectrometry. Chromatography 31:399–403

Dong P, Xue C, Sheng W, Xu J, Li Z (2008) Study on the determination of total saponin content in sea cucumber. China Mar Drugs 27:28–32

Fan S, Hu C, Zhang L, Wen J (2010) Morphological characteristics of bone fragments of eight species of tropical sea cucumbers from China Xisha Islands. J Trop Oceanogr 29:148–153

Fong B, Norris C, Lowe E, McJarrow P (2009) Liquid chromatography–high-resolution mass spectrometry for quantitative analysis of gangliosides. Lipids 44:867–874

Gao Y, Lin Y, Shu-Min H, Li Z, Hu Y, Lili S, Zhao Q (2016) Comparison of trace element contents of spiny ginseng from different origins and farming methods. J Dalian Ocean Univ 29:498–501

Groener JE, Poorthuis BJ, Kuiper S, Helmond MT, Hollak CE, Aerts JM (2007) HPLC for simultaneous quantification of total ceramide, glucosylceramide, and ceramide trihexoside concentrations in plasma. Clin Chem 53:742–747

Hayakawa T, Hirai M (2003) An assay of ganglioside using fluorescence image analysis on a thin-layer chromatography plate. Anal Chem 75:6728–6731

Hebert PD, Ratnasingham S, De WJR (2003) Barcoding animal life: cytochrome c oxidase subunit 1 divergences among closely related species. Proc R Soc Lond B Biol Sci 270:S96–S99

Hu R, Xing R, Wang N, Ge Y, Chen Y (2019) Identification of sea cucumber species based on DNA barcoding technology. Sci Technol Food Ind 40:145–151

Idris MY, Ahmad HF, Hashim R, Usup G, Kamarudin SK (2011) Phylogeny of *Holothuria leucospilota* (*Echinodermata*: *Holothuroidea*) as inferred from cytochrome c oxidase i mtdna gene sequences. BJRST 1:52–54

Jia Z, Xu J, Song Y, Li Z, Xue C (2015) Analysis of sea cucumber ceruloplasmin and ceramide by high performance liquid chromatography-evaporative light scattering. J Food Saf Qual Inspect 2:602–608

Jiang B, Zhang X, He X, Zhang J, Zhang H, Li Z, Xue C, Feng T, Jiang X (2021) Research progress of traceability analysis technology of commercial sea cucumber. Food Sci 42:309–318

Jo J, Park C (2016) Phylogenetic analysis of the three color variations of the sea cucumber *Apostichopus japonics*. J Aquac Res Dev 7:2–4

Kalinin VI, Aminin DL, Avilov SA, Silchenko AS, Stonik VA (2008) Triterpene glycosides from sea cucumbers (*Holothurioidea*, *Echinodermata*). Biological activities and functions. Stud Nat Prod Chem 35:135–196

Kamarudin KR (2015) 16S rRNA Barcoding technique for species identification of processed sea cucumbers from selected Malaysian markets. J Sci Math Lett 4:10–23

Kamarudin KR, Rehan AM, Hashim R, Usup G, Rehan MM (2016) Phylogenetic relationships within the genus *Holothuria* inferred from 16S mitochondrial rRNA gene sequences. Sains Malays 45:1079–1087

Kamarudin KR, Rehan MM, Bahaman NA (2017) Morphological and molecular identification of sea cucumber species *Holothuria scabra, Stichopus horrens* and *Stichopus ocellatus* from Kudat, Sabah, Malaysia, Pertanika. J Trop Agric Sci 40(1):161–172

Kang X, Zhao Y, Shang D, Zhai Y, Ning J, Ding H, Sheng X (2020) Identification of the geographical origins of sea cucumbers in China: the application of stable isotope ratios and compositions of C, N, O and H. Food Control 111:107036.1–107036.12

Kashima M, Nakagawa K, Sugawara T, Miyazawa T, Murakami C, Miyashita R, Ono J, Deschamps F, Chaminade P (2002) Method for quantitative determination of cerebroside in "plant ceramide" foodstuffs by high performance liquid chromatography with evaporative light scattering detection. J Oleo Sci 51:347–354

Kress WJ, Erickson DL (2012) DNA barcodes: methods and protocols. In: DNA barcodes, vol 858. Springer, Cham, pp 3–8

Kress WJ, Wurdack KJ, Zimmer EA, Weigt LA, Janzen DH (2005) Use of DNA barcodes to identify flowering plants. Proc Natl Acad Sci 102:8369–8374

Lacey KM, McCormack GP, Keegan BF, Powell R (2005) Phylogenetic relationships within the class *Holothuroidea*, inferred from 18S rRNA gene data. Mar Biol 147:1149–1154

Lee T, Shin S, Kim D (2017) A newly recorded sea cucumber of the genus *Psolus* (*Holothuroidea*: *Dendrochirotida*: *Psolidae*) from the East Sea of Korea. Anim Syst Evol Divers 33:195–199

Levin V, Kalinin V, Fedorov S, Smiley S (1986) Structure of triterpene glycosides and taxonomical position of two species of the family *Stichopodidae*. Biol Morya 4:72–77

Li Y, Fei L, Chen J (2008) Morphological study of fifteen species of sea cucumber bone fragments. J Ocean Univ China (Nat Sci Ed) 2:211–216

Li Y, Zhang X, Li Y, Wang H (2017) The application of compound-specific isotope analysis of fatty acids for traceability of sea cucumber (*Apostichopus japonics*) in the coastal areas of China. J Sci Food Agric 97:4912–4921

Li W, Wu G, Wang M, Yue A, Du W, Liu D, Zhao J (2020) Colorimetric detection of class A soybean saponins by coupling DNA zyme with the gap ligase chain reaction. Anal Methods 12: 3361–3367

Li Z, Song L, Wang H, Li Y, Yan Y, Li P (2022) Characterization of inorganic elements in different origins of silver cohosh by ICP-MS combined with multivariate statistics. China J Trad Chin Med:1–7. In Chinese

Liang J (2008) Application of mitochondrial sequences to the preliminary identification of several sea cucumbers and their application to deep processed sea cucumber products. Dissertation, Ocean University of China

Liang J, Xue C, Zhao Y, Tang Q, Xu J, Yang W (2008) Efficient and rapid extraction of DNA from sea cucumber nutrient solution and species identification method. Food Sci 29:269–272

Liao Y (2001) Chinese Sea Cucumber. Biol Bull 9:1–3+47

Ling Z, Jing W, Fan S, Chen Z et al (2018) Identification of sea cucumber species in processed food products by PCR-RFLP method. Food Control 90:166–171

Liu X, Xue C, Wang Y, Li Z, Xue Y, Xu J (2011) Cluster analysis and principal component analysis of inorganic elements in spiny ginseng. Spectrosc Spectr Anal 31:3119–3122

Liu Y, Xu J, Zhou Q, Xue Y, Xue C, Wang J (2015) Rapid determination of total and organic vanadium in sea cucumbers by ultrasonic-assisted digestion with graphite furnace atomic absorption. Food Sci 4:126–130

Liu Y, Zhao X, Zhang X, Li Y, Fang S (2018) Characterization of the carbon stable isotope composition of amino acids in the spiny ginseng of Dalian. Environ Chem 37:239–248

Liu Q, Du Z, Li J, Wang Y, Gu X, Cui X (2021) Analysis and evaluation of inorganic elements in *Salvia miltiorrhiza* and inter-root soils of different origins. Spectrosc Spectr Anal 41:3618–3624

Lyv Y (2012) A molecular biology method based on DNA barcoding for identification of sea cucumber species. Dissertation, Ocean University of China

Lyv Y, Zuo T, Tang Q (2011) Construction and application of DNA barcoding in sea cucumber *Holothurians*. J Fish Sci China 18:782–789

Ma L, Fong BY, MacGibbon AK, Norris G (2020) Qualitative and quantitative study of glycosphingolipids in human milk and bovine milk using high performance liquid chromatography–data-dependent acquisition–mass spectrometry. Molecules 25: 4024.1–4024.24

Maggi C, Gonzalez-Wanguemert M (2015) Genetic differentiation among *Parastichopus regalis* populations in the Western Mediterranean Sea: potential effects from its fishery and current connectivity. Mediterr Mar Sci 1:489–501

Pan C, Sun D, Che M, Liu R, Hu M (2021) National standard for food safety dried sea cucumber (GB 31602-2015) standard interpretation. Food Saf Guide 19:43–46

Peng K, Liu H, Liu G, Zhang X, Li F, Tian X, Gong X, Wang B, Sun C, Xu Y (2021) Study on the application of stable isotope technology in food traceability of plant and animal origin. Sci Technol Food Ind 42:338–345

Rrx A, Rrha B, Nan WA, Jkz A, Yqgb C, Ying C (2021) Authentication of sea cucumber products using NGS-based DNA mini-barcoding. Food Control 129:108199.1–108199.9

Shan Y, Xue C, Li Z (2005) Extraction and purification of mitochondria from dried sea cucumber. Food Sci 26:111–112

Sisu E, Flangea C, Serb A, Rizzi A, Zamfir AD (2011) High-performance separation techniques hyphenated to mass spectrometry for ganglioside analysis. Electrophoresis 32:1591–1609

Song Y, Xing P, Jia Z, Li Z, Xue C, Xu J (2016) Analysis of glucose cerebrosides and galactose cerebrosides by o-phthalaldehyde derivatized high performance liquid chromatography. J Nutr 38:75–80

Sørensen LK (2006) A liquid chromatography/tandem mass spectrometric approach for the determination of gangliosides GD3 and GM3 in bovine milk and infant formulae. Rapid Commun Mass Spectrom 20:3625–3633

Sun Y, Yuan Z, Xue X (2018) Research on the development of China's sea cucumber industry and countermeasures. World Agric 3:186–191

Svobodova Z, Novotny J, Ospalkova B, Slovakova M, Bilkova Z, Foret F (2021) Affiblot: a dot blot-based screening device for selection of reliable antibodies. Anal Methods 13:3874–3884

Taberlet P, Coissac E, Pompanon F, Gielly L, Miquel C, Valentini A, Vermat T, Corthier G, Brochmann C, Willerslev E (2007) Power and limitations of the chloroplast trn L (UAA) intron for plant DNA barcoding. Nucleic Acids Res 35:e14–e14

Tanabe A, Matsuda M, Fukuhara A, Miyata Y, Komuro R, Shimomura I, Tojo H (2009) Obesity causes a shift in metabolic flow of gangliosides in adipose tissues. Biochem Biophys Res Commun 379:547–552

Wang X, Cong P, Chen Q, Li Z, Xu J, Xue C (2020) Characterizing the phospholipid composition of six edible sea cucumbers by NPLC-Triple TOF-MS/MS. J Food Compos Anal 94:1–12

Wang X, Wang Y, Xu J, Xue C (2021) Sphingolipids in food and their critical roles in human health. Crit Rev Food Sci Nutr 61:462–491

Wen J, Zhang L, Hu C, Shen Q (2009) Species and morphology of bone fragments of six species of sea cucumbers from tropical China. J Oceanogr 2:139–145

Wen J, Hu C, Zhang L, Luo P, Zhao Z, Fan S, Su T (2010) The application of PCR–RFLP and FINS for species identification used in sea cucumbers (*Aspidochirotida: Stichopodidae*) products from the market. Food Control 21:403–407

Wen J, Hu C, Zhang L, Fan S (2011) PCR-RFLP identification of 16S rRNA in 16 species of commercial sea cucumbers. China Fish Sci 18:451–457

Wen J, Hu C, Zhang L, Fan S (2012) Identification of 11 sea cucumber species by species-specific PCR method. Food Control 27:380–384

Wilwet L, Jeyasekaran G, Shakila RJ, Sivaraman B, Padmavathy P (2018) A single enzyme PCR-RFLP protocol targeting 16S rRNA/tRNAval region to authenticate four commercially important shrimp species in India. Food Chem 239:369–376

Wu N, Chen S, Ye X, Li G, Yin LA, Xue C (2014) Identification of fucans from four species of sea cucumber by high temperature 1H NMR. J Ocean Univ China 13:871–876

Xu X, Wang S, Wang H, Hu W, Han L, Chen B, Li X, Wang H, Li H, Gao X (2021) Simultaneous quantitative assays of 15 ginsenosides from 119 batches of ginseng samples representing 12 traditional Chinese medicines by ultra-high performance liquid chromatography coupled with charged aerosol detector. J Chromatogr A 1655:462504.1–462504.12

Yang B, Chen Z, Zhang Y, Song Y, Zhao S (2019) Development of a duplex PCR assay for simultaneous detection of Mycoplasma gallisepticum and Mycoplasma synovium. Anim Husb Feed Sci (Inner Mongolia) 40:12–15

Yao W, Gu Y, Wu G, Zeng X (2017) Ultramicrostructure of bone fragments of six sea cucumbers from China Xisha Islands. Hacettepe Bull Med Surg 47:76–81

Yu L, Xue C, Dong P, Zou B, Li Z, Xu J, Wang Y, Wang J (2011) Comparative SPE-HPLC analysis of saponins in eight species of sea cucumbers. Anal Lab 30(1):13–15

Zeng L, Wen J, Fan S, Chen Z, Xu Y, Sun Y, Chen D, Zhao J, Xu L, Li Y (2018) Identification of sea cucumber species in processed food products by PCR-RFLP method. Food Control 90:166–171

Zhang X, Liu Y, Li Y, Zhao X (2017) Identification of the geographical origins of sea cucumber (*Apostichopus japonics*) in northern China by using stable isotope ratios and fatty acid profiles. Food Chem 218:269–276

Zhang H, Zhang X, Zhao X, Xu J, Lin C, Jing P, Hu L, Zhao S, Wang X, Li B (2019) Discrimination of dried sea cucumber (*Apostichopus japonics*) products from different geographical origins by sequential windowed acquisition of all theoretical fragment ion mass spectra (SWATH-MS)-based proteomic analysis and chemometrics. Food Chem 274:592–602

Zhao Y, Wang Z, Dai J, Chen L, Huang Y (2014) Preparation and quality assessment of high-purity ginseng total saponins by ion exchange resin combined with macroporous adsorption resin separation. Chin J Nat Med 12:382–392

Zhao Y, Zhang C, Qin X, Zheng H, Cao W (2015) A preliminary investigation on the traceability technology of Hong Kong oyster origin based on inorganic element analysis. J Guangdong Ocean Univ 35:94–99

Zhao X, Liu Y, Li Y, Zhang X, Qi H (2018) Authentication of the sea cucumber (*Apostichopus japonics*) using amino acids carbon stable isotope fingerprinting. Food Control 91:128–137

Zhao Z, Jiang J, Zheng J, Pan Y, Dong Y, Chen Z, Gao S, Xiao Y, Jiang P, Wang X, Zhang G, Wang B, Yu D, Fu Z, Guan X, Sun H, Zhou Z (2022) Exploiting the gut microbiota to predict

the origins and quality traits of cultured sea cucumbers. Environ Microbiol. Online March 16. https://doi.org/10.1111/1462-2920.15972
Zhou Y (2016) Study on the stable isotope composition of spiny ginseng and its environmental factors. Dissertation, Dalian Maritime University
Zhou Y, Zhang J, Li L (2012) Factor analysis and cluster analysis of inorganic elements in beers of different origins. Phys Chem Test (Chemistry) 48:49–52, 55
Zhu W, Rong X, Lin H, Jiang Y, Yao L, Wang L (2011) Current situation and development direction of Chinese sea cucumber standard system. Aquat Sci Technol Inf 38:52–54
Zhu L, Wu J, Zhou D (2015) Research progress on processing and utilization of sea cucumber and quality evaluation. Agro Process 24:60–63
Zhu Z, Zhu B, Ai C, Lu J, Wu S, Liu Y, Wang L, Yang J, Song S, Liu X (2018) Development and application of a HPLC-MS/MS method for quantitation of fucosylated chondroitin sulfate and fucoidan in sea cucumbers. Carbohydr Res 466:11–17
Zuo T, Li Z, Lyv Y, Duan G, Wang C, Tang Q, Xue C (2012) Rapid identification of sea cucumber species with multiplex-PCR. Food Control 26:58–62

Food Safety Issues and Regulatory Requirements of Sea Cucumber Products and Their Internationalization

Zhaojie Li, Wei Zhang, and Changhu Xue

Abstract The sea cucumber is not only a precious food but also a valuable traditional Chinese medicine. It contains a variety of bioactive compounds, such as bioactive peptides, saponins, lectins, sulfated polysaccharides, etc. The sea cucumber has antioxidant, anticancer, anti-inflammatory, immunity modulation, and many other good effects, making it widely used and consumed by people in China and Southeast Asia. However, with the development of the sea cucumber industry, a number of safety issues have arisen. Therefore, this chapter discusses current safety issues in the sea cucumber market. These issues mainly include sea cucumber diseases, illegal additives, microplastics, and crude oil pollution. These can confuse consumers and processors and disrupt the market. To ensure the healthy development of the sea cucumber industry, it is suggested to formulate high standards for sea cucumber breeding and processing that help to guarantee food safety. In addition, with the improvement of living standards, diversified and convenient sea cucumber products are becoming increasingly popular. To meet the demand of the international market for processed sea cucumber products, the development trends of sea cucumber product processing will focus on new sea cucumber products (freeze-dried, ready-to-eat and liquified sea cucumber, etc.) and standardized production technology for sea cucumber and its functional components (collagens, polysaccharides, etc.).

Keywords Sea cucumber · Edible safety · Antibiotics · Heavy metals · Microplastics · Illegal additives · Development trends

---

Z. Li · C. Xue (✉)
College of Food Science and Engineering, Ocean University of China, Qingdao, Shandong, P. R. China
e-mail: lizhaojie@ouc.edu.cn; xuech@ouc.edu.cn

W. Zhang
Centre of Marine Bioproducts Development, Medical Biotechnological, College of Medicine and Public Health, Flinders University, Bedford Park, SA, Australia
e-mail: wei.zhang@flinders.edu.au

C. Xue (ed.), *Advances in Sea Cucumber Processing Technology and Product Development*, Advances in Marine Bioprocesses and Bioproducts,
https://doi.org/10.1007/978-3-031-16512-2_12

1 Introduction

With the increasing demand for healthy food such as sea cucumbers, food safety has increasingly become a common public health issue. The sea cucumber belongs to benthic marine organisms, which inhabit the seabed, and the surface or interior of sediments and consumes the microorganisms in the sediments and the organic debris of animals and plants. Offshore waters are greatly affected by human activities. For example, traces of antibiotics and drugs from land animal breeding enter the sea through industrial sewage discharge. Similarly, oil pollution caused by ports and offshore drilling and microplastics (MPs) produced by the long-term crushing of plastic particles produced in human daily life and industrial production can enter offshore waters. For these reasons, farmed sea cucumbers may be polluted. At the same time, due to the high breeding density, the sea cucumber can be very easily infected with bacteria in the breeding process. The existence of a large number of microorganisms threatens the edible safety of sea cucumbers. In order to prevent diseases in sea cucumbers, antibiotics are widely and intensively used in their cultivation, and as a result, the antibiotics in sea cucumbers often exceed the standard. In the breeding process, some farmers also use hormones and illegal drugs to make sea cucumbers grow rapidly in order to increase their profits. After eating this kind of sea cucumber, both the human hormone level and the normal physiological function will be affected. Some chemicals harmful to the human body may also be added during processing. All of these factors make the food safety of sea cucumbers an increasingly pressing concern for consumers and government authorities. In addition, diversified and convenient sea cucumber products are increasingly sought after by consumers. At present, the forms of sea cucumber products on the market (mainly dried and salted sea cucumbers) and processing techniques (cooking, salting, etc.) are relatively monotonous. Therefore, the development trend of sea cucumber product processing will include building a modern and standardized sea cucumber processing technology system and enriching the types of sea cucumber products to meet the demands in the international market.

2 Edible Safety of Sea Cucumber Products

2.1 Microorganisms and Diseases

The sea cucumber is rich in water and protein, which makes it very vulnerable to microbial infection, and the existence of diverse microorganisms threatens the edible safety of sea cucumber. *Vibrio* infection is usually the main cause of diseases, such as skin ulcer syndrome and bacterial ulcer syndrome. Dang et al. (2006) found that the sea cucumber culture pool contained large amounts of antibiotic-resistant bacteria, mainly marine *Vibrio*. *Vibrio parahaemolyticus* is a halophilic *Vibrio*, which is widely distributed in the marine environment, such as in seawater, seabed sediments, seafood, and salted food. It was the main *Vibrio* causing the reduction of adhesion, body contraction, and collapse in juvenile sea cucumbers (Wang et al. 2012). It was reported that *Vibrio parahaemolyticus* was resistant to ampicillin, streptomycin, kanamycin, tetracycline, and ciprofloxacin (Devi et al. 2009). The sea cucumber can be used as a carrier to spread drug-resistant pathogens to the human body or other environments during eating or processing, which poses a threat to public health. Human acute gastroenteritis can very easily occur after aquatic products contaminated by this fungus are eaten. Cases of food poisoning caused by *Vibrio parahaemolyticus* have become the highest among all forms of microbial food poisoning.

In addition, skin ulcers, rotten skin, edema, and other diseases in the process of sea cucumber breeding are caused by bacteria, molds, parasites, and other pathogens. Becker et al. (2004) found an outbreak of an infectious disease known as "skin ulcer syndrome" during the raising of sea cucumber larvae seedlings. The lesions first appeared around the oral cavity or cloacal opening of sea cucumber and then spread to both sides of the individual, thus covering the whole body. The infection was accompanied by the loss of epidermal pigmentation and the presence of a large amount of viscous mucus. The disease spreads rapidly from diseased individuals to healthy ones, making it difficult to control. If the sea cucumber infected by the disease was not removed from the population, it would lead to the death of up to 95% of the breeding individuals (Morgan 2000). The "skin ulcer syndrome" has become the most serious and common disease in the sea cucumber breeding industry, because of its high infection rate and high mortality. The skin ulcer syndrome has been observed in various cultured sea cucumbers in different regions of the world. The disease mostly occurred in *Stichopus japonicus* during the seedling preservation period in the breeding area. The focus in the early stage of the disease teemed with bacteria. Bacteria, molds, or parasites were found in the focus in the middle and late stages of the disease, with bacteria as the main infectious substances (Wang et al. 2006). The sea cucumber is the species of echinoderms most susceptible to parasitic infection. It was reported that more than 1/3 of echinoderm parasites were parasitic on sea cucumbers. Most protozoa parasitized in sea cucumbers were sporidia, which parasitized in the body cavity, blood circulation system, or gonads of the sea cucumbers. The research conducted by Doignon et al. (2003) showed that planarians generally entered the body cavity and digestive tract through the mouth and respiratory tree of sea cucumber, causing a significant amount of sea cucumber infections. Among the collected sea cucumbers, 63% contained 6–10,000 egg sacs of planarians. Wang et al. (2005) observed many spherical, virus-like particles in the tissue extract of diseased cultured *Stichopus japonicus* by electron microscope negative staining technique. After negative staining and ultrathin section observation under the electron microscope, it was confirmed that a lot of globular viruses existed in the sea cucumber (Deng 2006).

2.2 Antibiotics or Veterinary Drugs

In the process of sea cucumber breeding, numerous antibiotics (such as fluoroquinolones and tetracyclines) are added to sea cucumber feed. They are widely and intensively used in aquaculture systems for the prevention of bacterial diseases due to their broad-spectrum activity and low toxicity. At the same time, antibiotics used in the terrestrial animal breeding process also entered the terrestrial environment through wastewater discharge and land application of feces and sludge and finally entered the marine environment through river input and sewage discharge, eventually making antibiotics a new type of pollutant in the marine environment and posing a threat to human health and the marine ecological environment (Jia et al. 2011). Detection methods for antibiotic residues in sea cucumbers include high-performance liquid chromatography (HPLC), liquid chromatography-tandem mass spectrometry (LC-MS-MS), solid phase extraction-high-performance liquid chromatography (SPE-HPLC), high-performance liquid chromatography-atomic fluorescence spectrometry (HPLC-AFS), and high-performance liquid chromatography (HPLC) with fluorescence detector. At present, the presence of antibiotics has been detected in natural water environments in many parts of the world. More than 110 drugs and their metabolites were detected in global offshore waters in 2014, of which more than 40 belonged to the antibiotic class, mostly in the range of 1 ng/L to several thousand ng/L (Gaw et al. 2014). Four antibiotics were detected in the Mediterranean Sea at concentrations up to 217 $\mu g \cdot L^{-1}$ (Nödler et al. 2014). The continued input of high concentrations of antibiotics from different sources into seawater has led to an increased potential for bioaccumulation and biomagnification of antibiotics in marine organisms, which could pose a high ecological risk to these organisms. Many studies found that antibiotic contamination in sea cucumbers was geographical region-dependent, with high concentrations of antibiotics found in sea cucumbers from the South China Sea. The maximum concentration of antibiotic residues was 32.8 $\mu g\text{-}kg^{-1}$ (dry weight), and the levels of antibiotics in ready-to-eat sea cucumbers were slightly higher than those in dried sea cucumbers (Zhu et al. 2018). In addition, the MCT model of aqueous organic chemicals in sea cucumbers was developed based on the passive diffusion principle, which was used to estimate the residual concentrations of antibiotics in different tissues of sea cucumber, indicating that sea cucumbers had a high potential to accumulate antibiotics (Zhu et al. 2020). These drug residues could enter the human body through the food chain and accumulate in the human body.

In order to make sea cucumbers grow quickly, some unscrupulous traders use hormones (such as steroidal anabolic hormones and non-steroidal anabolic hormones) to reduce the activity of sea cucumbers, so that sea cucumbers became dumber and did nothing but eat to grow. Long-term consumption of these sea cucumbers will cause drug accumulation in the body, thus affecting human hormone levels, disrupting normal physiological functions, and causing potential developmental toxicity (premature maturation in children) and feminization or masculinization (Wang 2012). Compounds such as estrogens (e.g., estradiol and estriol), ibuprofen, diclofenac, and acetylsalicylic acid are deposited into the ocean every year (Miller et al. 2015). Those compounds will persist in the marine environment, where they can be toxic to many organisms. Relevant studies have found that those drugs have harmful effects (such as chromosome damage and induction of DNA mutations) on the growth and fertility of echinoderms and invertebrates such as sea cucumbers, while they may be carcinogenic to other marine vertebrates and humans (Adeel et al. 2017; Lecomte et al. 2017). In addition, many sea cucumber farms use cephalosporins (such as ceftriaxone and cefotaxime) to improve the survival rate of sea cucumber seedlings and prevent diseases. However, the massive use of cephalosporin antibiotics will lead to the pollution of aquaculture waters, subjecting the ecosystem to a sub-health status; bacterial resistance increases, causing harm to other aquaculture species. Yang et al. (2021) used liquid-phase mass spectrometry to detect veterinary drug residues in commercially available dried sea cucumbers, including 11 kinds of residual substances such as malachite green, chloramphenicol, and furazolidone metabolites. The overall exceedance rate of veterinary drug residues in 100 parts of dried sea cucumber was 1.0%. Malachite green was detected in only one batch of samples, and the rest of the veterinary drugs were not detected. Malachite green has long been widely used in the prevention and treatment of various diseases affecting aquatic organisms such as water molds, gill molds, and sea cucumber diseases. However, some scholars discovered that malachite green had high toxicity, high residue, high carcinogenicity, and high teratogenicity. In view of the dangers of malachite green, many countries have prohibited its use in aquaculture, but the illegal use of the substance is still widespread. Zhang et al. (2012) used an optimized liquid chromatography-tandem mass spectrometry (LC-MS/MS) method to determine nitrofuran metabolite residues in sea cucumbers and sea cucumber seedlings. AOZ, AHD, SEM, and AMOZ were used as the standards for nitrofuran metabolites, using liquid chromatography-triple quadrupole tandem mass spectrometry, multiple reactions monitoring scan mode detection, and isotope internal standard method. It was found that the content of nitrofuran metabolites in sea cucumbers was lower than that in sea cucumber seedlings, where the detection rate of AOZ was 15%. AHD and AMOZ were not detected in 40 water samples. The results of the study also confirmed that nitrofurans were released during the seedling period. Therefore, nitrofurans in sea cucumber culture should be strictly controlled during the seedling period to ensure the quality and safety of sea cucumbers.

2.3 *Heavy Metals*

Heavy metals are natural components that exist in ecosystems. Some of these are essential nutrients, such as iron, zinc, copper, and manganese, while others, such as cadmium and lead, are highly toxic at high levels. Cadmium is a ubiquitous heavy metal, which was classified as a Class I carcinogen by the International Agency for Research on Cancer (IARC) in 2017 and is ranked as the sixth most hazardous toxic substance to human health by the Agency for Toxic Substances and Disease Registry

(ATSDR). Lead is considered an invisible killer in modern society, and lead poisoning can cause damage to the central and peripheral nervous systems in children, resulting in irreversible damage to intelligence and development. Globally, metal pollution is one of the most serious problems in marine aquatic ecosystems. As a result of intensive agricultural and industrial activities, increasing amounts of liquid wastewater with high heavy metal content are released into the environment, and some toxic elements tend to bioaccumulate in organisms (echinoderms, bivalves, crustaceans, etc.) and sediments, with harmful effects on organisms and human health through biomagnification in the food chain. Due to the serious heavy metal pollution in the water, and specifically in sea cucumbers, in recent years, scholars have investigated and studied the effects and extent of heavy metal pollution in the sea cucumber.

It was found that the heavy metal content in sea cucumbers of different origins differed significantly and that there were also differences in heavy metal levels among individual sea cucumbers in the same region. Moreover, there was a wide range of excessive heavy metal content in sea cucumbers. Studies conducted on two sea cucumber species commonly consumed by humans, namely *Holothuria leucospilota* and *Holothuria scabra*, showed tendencies for different metals to accumulate in different tissues of sea cucumber. Different heavy metals such as lead, cadmium, zinc, or copper were found in high concentrations in the body wall, dermis, respiratory tree, ventral fluid, and gonads of sea cucumber, particularly affecting their respiratory and reproductive functions. It was found that *Holothuria leucospilota* accumulated copper and cadmium in the gonads and was more likely to accumulate zinc and lead in the body wall. *Holothuria scabra* accumulated more copper and cadmium in the dermis and body wall (Mohammadizadeh et al. 2016). Warnau et al. (2006) showed that the variation in metal accumulation in sea cucumber (*Holothuria tubulosa*) depended on the sampling time, sampling area, and sampling depth. Marrugo-Negrete et al. (2020) found that copper, mercury, and lead levels in sea cucumber (*Holothuria floridana*) from the Cispata Bay on the Caribbean coast of Colombia were not expected to pose a risk to adults based on a health risk assessment of the maximum allowable daily consumption rate (CR_{lim}), but that excessive consumption of sea cucumber (e.g., more than three to five pieces per day, depending on their size) might pose a risk to human health, and that children should limit or avoid their consumption. The anthropogenic sources of metal pollution in that area were mainly agricultural activities (application of fertilizers and agrochemicals). Laboy-Nieves and Conde (2001) randomly selected 28 sample sites in Morrocoy National Park (Venezuela) to collect sediments and two sea cucumbers (*Holothuria Mexicana* and *Isostichopus badionotus*) and detected the six heavy metals Al, Cu, Mn, Ni, Pb, and Zn. It was found that the bioaccumulation effects of Cu, Ni, Pb, and Zn in the large individual specimen of sea cucumbers were obvious and that the presence of Cu, Ni, and Pb in sea cucumbers was significantly higher than in other marine invertebrates in the region and that these levels exceeded the local safety limits. In addition, tests on sea cucumbers sold in Kota Kinabalu, the capital of Sabah, Malaysia, found that As in *Holothuria leucospilota* and Cd in

Thelenota ananas exceeded the limits set by the Malaysian Food Safety Act and amendments (Hashmi et al. 2014).

Zhang et al. (2009) identified As, Cd, Hg, Pb, and other toxic elements in 10 batches of commercially available sea cucumbers in the Chinese market and found that the contents of As in all the 10 batches exceeded the standard, while some batches of sea cucumbers had excessive contents of Cu, Cd, and Pb, indicating that the commercially available sea cucumbers did have excessive toxic elements. Li et al. (1989) identified 11 trace elements including Pb and Cd in *Holothuria nobilis*, *Apostichopus japonicus*, and dried scallops by dry or wet digestion combined with atomic absorption spectrophotometry. The contents of Pb and Cd in the dry and wet digested samples were 2.055 and 2.904 mg/kg and 2.544 and 2.548 mg/kg, respectively, which all exceeded the limits specified in the *National food safety standard—Quantity of pollutants in food* (GB 2762). Ye et al. (2019) measured the heavy metal content in cultured sea cucumbers in different Chinese waters and found that the content of Cd and Hg in cultured sea cucumbers could meet the green food requirements, while Pb and As were enriched to a high degree in sea cucumbers, indicating that samples were slightly contaminated by Pb and As, which were the main risk factors for sea cucumber culture. Wang et al. (2009) examined the trace elements in sea cucumbers using autoclave ablation and inductively coupled plasma atomic emission spectrometry. The results showed that the sea cucumbers were rich in beneficial elements such as iron, zinc, selenium, and copper, but the contents of lead, mercury, and arsenic were 14.04, 57.20, and 27.98 mg/kg, respectively, which far exceeded the national safety limits. Gong and Que (2012) determined the contents of 15 elements in sea cucumber using the detection method of microwave digestion inductively coupled plasma mass spectrometry. The results showed that the contents of Cr, As, Hg, and Pb in the sea cucumbers were 0.74, 4.57, 1.89, and 2.37 mg/kg, respectively, which exceeded the limit standard specified in the Chinese industrial standard *Pollution-free food—Limits of toxic and harmful substances in aquatic products* (NY5073-2006).

In conclusion, the phenomenon of excessive heavy metals in sea cucumbers is widespread. The cumulative toxicity of heavy metals is considerable, with a long latent period and high enrichment, and it is difficult to excrete from the human body. For these reasons, the consumption of sea cucumbers contaminated with heavy metals, therefore, will pose a significant threat to human health.

2.4 Microplastics

In recent years, the presence of microplastics (MPs) in aquatic products has become an issue of increasing concern. With economic and social development, plastic products have become an important and convenient part of daily life. MPs are usually defined as plastic particles with a diameter of less than 5 mm. In 2004, Thompson et al. at the University of Plymouth first proposed the concept of "microplastics" (Thompson et al. 2004). Factory sewage, plastic products, industrial washing wastewater, and the development of tourism are closely related to the presence of microplastics and synthetic microfibers in the terrestrial ecosystem. In daily life, synthetic microfibers are produced by laundry and residue emissions from the catering industry, while widely used cosmetics are also major sources of microplastics in soil and water.

MPs can adsorb organic or inorganic pollutants around them, including polycyclic aromatic hydrocarbons, polychlorinated biphenyls, and heavy metals. Marine organisms that consume MPs will encounter increased mortality, reduced growth rate and fecundity, intestinal obstruction or injury, and other adverse effects. In addition to the physical properties of MPs, which will lead to physiological complications after ingestion by marine organisms, MPs may also carry various toxic chemicals that will penetrate animal tissues. Due to the non-degradability of MPs, most MPs will be enriched in biological species higher up the food chain, including humans. It has been found that microplastic pollution is common in the human diet. For example, the content of microplastics in marine shellfish can be as high as 20 pieces/g (Naji et al. 2018) and as high as 7–681 pieces/kg in salt (Yang et al. 2015). It can be seen that marine pollution caused by microplastics can be transmitted to humans through salt or aquatic products. Chen et al. (2021) pointed out that the microplastics in seawater or sediments could enter the body of sea cucumbers through the mouth or anus and enter the body cavity fluid through the respiratory tree. A high concentration of microplastics in seawater might damage the body cavity cell structure of *Halodeima atra*, reduce its immune defenses, and affect its normal physiological activities. Studies have found that, in samples of *Apostichopus japonicus* collected from eight locations in the Bohai Sea and the Yellow Sea, MPs cannot be discharged in the coelomic fluid of these specimens, therefore damaging their biological function and development (Mohsen et al. 2018). In addition, the fragments of PVC or nylon in MPs ingested by sea cucumber were high in toxic polychlorinated biphenyls, which made polychlorinated biphenyls enter sea cucumber tissue and accumulate continuously. When the sea cucumber is ingested by other organisms, pollutants will enter higher levels of the food chain (such as seabirds, whales or humans) and development bioaccumulation, which can lead to serious liver and nerve disorders (Graham and Thompson 2008).

2.5 Crude Oil Pollution

With the continuous increase in global crude oil production and consumption, oil pollution from ports, shipping, and offshore drilling may seriously threaten the marine ecosystem and marine economy. Harmful substances in oil will affect fish, shellfish, and human health through the food chain. Although crude oil contains thousands of organic compounds, polycyclic aromatic hydrocarbons (PAHs) are generally considered to be the main toxic component of crude oil to marine organisms (Hodson 2017). Due to their strong hydrophobicity and stable structure, polycyclic aromatic hydrocarbons can be distributed to non-aqueous phase and easily adsorbed on suspended particles and sediments in the water. Once they are bound to the sediments, it is difficult for polycyclic aromatic hydrocarbons to undergo photochemical degradation or oxidative decomposition by microorganisms. Previous studies have shown that PAHs, as persistent organic pollutants, have semi-volatility and bioaccumulation, especially polycyclic aromatic hydrocarbons with more than four rings, which have carcinogenic, teratogenic, and mutagenic effects (Zhong et al. 2014). Recent studies have shown that sea cucumber has a strong potential to absorb polycyclic aromatic hydrocarbons, and the concentration of PAHs in sea cucumber is positively correlated with the concentration of PAHs in seawater. PAHs will stimulate the production of reactive oxygen species (ROS) in the process of biotransformation, resulting in an increase in reactive oxygen species in marine organisms such as sea cucumbers. The high level of ROS may further induce oxidative damage in biological macromolecules (including DNA, protein, and lipid) and damage the antioxidant capacity of marine benthos (Li et al. 2020). 3-methylphenanthrene is a typical PAH with high content in oil. The physiological activities and metabolism of organisms living in the marine environment containing 3-methylphenanthrene would change (Liu et al. 2022).

2.6 Illegal Additives During Processing

Due to the high commercial value of sea cucumber, some merchants add formaldehyde, borax, and other illegal additives to aquatic products to increase their shelf life, water retention, and toughness. However, these also increase the toxicity of the aquatic products. Accidental ingestion of borax can cause cancer and death in severe cases.

Formaldehyde solution is commonly used for soaking rehydrated sea cucumber in aquafaba to increase weight, reduce cost, preserve freshness, improve appearance, and prolong the marketing time of the product (Xie 2013). Formaldehyde has general protoplasmic toxicity, which causes substantial harm to health. Consumers can inadvertently introduce residual formaldehyde into their bodies by eating sea cucumbers, which can damage the central nervous system; cause loss of appetite, vomiting, and diarrhea; endanger human health; and have carcinogenic effects. Formaldehyde was classified as an A1 carcinogen by the World Health Organization (WHO) in 2004. The US Environmental Protection Agency (EPA) states that the maximum recommended dose (RfD) of formaldehyde without posing a significant health risk is 0.2 mg/(kg·day).

Alkaline substances such as kerosene and borax are added to sea cucumbers by unscrupulous businessmen to increase their volume and preserve their freshness, but such alkaline substances seriously destroy their nutritional value. Caustic soda is highly corrosive to proteins, which can penetrate into every part of the sea cucumber, causing changes in its internal structure and resulting in the loss of nutrients and even changes in the structure of nutrients. If people frequently consume this type of sea cucumber, the caustic soda left in the sea cucumber will harm the digestive tract and damage the stomach mucosa, causing harm to the human body. Borax enters the body and reacts chemically with stomach acid to produce boric acid. Boric acid is a toxic substance with a lethal dose of 15–20 g for adults and 3–6 g for children.

In the past few years, a kind of dried sea cucumber product called "sugar-dried sea cucumber" appeared in the Chinese market. The processing method is basically the same as that of salted sea cucumber, except that the salt used in curing is replaced by sugar for shaping and weight gain. The purpose of salt-dried sea cucumber is to extend the shelf life, but dried sea cucumbers obtained by sugar treatment tend to absorb water during storage with shortened shelf life, and the caramelization caused by high temperature during processing can reduce the quality of the sea cucumbers (Pan et al. 2019).

To regulate and standardize the sea cucumber market, the National Health and Family Planning Commission of China issued *National Food Safety Standard—Dried Sea Cucumber* (GB31602-2015) on November 13, 2015, which has been implemented since November 13, 2016. This standard is the first mandatory national standard for dried sea cucumber products formulated by the Chinese government. The standard specifies strict regulations on the protein, salt, and total water-soluble sugar content in dried sea cucumbers, forbidding the addition of other ingredients other than edible salt during the processing of dried sea cucumbers. At the same time, the limits of contaminants and veterinary drug residues are also strictly specified to ensure the quality and safety of dried sea cucumber from the perspective of raw material production.

3 Development Trend of Sea Cucumber Processing

3.1 Development Trend of Processing Technology

Sea cucumbers have been consumed in China for nearly 2000 years since there was formal documentation during the Three Kingdoms period. At present, more than 60 species of sea cucumber have been widely cultivated and utilized all over the world. Sea cucumber fishing, processing, and trading are spread across more than 70 countries and regions around the world (FAO 2012), thus making it an aquatic product with a high degree of internationalization. Before the 1980s, the supply of sea cucumbers mainly depended on the fishing of wild sea cucumbers. Due to the sharp increase in demand, China carried out research on the artificial breeding technology for *Apostichopus japonicus* in the 1950s. In the 1980s, a breakthrough in this technology resulted in a rapid increase in the output of sea cucumbers and the emergence of new products. In the sea cucumber market, dried sea cucumbers, salted sea cucumbers, and ready-to-eat sea cucumbers are the main products that maintain the inherent shape of sea cucumber. These are supplemented by deep-processed products focusing on sea cucumber collagen (peptide), sea cucumber polysaccharide, sea cucumber lipid, and other functional components. However, there are some issues in the sea cucumber processing industry, such as a low degree of mechanization, the small market scale for the deep processed products, and a low level of product internationalization. The following innovations in processing technology are required to meet the needs of the development of the sea cucumber processing industry in the future:

1. Build a standardized modern processing technology system for sea cucumbers. Large-scale sea cucumber processors meet the production requirements of modern food companies in that they adopt a variety of modern processing equipment and technologies, including mechanized pretreatment equipment, as well as low-temperature curing, hot air drying, heat pump drying, freeze drying, back pressure sterilization, microwave-assisted rapid sterilization and other technologies. However, the traditional manual treatment, conventional cooking, salting treatment, and natural drying techniques (such as sun exposure) are still the main methods in the pretreatment and drying of fished sea cucumbers. During processing, there is a significant loss of nutritional components such as collagen and polysaccharide, which is still the key factor restricting the maintenance of product quality.
2. Further develop new forms of sea cucumber products. In addition to the traditional dried sea cucumber, salted sea cucumber, and rehydrated sea cucumber, a series of new products such as freeze-dried sea cucumber, ready-to-eat sea cucumber, semi-dried sea cucumber, sea cucumber capsules, sea cucumber oral liquid, sea cucumber liquor, sea cucumber punch, and sea cucumber milk have emerged in the market, but these products still only occupy a small share. The main reason is that the forms of these new products have not been acceptable or satisfactory to consumers, and the production cost is considerable. Huang and Yuan (2022) analyzed the development patents of sea cucumber resources and found that the utilization of sea cucumber mainly focused on food processing, while those related to the extraction and utilization for medicinal purposes were limited, mainly focusing on components such as polypeptides and polysaccharides. Therefore, the development of functional components in sea cucumber should be strengthened and new products with international appeal could be developed.
3. Develop commercialized production technologies for effective components of sea cucumber. Collagens, polysaccharides, and lipids (phospholipids and cerebrosides, etc.) in sea cucumber are important raw materials for the development of functional foods, which have a variety of physiological functions. There are many methods to extract, prepare, and refine the functional components in sea cucumber, and some products have been developed with these methods. However, most of the methods are still in the laboratory stage, and standardized and commercialized large-scale processing technologies have not yet been established. It is necessary to develop green and safe pretreatment methods for sea cucumber raw materials and industrialized extraction technology for the functional components (especially those in by-products, including cooking liquid and viscera) to realize the commercialized production of sea cucumber functional components, increase the economic value of the sea cucumber food sector, and

bring real health benefits to consumers. The traditional processing methods of sea cucumber also need to be further improved to adapt to the commercial production and preparation of functional components.

4. Further improve the quality standard system of sea cucumber products. China is the major consumer market of sea cucumbers. Government authorities in China have formulated relatively high-quality standards for sea cucumber related products. However, with the in-depth development of the sea cucumber industry, new quality and safety issues continue to emerge, which pose problems for consumers and processors, disrupt the sea cucumber market, and limit the healthy development of the sea cucumber industry. For example, the origin tracing of sea cucumber, the illegal additives used in the processing of dried sea cucumber, the quality differences resulting from various sea cucumber breeding methods, the control of hazard factors in sea cucumber breeding, and the identification of counterfeit sea cucumber products all still require further research, so as to guide consumers' consumption and standardize the market behavior of sea cucumber businesses. At the same time, the health-care products on the market, which use collagen (peptide), polysaccharides, and active lipid in sea cucumber as functional components, need to strengthen the formulation of their quality standards to ensure the healthy development of the industry.
5. Strengthen the development of the nutritional value and active efficacy of sea cucumber. Based on the nourishing and health function of sea cucumber under the theory of traditional Chinese medicine, sea cucumber products are much favored by consumers in China and Southeast Asia. Therefore, the product research of sea cucumber and other related food materials and materials for both food and medicinal purposes should be carried out under the guidance of the structure-activity relationship research for the effective components of sea cucumber under the theory of traditional Chinese medicine. New sea cucumber products with significant nourishing effects should be developed to meet the needs of the international development of sea cucumber products.

3.2 International Prospects of Sea Cucumber Products

Sea cucumber fishing, processing, and trading are spread across more than 70 countries and regions around the world (FAO 2012). The sea cucumber is one of the aquatic products with a high degree of internationalization. However, consumers of aquatic products are mostly concentrated in China, Japan, South Korea, and Southeast Asian countries. Asia is the largest consumer market of dried sea cucumber in the world, with a market share of more than 70%. North America has a market share of dried sea cucumber of about 15%, where it is mainly consumed by consumers of Asian ethnicity. The main reasons for the high degree of international trade and low degree of international consumption of sea cucumber are as follows:

1. Dried sea cucumber and salted sea cucumber are the main processed products of sea cucumber. These require lengthy pretreatment and complex cooking processes in order to make appealing food.
2. The nutritional and health functions of sea cucumber have not been fully recognized by consumer groups other than the Chinese.
3. The processing of ready-to-eat sea cucumber products that can be stored at room temperature or low temperature is in its early stages with only a tiny market share.
4. There are few functional foods with sea cucumber functional components and clear-cut health effects, and these have not yet found a strong role as a supplement to the functional food industry.

Therefore, to expand the global consumption of sea cucumber products, processed sea cucumber products that meet the demands of the international market need to be vigorously developed. The development of sea cucumber functional foods with clear efficacy should therefore be prioritized. The functional food industry in developed nations has reached a large industrial scale, and the health function of functional foods has been widely recognized there, making the promotion of sea cucumber functional foods easy and feasible. Second, develop vigorously ready-to-eat sea cucumber products that can be stored at room temperature or low temperature, eliminating complicated cooking processes that always require a high level of skill and meeting the demands of busy consumers. Third, science-based promotion of the nutritional and health functions of sea cucumber is recommended, alongside the development of prefabricated sea cucumber dishes with clear health benefits. Fourth, the construction of international brands should be strengthened, while adhering to the green development of the sea cucumber processing industry and enhancing international competitiveness.

To sum up, there are many unsafe factors in the breeding process of sea cucumber in terms of environment, feed, storage, and transportation and man-made non-standard operations, resulting in various issues such as deterioration of the ecological environment and diseases. This reveals that several disadvantages accompany the sea cucumber breeding industry, such as insufficient awareness of hygiene and disease prevention, lack of disease inspection procedures and corresponding treatment measures, etc. Therefore, to improve the food safety of sea cucumbers, prevention should be carried out in the following aspects:

1. Strengthen the publicity of and prevention against marine micro pollution. First, science-based publicity can be used to improve environmental awareness among the public and reduce plastic pollution at its source. Second, standard microplastic treatment methods should be formulated, the treatment of plastic products should be normalized, and a unified standard system (including monitoring, analysis, evaluation standards and treatment measures) should be established.
2. Formulate strict residue and medication specifications on heavy metals. Publicity and education on healthy breeding should be conducted for sea cucumber farmers, and monitoring systems for heavy metals and agricultural and veterinary drug residues should be established, so as to effectively control the quality and

safety of sea cucumber products and ensure the consumption safety and health of consumers.

3. Improve the laws and regulations on the quality of sea cucumber and other aquatic products. Bring green aquatic product standards into the legislature. Green aquatic products (also known as pollution-free aquatic products) refer to pollution-free, safe, and high-quality nutritional aquatic products approved by special authorities and licensed to use green food signs. The requirements include that the raw materials and processing of the products must meet the ecological environment standards and production and operation specifications for green food.
4. Adapt to changes in the international market. Countries differ in legislative systems, certification systems, and import and export management methods. Therefore, to reduce food issues, the product quality standards of import and export countries need to be combined to reach an agreed production requirement.

4 Conclusion

With the increasing demand for healthy foods such as sea cucumbers, food safety has increasingly become a widespread public health issue. The food safety of sea cucumber is affected by diseases, antibiotics, residues of drugs illegally used in aquaculture, oil pollution from offshore drilling, and plastic particles produced in human daily life and industrial production. These issues still pose challenges for consumers and processors, disrupt the sea cucumber market, and limit the healthy development of the sea cucumber industry. Therefore, to ensure the healthy development of the sea cucumber industry, it is suggested to establish universal breeding, processing, and safety monitoring standards in the sea cucumber market, such as breeding standards, quality standards, traceability and counterfeiting, monitoring of illegal additives, etc., thereby guaranteeing food safety. In addition, with the improvement of people's living standards, consumers are pursuing diversified sea cucumber products. The development trend of sea cucumber product processing in the future therefore mainly lies in the development of new sea cucumber products. Standardized and commercialized production technology for sea cucumber and its functional components (collagen, polysaccharide, etc.) can assist in vigorously developing processed sea cucumber products that meet the demands of the international market.

References

Adeel M, Song XM, Wang YY, Francis D, Yang YS (2017) Environmental impact of estrogens on human animal and plant life: a critical review. Environ Int 99:107–119

Becker P, Gillan D, Lanterbecq D, Jangoux M, Rasolofonirina R, Rakotovao J, Eeckhaut I (2004) The skin ulceration disease in cultivated juveniles of *Holothuria scabra* (*Holothuroidea echinodermata*). Aquaculture 242:13–30

Chen ML, Gao F, Wang XY, Wei YF, Xu Q, Liu CS (2021) Effects of microplastics on immune and digestive physiology of *Holothuria* ATRA. Mar Sci 45:126–1359

Dang HY, Song LS, Chen MN, Chang YQ (2006) Concurrence of cat and tet genes in multiple antibiotic-resistant bacteria isolated from sea cucumber and sea urchin mariculture farm in China. Microb Ecol 52:634–643

Deng H (2006) A new pathogen in *Stichopus japonicus*—a globular virus. Fish Sci 1:30–31

Devi R, Surendran PK, Chakraborty K (2009) Antibiotic resistance and plasmid profiling of vibrio parahaemolyticus isolated from shrimp farms along the southwest coast of India. World J Microbiol Biotechnol 25:2005–2012

Doignon G et al (2003) Seasonal release of the egg capsules of Anoplodium parasita Schneider, 1858, intracoelomic turbellarian (Platyhelminthes, Rhabdocoela) symbiotic of the sea cucumber Holothuria tubulosa gmelin, 1788 (*Echinodermata holothuroida*). Echinod Res 2001:261–264

FAO, WFP and IFAD (2012) The state of food insecurity in the world. Economic growth is necessary but not sufficient to accelerate reduction of hunger and malnutrition. FAO, Rome

Gaw S, Thomas KV, Hutchinson TH (2014) Sources impacts and trends of pharmaceuticals in the marine and coastal environment. Philos Trans R Soc Lond B Biol Sci 369:20130572

Gong S, Que F (2012) Determination and analysis of 15 elements in sea cucumber. Mod Agric Sci Technol 2:347–357

Graham E, Thompson J (2008) Deposit—and suspension-feeding sea cucumbers (Echinodermata) ingest plastic fragments. J Exp Mar Biol Ecol 368:22–29

Hashmi MI, Thilakar R, Ali M, Hussein S, Hoque Z (2014) Determination of seven heavy metals in eight species of sea cucumbers. Sci Int 26:261–262

Hodson PV (2017) The toxicity to fish embryos of PAH in crude and refined oils. Arch Environ Con Tox 73:12–18

Huang JS, Yuan LH (2022) Development and utilization of holothurian biological resources. Aquaculture 43(01):76–80

Jia A, Hu JY, Wu XQ, Peng H, Wu SM, Dong ZM (2011) Occurrence and source apportionment of sulfonamides and their metabolites in Liaodong Bay and the adjacent Liao River basin North China. Environ Toxicol Chem 30:1252–1260

Laboy-Nieves EN, Conde JE (2001) Hydrobiologia, metal levels in eviscerated tissue of shallow-water deposit-feeding holothurians. Hydrobiologia 459:19–26

Lecomte S, Habauzit D, Charlier TD, Pakdel F (2017) Emerging estrogenic pollutants in the aquatic environment and breast cancer. Genes 8:229

Li J, Zhou ZH, Kang JT et al (1989) Analysis of trace elements in sea cucumber. Mar Sci 13:38–42

Li XS, Liao GX, Ju ZL, Wang CY, Li N, Xiong DQ, Zhang YL (2020) Antioxidant response and oxidative stress in the respiratory tree of sea cucumber (*Apostichopus japonicus*) following exposure to crude oil and chemical dispersant. J Mar Sci Eng 8:547

Liu S, Che J, Zhao YM et al (2022) Toxicity of 3-methylphenanthrene on Japanese spiky sea cucumber (*Apostichopus japonicus*). Appl Ecol Environ Res 20:69–77

Marrugo-Negrete J, Pinedo-Hernández J, Marrugo-Madrid S, Navarro-Frómeta E, Díez S (2020) Sea cucumber as bioindicator of trace metal pollution in coastal sediments. Biol Trace Elem Res 199(5):2022–2030

Miller T, McEneff G, Brown R, Owen S, Bury N, Barron L (2015) Pharmaceuticals in the freshwater invertebrate Gammarus pulex determined using pulverised liquid extraction solid phase extraction and liquid chromatography-tandem mass spectrometry. Sci Total Environ 511: 153–160

Mohammadizadeh M, Bastami K, Ehsanpour M, Afkhami M, Mohammadizadeh F, Esmaeizadeh M (2016) Heavy metal accumulation in tissues of two sea cucumbers *Holothuria leucospilota* and *Holothuria scabra* in the northern part of Queshm Island. Mar Pollut Bull 103:354–359

Mohsen M, Wang Q, Zhang LB, Sun LN, Lin CG, Yang HS (2018) Microplastic ingestion by the farmed sea cucumber *Apostichopus japonicus* in China. Environ Pollut 245:1071–1078

Morgan AD (2000) Aspects de la gestion des stocks ge´niteurs d'holothuries de sable (*Echinoderme: Holothurides*). La Be^che-de-mer Bulletin d'information 13:2–8

Naji A, Nuri M, Vethaak AD (2018) Microplastics contamination in molluscs from the northern part of the Persian Gulf. Environ Pollut 235:113–120

Nödler K, Voutsa D, Licha T (2014) Polar organic micropollutants in the coastal environment of different marine systems. Mar Pollut Bull 85:50–59

Pan QL, Che MX, Li SL, Hu MY, Wang J (2019) Research status of standard methods for determination of exogenous sugars in dried sea cucumber. J Food Saf Qual 10:5282–5287

Thompson RC, Olsen Y, Mitchell RP et al (2004) Lost at sea: where is all the plastic. Science 304: 838–838

Wang YT (2012) Aquaculture drugs and aquatic product quality and safety. China Fish 5:56–57

Wang PH, Chang YQ, Xu GR, Song LH (2005) Isolation and ultrastructure of an enveloped virus in cultured sea cucumber *Apostichopus japonicus* (Selenka). China Fish Sci 6:100–105

Wang YG, Fang B, Zhang CY, Rong XJ (2006) Pathogen and infection source analysis of "skin ulcer syndrome" in seedling preservation period of cultured *Stichopus japonicus*. Chin Fish Sci 4:610–616

Wang Y, Kang WL, Xin SG et al (2009) Analysis of trace elements in abalone and sea cucumber. Spectrosc Spect Anal 29:511–514

Wang YG, Guo WL, Rong XJ, Li B, Liao MJ, Zhang Z, Chen X, Liu SP (2012) Isolation and identification of a pathogenic bacterium associated with off-plate syndrome from *Apostichopus japonicus* cultured in Liaoning Province. Prog Fish Sci 33:81–86

Warnau M, Dutrieux S, Ledent G, Rodriguez-Baena A, Dubois P (2006) Heavy metals in the sea cucumber *Holothuria tubulosa* (*Echinodermata*) from the Mediterranean Posedonia oceanica ecosystem: body compartment seasonal geographical and bathymetric variations. Environ Bioindic 1:268–285

Xie WW (2013) Residue trand conol of formaldehyde in aquatic products. Agric Products Process 1:20–23

Yang D, Shi H, Li L, Li J, Jabeen K, Kolandhasamy P (2015) Microplastic pollution in table salts from China. Environ Sci Technol 49(22):13622–13627

Yang ZD, Pan QL, Hu MY et al (2021) Investigation on veterinary drug residues and heavy metal pollutants in dried sea cucumber in Shandong Province. J Food Saf Qual Inspect 12:3055–3059

Ye H, Zhu PL, Zhong MS, Xiao Y, Chen SY (2019) Analysis and evaluation of heavy metals in cultured sea cucumber in Fujian sea area. Fish Res 41:234–240

Zhang DL, Zhou M, Chen JH, Cui WG, Cheng HY, Zhao HQ, Wang XR, Li GQ (2009) Determination of toxic elements in sea cucumber by microwave coupled plasma mass spectrometry. Res Trace Elem Health 26:41–43+52

Zhang XY, Ma L, Wang HL (2012) Determination of nitrofuran metabolites residues in sea cucumber and sea cucumber seedlings by liquid chromatography—mass spectrometry. Port Health Control 17:24–28

Zhong LY, Xie YP, Lai JP et al (2014) Toxic effects of 3, 4-benzopyrene exposure on the growth and development of mosquito eating fish Jiangxi. Agric J 26:94–97

Zhu MH, Zhao HX, Chen JW, Xie HJ, Du J (2018) Investigation of antibiotics in sea cucumbers: occurrence pollution characteristics and human risk assessment. Environ Sci Pollut Res 145:1–7

Zhu MH, Wang ZY, Chen JW, Xie HJ, Zhan HX, Yuan XT (2020) Bioaccumulation biotransformation and multicompartmental toxicokinetic model of antibiotics in sea cucumber (*Apostichopus japonicus*). Environ Sci Technol 54:13175–13185. https://doi.org/10.1021/acs.est.0c04421

The manufacturer's authorised representative in the EU is Springer Nature Customer Service Centre GmbH, Europaplatz 3, 69115 Heidelberg, Germany. If you have any concerns regarding our products, please contact ProductSafety@springernature.com

Printed and bound by CPI Group (UK) Ltd, Croydon, CR0 4YY

30/09/2026

02225228-0014